T0076574

Exotic Ants

Exotic Ants

Biology, Impact, and Control of Introduced Species

EDITED BY

David F. Williams

NEW YORK AND LONDON

First published 1994 by Westview Press, Inc.

Published 2021 by Routledge
605 Third Avenue, New York, NY 10017
2 Park Square, Milton Park, Abingdon, Oxon OX14 4RN

Routledge is an imprint of the Taylor & Francis Group, an informa business

Library of Congress Cataloging-in-Publication Data
Exotic ants : biology, impact, and control of introduced species /
edited by David F. Williams
p. cm. — (Westview studies in insect biology)
Summaries also in Spanish.
Includes bibliographical references and indexes.
ISBN 0-8133-8615-2
1. Ants. 2. Insect pests. 3. Ants—Ecology. 4. Ants—Control.
5. Pest introduction. I. Williams, David F., 1938–
II. Series.
QL568.F7E97 1994
628.9'657—dc20

93-33266
CIP

ISBN 13: 978-0-3670-1093-5 (hbk)
ISBN 13: 978-0-3671-6080-7 (pbk)

Contents

Foreword

Most ants are relatively benign residents of natural arthropod communities. A few exotic species, however, have become serious environmental or human pests. This book is about those ants. Readers will find a veritable rogues gallery, a who's who of exotic pest ants. Within this book is a scholarly reservoir of information concerning the "Al Capones" and "Blackbeards" of the ant world. Read on, because these are the ants with which you are most likely to share your picnics, pantries, parties and pants.

The irony of this often vexing relationship is that humans are clearly a tramp ant's best friend. Without our houses, fields, roads and garbage dumps, these exotic pests would find few places to live. Not only do we provide them with ideal living space, but we also chauffeur them around the world during the course of international trade. Virtually every city on the globe has some of these intriguing social tramps gracing the doorsteps of its honorable and less honorable citizens.

By and large, people would prefer to see their ant guests permanently eliminated. Unfortunately, eradication attempts are almost never successful once exotic species become well established. Fortunately, most ants can be successfully controlled in selected areas, given some persistence and the right information. Readers will find this book filled with the right information—specifically, it contains many of the technical and biological details that form the basis for reasoned control and management of pest ants. A major strength of this book is that it combines articles from many authors about numerous species of exotic ants from all over the world in one volume.

Sanford D. Porter

Preface

Major problems caused by ants are usually a result of exotic species that have been introduced into areas where they have been released from any natural control. The imported fire ant, *Solenopsis invicta*, in the southern United States; leaf-cutting ants, *Atta* and *Acromyrmex* species, on several islands in the tropics; Pharaoh ants, *Monomorium pharaonis*, and Argentine ants, *Linepithema humile* (formerly *Iridomyrmex humilis*), in urban environments; the big-headed ant, *Pheidole megacephala*, in Hawaii; the crazy ant, *Anoplolepis longipes*, in the Seychelles; and, more recently, the little fire ant, *Wasmannia auropunctata*, in the Galapagos Islands are examples of introduced ants that have caused severe problems. In fact, it was because of the problems with the little fire ant in the Galapagos Islands that a conference on the biology, impact and control of introduced ant species was held in 1991 at the Charles Darwin Research Station in the Galapagos Islands to address the issue of exotic ants. The need for help in the Galapagos was such that the editor and Dr. Daniel Cherix (Museum of Zoology, Lausanne, Switzerland) decided to convene a conference, with the Charles Darwin Research Station acting as host. This conference was to provide a forum for addressing the problems caused by introduced ant species in many parts of the world. In addition, bringing together researchers in the Galapagos Islands who work on pest ants would allow active dialogue on the problems caused by introduced ants in these islands and how to deal with them. Because of illness, Dr. Cherix had to withdraw as an organizer. In his place, Drs. R. S. Patterson (USDA-ARS) and P. G. Koehler (University of Florida) were added to the organizing committee. The conference planning continued, a program was developed, funds were raised, and local arrangements were made. These efforts concluded in the Galapagos Ant Conference, titled "Exotic Ants: Biology, Impact and Control of Introduced Species," held at the Charles Darwin Research Station, October 14-17, 1991.

At the end of this very successful conference, a day-long workshop was held on the problems that pest ants were causing in the Galapagos Islands, along with deliberations on future research needs to develop management programs and control methods for these ant pests.

Most of the contributing authors were invited speakers at the conference. In addition, several other scientists were each asked to submit a chapter relating to their work with pest ants. This volume presents the latest research and ideas from these experts on introduced pest ant species and their control. Topics covered include current research on basic and applied aspects of the distribution, biology, ecology and behavior of several pest ants; the ant fauna of the Galapagos Islands and islands in the Caribbean; the ecology of an exotic ant in a tropical rainforest and the impact of some introduced ants on other species in several island habitats, with a discussion of successful and unsuccessful control strategies.

This book should be useful to individuals conducting basic and applied research on ants as well as other social insects and for scientists responsible for managing pest ant populations. The royalties from the sale of this book will be used to further research efforts at the Charles Darwin Research Station in the Galapagos Islands.

David F. Williams

Acknowledgments

The editor wishes to thank all of the contributors to the book. Their patience and response to editorial requests is greatly appreciated. A special thanks goes to Clifford S. Lofgren for his invaluable assistance in reviewing and editing the manuscripts and to Mary Chichester for the many hours spent typing, formatting and assisting in the editing of the publication. I also thank William A. Banks for providing a critical review of each chapter, Daniel P. Wojcik for contributing valuable editorial assistance on references and scientific names and Sanford D. Porter for writing the Foreword in addition to reviewing most of the chapters. Thanks also to Jorge Peña and Andrea Williams for the Spanish translations, and to Margaret Haile, Bonnie Bayer and Karen Vail for assistance in typing, copying and mailing manuscripts. I thank the U. S. Department of Agriculture, Agricultural Research Service, for allowing me to work on the book and for the use of the facilities. A special thanks to the following companies for their financial contributions to the Galapagos Ant Conference and to the publication of this book: Griffin Corporation; ICI Professional Products; McLaughlin Gormley King Company; S. C. Johnson & Son, Inc.; Sumitomo Chemical Company, Limited; Valent U.S.A. Corporation and Zoecon Corporation. Finally, I thank the staff at Westview Press, especially Ellen McCarthy, associate editor, and Mike Breed, scientific adviser, for their encouragement and advice, and for making a difficult job easier.

D.F.W.

Editorial Note

During the publication of this book, a revision was made of the genus *Iridomyrmex* by S. O. Shattuck, 1992, "Review of the dolichoderine genus *Iridomyrmex* Mayr with descriptions of three new genera (Hymenoptera: Formicidae)," *J. Australian Entomol. Soc.* 31: 13-18. Therefore, the scientific name of the Argentine ant, *Iridomyrmex humilis*, was changed to *Linepithema humile* in this book.

D.F.W.

1

The Galapagos Ant Fauna and the Attributes of Colonizing Ant Species

Carlos Roberto F. Brandão and Ricardo V. S. Paiva

The natural history of these islands is eminently curious, and well deserves attention. Most of the organic productions are aboriginal creations, found nowhere else; there is even a difference between the inhabitants of the different islands; yet all show a marked relationship with those of America, though separated from the continent by an open space of ocean, between 500 and 600 miles in width. The archipelago is a little world within itself, or rather a satellite attached to America, whence it has derived a few stray colonists, and has received the general character of its indigenous productions. Considering the small size of these islands, we feel the more astonished at the number of their aboriginal beings, and at their confined range. Seeing every height crowned with its crater, and the boundaries of most of the lava-streams still distinct, we are led to believe that within a period, geologically recent, the unbroken ocean was here spread out. Hence, both in space and time, we seem to be brought somewhere near to that great fact--that mystery of mysteries--the first appearance of new beings on this earth.

Charles Darwin, 1860

Introduction

Charles Darwin first realized the wealth of issues that could be investigated by consideration of the terrestrial faunas of oceanic islands. For instance, the establishment of a population in an area not previously occupied by a species often has dramatic effects in isolated environments.

Oceanic islands may be regarded as open sky laboratories for studying biological attributes of species, such as their differential abilities to colonize new environments, the genetics of colonizing propagules in relation to populations in nearby continents, and the effect of taxon cycles on biogeography. It is also possible to determine behavioral and other syndromes that allow tramp species to be, according to Carson (1987) "preadapted to movements over long distances and to quite literally establish beachheads on isolated and distant islands in the course of their normal dispersion".

Inspiration for studies of this type in the Galapagos archipelago can be obtained from research in the chronologically and spatially linear series of highly isolated islands of Hawaii. Here, ecological studies integrated with a multidisciplinary investigation of the *Drosophilidae,* made possible the tracing of both the continental and the inter-island origin of certain populations and species (Carson and Kaneshiro 1976).

The often quoted citation of Darwin which opens this article, does not hold true for all instances and groups. The Hawaiian Islands and Galapagos archipelago house very poor ant faunas, which are by no means specialized. In Hawaii, notwithstanding, the *Drosophilidae* attains its maximum diversity, while in Galapagos the seabirds are very much specialized in relation to the nearby continental fauna, and are more highly endemic than those of any other archipelago (Snow and Nelson 1984). The most important problem to understand is the significance of the poverty of species and the differentiation of so many forms that inhabit the islands (Patton 1984).

Thirty-six ant species have been recorded in Hawaii (none are endemic and most are widely distributed in other tropical places). For the Galapagos archipelago, we can not consider the ant fauna fully recorded. A better knowledge of the species occurring on each island must be obtained in a systematic and standardized survey. This will permit the monitoring and management of the species considered to be introduced on each island as well as enable us to make comparisons between other islands and with nearby areas in continental South America.

Historical review of the ant collections in Galapagos

According to Wheeler (1919), the first ants recorded from Galapagos were three species of *Camponotus* (*C. planus, C. macilentus* and *C. senex*) that were collected by Charles Darwin on the "Beagle" voyage and by Cook in 1875. The species were identified by Frederick Smith in 1877. The record of *C. senex,* however, has been challenged by Wheeler since the species has not been recorded in the islands since.

Wheeler (1919) identified some *Camponotus* males collected by the Albatross Expedition to Albemarle and Charles Islands. An expedition by George Baur in 1891 resulted in the collection of ant species that were identified by Emery (1893) as *Solenopsis geminata, Tetramorium guineense,*

Tapinoma melanocephalum, Odontomachus bauri and *Camponotus peregrinus* (considered later by him as a "variety" of *C. planus*). All of them were collected on Chatham Island, however with the exception of the first, they were also recorded in the ship.

Wheeler (1919) published a list of the known Galapagos ants, including those collected during the Expedition of the California Academy of Sciences in 1905, and later voyages of the Albatross. The list includes material collected on 12 islands of the archipelago, and records 36 "forms", representing 12 species. At least six of them (*Monomorium pharaonis, T. guineense, T. simillimum, Tapinoma melanocephalum, Paratrechina longicornis* and *P. vividula itinerans*) are well-known "tramp" species. Others, like *Solenopsis geminata, S. saevissima* and *C. senex*, may have been introduced also.

The first ant specialist who collected in the Galapagos was William Morton Wheeler in 1923 as a member of the Williams-Beebe Expedition. In 1924 he published a list of the species collected, including 36 forms representing 18 species, 9 of which were considered to be endemic to the islands. Some years later Wheeler (1933) recorded the first occurrence of *Crematogaster* in the Galapagos. It was collected by the Templeton Crocker Expedition in 1932.

Clark et al. (1982) published the results of the first systematic collection of ants from one island of the archipelago, based on attraction to baits within a given area and for a definite time. They recorded 17 species on Santa Cruz Island (Indefatigable) and considered 4 of them endemic. Finally, Lubin (1984), published a list which included the ancient collections cited above plus her own collections. The list includes *Cyphomyrmex* and *Leptogenys*, both genera never recorded in the Galapagos before.

A comparison of the lists from Santa Cruz Island by Wheeler (1924) and Clark et al. (1982) revealed some interesting contrasts. All endemic species appear in both lists and, surprisingly, 4 introduced ones recorded by Wheeler are not found in Clark's list. On the other hand, Clark et al. recorded 8 new records on Santa Cruz Island, 3 of them new to the archipelago. Possibly some introduced species have disappeared or their populations have been drastically reduced and Clark's list reveals new introductions. It is altogether more plausible, that both lists represent subestimations of the fauna. According to Beebe (1924), the expedition in which Wheeler was the ant specialist, spent only 100 hours in the archipelago. Clark and his collaborators worked there nearly a month. Lubin collected ants with sticky traps and visual searches on many islands and had access to other collections as well. The techniques used by Wheeler were not made clear in his paper. Clark et al. used baits of sugar/water solutions on transects along altitudinal gradients for quantitative studies on Santa Cruz Island. In fact, consideration of any aspect involving the Galapagos ant fauna must bear in mind that our knowledge is still fragmentary. There are many places on the islands where the ant fauna has never been studied. This may be true for most animal groups we call "Invertebrates" (Kramer 1984).

The case of **Wasmannia auropunctata**

The notion that this species had been recently introduced is supported by the fact that it does not appear in old surveys, although we have already questioned their use as evidence. Silberglied (1972) was the first to describe the recent problems caused by *W. auropunctata* and dated the introduction at "sometime in the early part of this century". According to Lubin (1984), the original site of *W. auropunctata* in the Galapagos was la Trágica at San Salvador Island (1700m) around 1967. On Isabela it was "introduced between 1966 and 1967" in a shipment of clumps of elephant grass *(Pennisetum purpurea).*

The problems posed by the reported spreading of *W. auropunctata,* in at least two of the islands forming the Galapagos archipelago, is that it is still in the process of expanding its range and reducing the population densities of most other ant species and some other animals as well (Lubin 1984). We do not know, however, if the species is enlarging its distribution due to its own colonizing ability (which seems to be the case) or the environment is changing as to permit its expansion.

The case of *Wasmannia* seems to be one of primary colonization by this widespread species followed by naturalization to the Galapagos. Other ant species with large distributions occurring in the archipelago (Clark et. al. 1982) belong to the same category: *Solenopsis geminata, S. saevissima, Monomorium pharaonis, Tetramorium simillimum, T. caldarium, Tapinoma melanocephalum, Paratrechina vividula,* and *P. longicornis*. These colonizers do not seem to have undergone speciation. Other species, however, belong to lineages that became species rich after establishment of an ancestral immigrant. They are: *Camponotus (Pseudocolobopsis) macilentus* and *C. (Myrmocladoecus) planus,* both with many described varieties, each roughly corresponding to a major island with a related species occurring on the continental South America. Native and introduced species most likely have had more than one episode of introduction from already differentiated mainland stocks.

Biological attributes of introduced species

Behavioral and genetic studies in the *grimshawi* species complex of the picture-winged *Drosophila* of Hawaii, for instance, suggested that a colonizing species may arise as an "escape from specialism" and that the genetic basis which initiates the shift may in some cases be relatively simple (Carson and Ohta 1981).

A spreading population may be genetically constituted so that it can exploit many ecological opportunities or it may be rather specialized genetically but somehow, its niche becomes disseminated and the species follow it without much genetic change.

Although the definitions are rather imprecise and continuums of conditions certainly exist, primary colonization has been defined as the establishment of a population in an area not previously occupied by the species. In general, these areas have a recent origin or have suffered a recent catastrophe. Secondary colonization occurs in small confined areas usually less altered and less isolated from colonizing propagules, for instance, successions, forest clearings, etc.

Most ant species that are called tramps by biologists have a genome that permits expression of many phenotypic possibilities. It seems, in contrast, that many ant species with smaller distributions may be composed of several morphologically indistinguishable cryptic species, but genetically differentiated.

Tramp ants are characterized by several biological attributes: (1) polygyny (with queens, many times brachipterous or even wingless), (2) unicolonial (occupying patchy, species-poor or manmade environments), (3) relaxed discrimination among nestmates (and thus often non-territorialists), (4) opportunistic in regards to nest sites preferences, (5) omnivorous, (6) colonies founded by fission or budding (queens following groups of workers by foot) and (7) always aggressive in competitive encounters. The following example describes this last attribute.

In November 1990, J. Diniz and the senior author (unpublished observations) observed in a semi-arid locality in Northeastern Brazil a group of *W. auropunctata* workers interfering with the foraging of unidentified medium-sized *Pheidole sp.* workers. The *Pheidole sp.* nested in the soil and were collecting dead or actively searching and stinging small arthropods. When carrying their prey back to the nest entrance (a circular hole of approximately 0.5cm) the workers and soldiers encountered groups of 10-20 *W. auropunctata* workers encircling the entrance. The *W. auropunctata* workers began biting the legs of the *Pheidole* "forcing" them to release the prey, which were readily taken and carried away. *Pheidole* ants, including large soldiers (probably recently recruited), who tried to leave the nest were bitten on the antennae by *W. auropunctata* workers in the circle who were facing inwards. The *Pheidole* finally retreated to the nest. During the one hour of observation, the *W. auropunctata* workers robbed all of the items the *Pheidole* were trying to collect.

In the case of species invading oceanic islands in Melanesia, Wilson (1961) also recognized three general qualities associated with colonizing success: (1) behavioral traits which presumably reduce interspecific competition, (2) the ability to penetrate marginal habitats, and (3) the ability to disperse across water gaps. He suggested that these attributes are casually related in the sequence listed. Success in the marginal habitats gives expanding species the advantage needed to encompass and progressively replace older resident taxa. Propagules may also differ in their potential as colonizers and may differ in other characters (genetically) as a result of the founding process. Repeated introductions or the buildup of the populations may further alter the original populations.

It seems that colonizing species may change their behavior over time, being much more aggressive while invading a new territory than when they are well established. The "taming" of an exotic ant was reported by Reaumur in the following way (Hölldobler and Wilson 1990) "in early 1500s a stinging ant appeared in such huge numbers as to cause the near abandonment of the early settlements on Hispaniola and Jamaica. The colonists of Hispaniola called on their patron saint, St. Saturnin, to protect them from the ant and conducted religious processions through the village streets to exorcise the pest. What was evidently the same species, which came to be known by the Linnaean name *Formica omnivora*, appeared in plague proportions in Barbados, Grenada, and Martinique in the 1760s and 1770s. The legislature of Grenada offered a reward of £20,000 for anyone who could devise a way to exterminating the pest, to no avail. It now appears that *F. omnivora* was none other than the familiar "native" fire ant, *S. geminata*, which at the present time is a peaceable and only a moderately abundant member of the West Indies fauna".

Linepithema humile [=Iridomyrmex humilis] was discovered in Bermuda in 1953 and recognized as a major economic pest by the local authorities in 1957. Thirty years after its introduction, it still coexists with another invading pest ant, *Pheidole megacephala* which has been established there at least since 1902. Their patterns of distribution now form an interdigiting mosaic which even includes other species in some places in a situation similar to that found for most ant species communities in continental areas. Although one of the two competing species may have had "hold possession" of some sites throughout the whole time period "the situation would seem to be one of equilibrium rather than slow replacement, at least at the time-scale involved" (Haskins and Haskins 1989).

It is interesting to note that the rate of expansion of *L. humile* on the island since its establishment slowed as the territory became saturated. Also, other ants have been regularly reported, some now as "cryptic rarities", but a significant number in a long-continuing basis, despite the heavy environmental disturbance around them and the two invaders *P. megacephala* and *L. humile*. One of these species is *W. auropunctata*.

Within continents it is also known that a given area may suffer successive invasions by new species who replace the former invader in a relatively short time. It is altogether possible that on isolated oceanic islands cyclic invasions followed by the "taming" of the former invading species may be the rule.

According to Wilson (1961), however, expanding species play a major role in the fragmentation and speciation of older taxa. By dominating the faunas of smaller islands, they maintain hiatuses in the ranges of disjunct taxa. By saturating the marginal habitats they restrict the distribution of older taxa.

A Proposal for a Survey of the Galapagos Ant Fauna

Any survey of the ant fauna in the Galapagos Islands aimed at obtaining qualitative and quantitative data, will have to observe the diversity of habitats of each island, as they vary considerably in size, terrain, availability of fresh water and degree of vegetation. The survey should include transects or grids with a known number of baits, or collecting points, thus allowing for statistical analysis. The procedures should be conducted during the day and at night, since the fauna seems to be clearly different from day to night. The number and location of the collecting sets depends upon many variables like topography of the islands, vegetation types, elevation, etc.

Qualitative collections are also important. Twig hollows, dead seeds, rotten wood, and litter are good places for finding criptic species. Techniques like Tulmgrem-Berleze funnel and pitfall are also useful to collect inconspicuous species.

The great number of islands forming the archipelago, and their huge differences along with the logistic problems of working on some of them, may be regarded as the main problems to be faced by a survey group. On the other hand the collections are favored by the meagerness of the fauna and the conservation of the islands environments.

The information that could be theoretically obtained in this study includes richness of the ant fauna of the archipelago and each island, relative abundance of individuals, and frequency of species. Some correlations, like number of species x area of the island, and vegetation type or altitude x species occurrence could be made. Inferences on niche characteristics and knowledge on resource partitioning among species could be obtained. Nearby continental areas should be surveyed similarly to afford measures of comparison.

To address some of the questions outlined above, it is expected that this study would reveal patterns that may assist in distinguishing between alternative hypotheses of island biogeography, evolutionary processes of differentiation, and models of speciation.

With the survey information in hand, it will also be possible to investigate hypotheses on community structure and dynamics of the Galapagos archipelago ant fauna.

Conclusion

Even these preliminary considerations are very much affected by the poor state of knowledge on the taxonomy of Neotropical ants. This is especially true for some taxa cited in this text. The "necessity" to apply names to individuals or samples has resulted in a long list of names, often not associated with the original specimens anymore. This prevents us from determining whether other species reported from the Galapagos (Clark et. al. 1982 and references, Kempf 1972, Brandão 1991) belong to continental species or are restricted to the archipelago. The concepts of native versus exotic species may not be readily applicable yet to the Galapagos ant fauna.

Acknowledgments

We thank David F. Williams, Harry G. Fowler and everybody involved in the organization of the Galapagos meeting in Santa Cruz for making possible our participation, and for making the time we spent there so agreable. Further support came from FAPESP grant 90/2775-6. We thank Eliana Cancello for advice on the manuscript.

Resumen

En este capitulo, discutimos el concepto de las hormigas exoticas en comparación con la fauna de hormigas nativas, el concepto de invasión, el cual debe ser valorado de acuerdo a nuestro conocimiento de la dinámica de las comunidades y los ciclos de los taxon en una base continental y los sindromes ecologicos y de comportamiento relacionados con la así llamada trampa o especies de hormigas exoticas. Comparamos sus atributos con los de las especies que consideramos nativas. Las especies trampa en Suramerica son las mismas que se encuentran distribuídas en todo el mundo. Para poder encontrar sus centros de origen va a ser necesario el utilizar una tecnica genetica moderna la cual podría llevarnos a descubrir eventualmente varias especies cripticas las cuales no pueden ser distinguidas morfologicamente.

References

Beebe, W. 1924. *Galapagos: Worlds End.* The Knickerbocker Press. G. P. Putnam's Sons. New York and London.

Brandão, C. R. F. 1991. Adendos ao Catálogo Abreviado das Formigas da Região Neotropical. *Rev. Brasil. Entomol.* 35: 319-412.

Carson, H. L. 1987. Colonization and Speciation. Pp. 187-206. In: A. J. Gray, M. J. Crawley, and P. J. Edwards [eds.]. Colonization, Succession and Stability. Oxford, 26th Symposium of the British Ecological Society, Linnean Society of London. Blackwell Sci. Publ.

Carson, H. L. and K. Y. Kaneshiro. 1976. *Drosophila* of Hawaii: systematics and ecological genetics. *Ann. Rev. Ecol. and Syst.* 7: 311-345.

Carson, H. L. and A. T. Ohta. 1981. Pp. 365-370. Origin of the genetic basis of colonizing ability. In: G. G. EScudder and J. L. Reveal [eds.]. Evolution Today. Proc. of the 2nd International Congress of Systematics and Evolutionary Biology. Carnegie-Mellon Univ.

Clark, D. B., C. Guayasamín, O. Pazmiño, C. Donoso, and Pãez de Villacís, Y. 1982. The tramp ant *Wasmannia auropunctata*: autoecology and effects on ant diversity and distribution on Santa Cruz Island, Galapagos. *Biotropica* 14: 196-207.

Darwin, C. 1860. Galapagos archipelago, chapter 17. In: J. M. Dent. Journal of researches into the Geology and Natural History of the various countries visited during the voyage of H. M. S. Beagle round the world. London. (revised edition 1904).

Emery, C. 1893. Notices sur qualques fourmis des Iles Galapagos. *Ann. Soc. Entomol. France* 63: 89-92.

Haskins, C. P. and E. F. Haskins. 1989. Final observations on *Pheidole megacephala* and *Iridomyrmex humilis* in Bermuda. *Psyche.* 95: 177-184.

Hölldobler, B. and E. O. Wilson. 1990. *The Ants.* The Belknap Press of Harvard Univ. Press, Cambridge, MA.

Kempf, W. W. 1972. Catálogo Abreviado das Formigas da Região Neotropical. *Studia Entomologia.* 15: 3-344.

Kramer, P. 1984. Man and other introduced animals. *Biol. J. Linn. Soc.* 21: 253-258.

Lubin, Y. D. 1984. Changes in the native fauna of the Galapagos Islands following invasion by the little red fire ant, *Wasmannia auropunctata. Biol. J. Linn. Soc.* 21: 229-242.

Patton, J. L. 1984. Genetical process and the Galapagos. *Biol. J. Linn. Soc.* 21: 97-111.

Silberglied, R. 1972. The little fire ant *Wasmannia auropunctata* a serious pest in the Galapagos Islands. *Not. Galápagos* 19/20: 13-15.

Snow, D. W. and J. B. Nelson. 1984. Evolution and adaptations of Galapagos sea-birds. *Biol. J. Linn. Soc.* 21:137-155.

Wheeler, W. M. 1919. The ants of the Galapagos Islands. Proc. California Acad. Sci. 2: 259-297.

______ . 1924. The Formicidae of the Harrison Williams Galapagos Expedition. *Zoologica* 5: 101-122.

______ . 1933. The Templeton Crocker Expedition of the California Academy of Sciences 1932, no. 6. Formicidae of the Templeton Crocker Expedition. Proc. California Acad. Sci. ser. 4, 21: 57-64.

Wilson, E. O. 1961. The nature of the taxon cycle in the Melanesian ant fauna. *Am. Nat.* 95: 169-193.

2

Distribution and Impact of Alien Ants in Vulnerable Hawaiian Ecosystems

Neil J. Reimer

Introduction

The Hawaiian Islands emerged from the ocean floor more than 70 million years ago (Dalrymple et al. 1973) in one of the most secluded spots on earth: the middle of the Pacific Ocean. They are a chain of 132 islands, reefs, and shoals stretching over a distance of 2,466 km (Armstrong 1973). However, the eight largest islands, Hawaii, Maui, Oahu, Kauai, Molokai, Lanai, Niihau, and Kahoolawe make up 99% of the land area (Armstrong 1973). They are the most isolated islands in the world, occurring ca. 4,000 km from the nearest continent and 1,600 km from the nearest island group (Simon 1987). They range in altitude from sea level to 4,205 m. The mean monthly temperature ranges from 27°C at sea level to 0°C at the summit of the highest mountain, Mauna Kea. These altitude and temperature extremes along with varying soil conditions, precipitation gradients, and topography have allowed for the development of all of the worlds major plant communities (Mueller-Dombois et al. 1981). Ecosystems found in Hawaii include: littoral, strand plant community, lowland dry scrub, desert, grassland, deciduous dry forest, mesic forest, lowland and montane rain forest, cloud forest, cool dry forest (above 1,500 m), alpine scrub (above 2,000 m), and a stone desert supporting an aeolian community (above 3,000 m) (Howarth 1990). The few pioneering animals and plants that survived the long journey across the Pacific to the islands have evolved into a fascinating assemblage of diverse and unique organisms (Zimmerman 1948).

All ant species occurring in Hawaii are alien (Zimmerman 1970). They were not among the original prehistoric colonizers of the Hawaiian islands but reached Hawaii after human occupation; the majority of them within the last 100 years (Reimer et al. 1990). Therefore, the presence of ants in Hawaii is a relatively recent phenomenon especially in light of the millions of years of evolution that occurred prior to their appearance. Indeed, no predatory social insects were present in the native fauna (Howarth 1985). The native terrestrial invertebrates had no need to evolve effective defenses against ants (Reimer et al. 1990) as have invertebrates which coevolved with ants (Holldobler and Wilson 1990). Attributes such as mimicry, appeasement glands, defensive secretions, etc., are rarely found in the native ant fauna. In fact, many native arthropods did not develop effective ant defenses, such as hard exoskeletons (Gillespie and Reimer 1992) and the ability to fly (Zimmerman 1970). In addition, no specialized predators of ants exist in Hawaii. When ants did reach Hawaii, they found ideal climatic conditions, abundant, easily captured prey, and no predators. They spread rapidly throughout the islands preying on the vulnerable native invertebrates with devastating consequences.

Distribution of Ants in Hawaii

Ants have been found on all of the eight main islands (Reimer et al. 1990) and each of the eight largest islands of the Northwest Hawaiian Islands chain (i. e. Kure Atoll, Midway, Pearl and Hermes Atoll, Lisianski, Laysan, French Frigate Shoals, Necker, and Nihoa). All of the known ant species occur or have been collected from the island of Oahu, probably because this is the major port of entry. Currently, only 5 of the known 40 species are restricted to Oahu. Three of the five species limited to Oahu, *Amblyopone zwaluwenbergi* (Williams), *Monomorium latinode* Mayr, and *Strumigenys lewisi* Cameron, have not been collected since 1945, 1925, and 1920, respectively, despite extensive surveys during the 1960s and 1980s, and may no longer be present. The other two species, *Iridomyrmex glaber* (Mayr) and *Pseudomyrmex gracilis mexicanus* (Roger), are more recent immigrants which arrived in the 1970s (Beardsley 1979, 1980). These two species are common on Oahu.

In general, the Hawaiian ant fauna is composed of vagile, mostly cosmotropical tramp species. The majority of the species occur in the warmer, dry (0 to 120 cm precipitation per year) to mesic (120 to 250 cm precipitation) lowland (0 to 900 m) areas of the state (Table 2.1). These lowlands, which have undergone the most human disturbance due to land clearing, fire, agriculture, grazing, and urbanization (Smith 1985) serve as the principal point of entry for most introductions to the islands. Few undisturbed native plant communities exist below 600 m (Wagner et al. 1985). The depletion of the native invertebrate fauna has also been the greatest in these lowland areas.

In response to this loss, over 415,000 ha in Hawaii (ca. 23% of the State) have been legally dedicated to the protection of native species and ecosystems (Holt and Fox 1985). Until recently, the ant fauna in most of this area had never been surveyed and was virtually unknown. We now have a better understanding of the ants in these areas because of recent surveys in Haleakala National Park (Fellers and Fellers 1981, Medeiros et al. 1986), Hawaii Volcanoes National Park (Medeiros et al. 1986), the U.S. Fish and Wildlife Service National Wildlife Refuges in the Northwest Hawaiian Islands, and in the State Forests and Natural Area Reserves (N. Reimer, unpublished data). Prior to these surveys, ant distribution studies had been concentrated in human-disturbed areas, primarily along roadsides and in agricultural and urban areas; areas usually depauperate in native fauna and flora.

The majority of the areas with undisturbed native plant communities and relatively intact native invertebrate fauna occur above 1,000 m in montane (900 to 1800 m), subalpine (1800 to 2700 m), and alpine (>2700 m) communities. Surveys in the national parks by Medeiros et al. (1986), and by the author in the Natural Area Reserves and State Forests demonstrate that few ant species have proliferated in these cooler montane ecosystems (Table 2.1). This is probably due to the fact that species adapted to these environments must first survive at their point of entry in the lowlands. The few that have proliferated above 1,000 m pose a serious threat to the remaining native biota.

Ants occurring in these areas are listed in the Table that follows this brief review of information available on the most common species:

***Leptogenys falcigera* (Roger).** This ant was collected only once in a montane community, at 1219 m, but is frequently collected in lowland areas. *L. falcigera* appears to be better adapted to the warmer lowlands. The species forms small colonies (<50 individuals). It may have little impact on the native invertebrate fauna due to its small colony size and uncommonness in habitats frequented by native invertebrates.

***Hypoponera punctatissima* (Roger).** This ponerine occurs occasionally in mesic montane areas below about 1200m. Single foragers of this uncommon, cryptobiotic species can be found under leaf litter. Its impact on the native fauna is probably minimal.

***Hypoponera opaciceps* (Mayr).** This ponerine occurs in lowland and montane ecosystems. It can typically be found in dry, mesic, and wet

(>250 cm precipitation) areas between 600 and 1200 m but has been collected as high as 2700 m (Medeiros et al. 1986) and as low as 150 m. As with the preceding species, *H. opaciceps* is also cryptobiotic; its workers foraging under leaf litter. It forms small colonies (<50 workers) in the soil and under rocks. This is one of the few species which has been able to invade the wet undisturbed rainforests. Single foragers and occasional nests can be collected quite commonly in these areas, but it never builds up large populations. It appears to be a relatively innocuous species which has an insignificant impact on the native fauna.

***Pheidole megacephala* (Fabricius).** The impact of this ant on native invertebrates in lowland areas at the turn of the century was described by Perkins (1913). It has not penetrated deeply into higher altitudes since that time, but is primarily restricted to dry and mesic lowland areas. However, it can occasionally be found in a few dry and mesic montane communities up to 1200 m.

This is a polydomous, polygynous, aggressive species. It is almost always the dominant ant in areas where it is found. Its only natural enemies in Hawaii are other unicolonial, polygynous, aggressive ant species such as *Linepithema humile* [=*Iridomyrmex humilis* (Mayr)], *Anoplolepis longipes* (Jerdon), *Iridomyrmex glaber* (Mayr), *Solenopsis geminata* (Fabricius), and *Pheidole fervens* Fr. Smith which can displace it in environments less suitable for its survival.

***Solenopsis papuana* (Emery).** This small, black myrmecine was first collected on Oahu and Hawaii islands in 1967 (Huddleston and Fluker 1968). It has since spread to Maui, Molokai, Lanai, and Kauai. It occurs primarily in mesic and wet forests between sea level and 1100 m, and has occasionally been collected in shaded, dry forests at about 1000 m.

This is the only aggressive, polygynous species which has been able to penetrate into undisturbed wet forests. Other ants found in this habitat have been species with small, scattered colonies such as *H. opaciceps* which appear to have relatively little impact on the environment. *S. papuana* is a very aggressive species which can reach very dense populations in mesic and wet forests. It most likely has a severe impact on the native biota. This species is very common in these communities, but has not yet moved into all mesic and wet forests at these altitudes. It still appears to be expanding its distribution.

***Cardiocondyla emeryi* Forel, *C. nuda* (Mayr), and *C. wroughtoni* (Forel).** These three species are all very similar in biology and ecology. They form small, polygynous colonies in lowland communities. They rarely form dense populations. Their impact on native invertebrates appears to be minimal.

***Cardiocondyla venustula* (Wheeler).** *C. venustula* is similar to the other *Cardiocondyla* species in Hawaii in that it tends to form small, polygynous colonies. It differs in that it has a much wider tolerance of environmental conditions. *C. venustula* occurs from sea level to 1900 m in dry to wet environments. It does not penetrate into undisturbed forests as is seen with *S. papuana* and *H. opaciceps*, but is limited to roads and trails in these areas. It does, however, occur commonly in disturbed areas such as agro-ecosystems and urban developments.

Tetramorium simillimum (Fr. Smith). This species occurs from sea level to about 1100 m in dry and mesic areas. It can form large, polygynous colonies. It is limited to disturbed areas similar to the preceding species. In forested areas it is found along trails (especially on hill tops) and roads. It has not been found in undisturbed forest.

Linepithema humile (Mayr). This ant has been displaced by other ants, such as *P. megacephala*, in many areas and is now limited to the cooler higher altitudes in Hawaii. It occurs from 900 to 2800 m but is found most frequently between 900 and 2000 m. It has spread into undisturbed montane native habitat in dry and mesic areas of the state but has not been able to penetrate the wet forests. *L. humile* can have a significant negative effect on the native biota (Medeiros et al. 1986).

Plagiolepis alluaudi (Forel). This ant species can be found occasionally in mesic montane environments up to about 1000 m, but occurs more commonly in dry, mesic, and, wet lowland communities. It rarely occurs in large numbers in montane habitats.

Camponotus variegatus (Fr. Smith). This ant has only been collected in dry and mesic areas below 500 m where it is quite common. *C. variegatus* may occur in other areas but surveys to date have been during the day. This nocturnal ant is rarely picked up in these faunal surveys. This is the only exclusively nocturnal species in Hawaii.

Anoplolepis longipes (Jerdon). This aggressive long-legged ant forms large unicolonial, polygynous colonies. It is generally found from sea level to 800 m but has been collected as high as 1200 m at Haleakala National Park (Medeiros et al. 1986). This species does penetrate into the wet rain forests in the lowlands. These forests are primarily composed of introduced plants. Native invertebrates do occur in these forests, but not as commonly as in undisturbed native forests. This species has been implicated in causing depletion of native fauna in riparian habitats in the lowlands (Hardy 1979).

Paratrechina bourbonica (Forel) and P. vaga (Forel). Both species have been collected in disturbed montane habitats (1000-1200 m) such as roadsides or urban development sites, never in undisturbed sites. They are more commonly found at lower elevations, below 1000 m.

TABLE 2.1. Distribution of ants in the Hawaiian Islands by ecological community.

	Lowland			*Montane*			*Subalpine*			*Alpine*		
	0-900m			*900-1800m*			*1800-2700m*			*>2700m*		
Species	d	m	w	d	m	w	d	m	w	d	m	w
Leptogenys falcigera	X	X	X		X							
Cerapachys silvestrii	X	X										
Amblyopone zwaluwenbergi	X											
Ponera swezeyi	X	X										
Hyponera zwaluwenbergi		X	X									
H. sinensis	X	X										
H. punctatissima	X	X			X							
H. opaciceps	X	X	X	X	X	X	X					
Pheidole megacephala	X	X		X	X							
P. fervens			X									
Solenopsis geminata	X	X										
S. papuana	X	X	X	?	?	?						
Monomorium floricola	X	X										
M. fossulatum	X	X										
M. pharaonis	X	X										
M. destructor	X											
M. minutum	X	X										
Cardiocondyla emeryi	X	X										
C. nuda	X	X	X		?							
C. wroughtoni	X	X	X		?							
C. venustula	X	X	X	X	X	X	X					
Tetramorium bicarinatum	X	X	X									
T. simillimum	X	X		X	X							
T. tonganum		X										
Quadristruma emmae		X										
Strumigenys godeffroyi		X	X									
S. rogeri		X	X									
Linepithema humile	?	?		X	X	X	X					
Iridomymex glaber	X	X										
Tapinoma melanocephalum	X	X										
Technomyrmex albipes	X	X	X									
Plagiolepis alluaudi	X	X	X		X							
Camponotus variegatus	X	X										
Anoplolepis longipes	X	X	X									
Paratrechina longicornis	X											
P. bourbonica	X	X	X		X							
P. vaga	X	X	X			X						
Pseudomyrmex gracilis mexicanus	X											

d=dry=0-120 cm; m=mesic=120-250 cm; w=wet= >250 cm

?= collected only on periphery of community

Impact of Ants on Native Hawaiian Fauna

The primary impact of ants on native Hawaiian fauna is as predators. Medeiros et al. (1986) showed that ant predation had a severe impact on *Nesoprosopis* sp., an endemic ground-nesting bee. *Nesoprosopis* nests could not be found in ant-infested plots but were commonly encountered in ant-free sites of the same habitat. Heavy predation on these endemic bees could have an indirect effect on *Nesoprosopis*-pollinated native plants such as *Dubautia* and the Haleakala silver-sword, *Argyroxiphium sandwicense* DC. In the same study, Medeiros et al. (1986) demonstrated that native carabid, spider, and noctuid moth populations were depleted or eliminated from the study sites by *L. humile*. One of the affected carabid species is *Mecyclothorax robustus* (Blackburn) which is one of 32 native species of this genus which exist in relatively small areas on the dormant volcano, Haleakala (Britton 1948). With such a localized distribution, predation on *M. robustus* by *L. humile* could result in the extinction of the species. In contrast, according to Medeiros et al. (1986), alien isopods and millipedes benefited from the presence of ants, since their populations were higher in ant-infested plots than ant-free plots.

Studies with native *Tetragnatha* spiders demonstrated that their distribution in mesic to wet forests was inversely related to the presence of ants, primarily *A. longipes* and *P. megacephala* (Gillespie and Reimer 1992). Laboratory confrontations between *A. longipes* or *P. megacephala* and native or alien spiders always resulted in death of native spiders and survival of alien spiders. The alien spiders survived due to their hard exoskelton or by leg autotomy; characteristics not present in the native species (Gillespie and Reimer 1992).

Gagne (1979) found the composition of invertebrate species inhabiting tree canopies changed along transects as they entered ant-infested areas. Only species resistant to ant predation such as borers, vagile species, gall-formers, and species with repugnatorial glands were present in ant-infested areas. These are primarily alien species. Gagne (1979) collected the highest diversity of native species in ant-free canopies.

Other evidence for the impact of ants on native Hawaiian invertebrates has been more observational rather than experimental. A number of authors have commented on the disappearance of native fauna in Hawaii after ants colonized an area. These authors have implicated ants in the extinction of endodontid snails (Solem 1976), the depletion of lowland native invertebrates (Perkins 1913), and the disappearance of endemic aquatic and semiaquatic Diptera in riparian habitats (Hardy 1979). Interestingly, a few unique native insects associated with dry lowland scrub vegetation have been able to survive on Nihoa island (J.

W. Beardsley, pers. comm.). This is the only Hawaiian island which does not contain any aggressive polydomous ant species.

Ants have been implicated in causing other negative effects on native biota besides those shown experimentally or observed in the field. Ants have been hypothesized to threaten the survival of the last known populations of the native damselfly, *Megalagrion pacificum* (McLachlan), by attacking teneral adults and depleting its food source, streamside fauna (Hardy 1979, Moore and Gagne 1982). In addition, Howarth (1985) stated that ants help spread alien honeydew-producing homopterans, deplete food sources of native predators and parasites, feed on plant seeds, and spread weed seeds. These accusations are plausible but have yet to be proven with experiments.

The impact of ants on native biota is not always negative. It is very likely that some native homopterans may benefit from ants tending them for honeydew. For example, the Argentine ant has been observed feeding on honeydew produced by an endemic mealybug, *Pseudococcus nudus* Ferris (Reimer et al. 1990).

Management and Control Strategies

The emphasis for management and control of ants in Hawaii emphasizes prevention since the use of other control methods, such as biocontrol or pesticides, is either not feasible or unavailable in Hawaii's native ecosytems. Prevention can best be accomplished through quarantines and minimal disturbance of intact ecosystems.

A number of ecosystems, primarily montane areas above 1200 m, remain ant-free in Hawaii. Hawaii's ant fauna has not been able to penetrate into intact ecosystems at these elevations, but ant species do exist throughout the world which are adapted to similar habitats. Continued vigilant quarantine restrictions are an essential barrier in preventing access of these species.

The species of ants collected above 1200 m are *L. humile, H. opaciceps, C. emeryi, C. venustula,* and *P. bourbonica.* Of these, *H. opaciceps* is the only species which has been able to survive in mesic to wet forests exposed to minimal human impact or in undisturbed habitat. The other four species move into human-disturbed sections such as along roads, trails, and clearings where they then forage into the intact forest. Minimal disturbance of the forest would prevent the present ant fauna from penetrating deeper into these areas.

Other management methods, such as the use of natural enemies or pesticides, also need to be investigated. No effective parasites or predators are known for ant species in Hawaii, except for other aggressive ants. The use of pesticides, however, may be useful in certain situations.

Ants are so widespread throughout the islands and often in such inaccessible areas that the use of spray or bait insecticides is usually not feasible. An exception may be in some situations where an ant species spreads by budding or where a small population has been accidentally introduced to an area by man. The invasion of *L. humile* into Haleakala National Park is a good example. The Argentine ant occurs in limited areas at some sites in the park. These spot infestations should be eliminated with an appropriate bait. A suitable bait may also be used along the edges of an ants' distribution to stop an invasion into critical ant-free areas. However, care must be taken that the bait is not picked up by non-target organisms.

Conclusions

Experimental and observational evidence has shown that the Hawaiian ant fauna, composed entirely of alien species, has had a devastating effect on the native terrestrial fauna. The species that have had the greatest impact have been aggressive, unicolonial, polygynous species such as *L. humile, P. megacephala, A. longipes, S. geminata,* etc. Their effects have been limited primarily to lowland ecosystems to which they are best adapted. A few of these aggressive species have been able to move into human-disturbed areas in montane habitats, but have not penetrated into undisturbed habitat in these areas. The reasons for this are unclear and need to be investigated. However, to prevent the devastation observed in the lowlands from occurring in these pristine native habitats, it is important to keep these habitats in an undisturbed state. Once these areas are disturbed by humans (eg. roads, clearing, etc.), alien ant fauna are able to become established. It is also important to prevent the introduction of new ant species into Hawaii which may adapt to these areas.

The above discussion reveals that few experimental studies on the effects of ants on native biota and ecosystems have been completed. However, much observational evidence and hypothetical impacts have been recorded; these should be substantiated with experiments when possible.

The fact that the Hawaiian ant fauna is composed entirely of introduced species and that native arthropods are not adapted to ant predation provides an excellent opportunity for research on biological invasions. This is especially true for ant species that are still expanding their distributions. The most conspicuous and possibly most damaging impacts appear to occur during the initial wave of the invasion of ants into previously ant-free areas. Experimental field studies to date have all compared fauna in ant-free vs. ant-infested habitats. These studies have given excellent information on the impacts of ants on native biota and should be continued, but studies investigating the invasion phase could better enhance our understanding of the processes involved in the invasion. Research studies on the rate of movement, invasion behavior, and nest requirements of ants as they move into previously uninfested areas and the responses of susceptible invertebrates would all add to our understanding of the invasion process. The effects of aliens on native ecosystems and the interactions among alien species are also fertile areas for research with the Hawaiian ant fauna. Such research would provide important information applicable to other biological invasions throughout the world, particularly those involving ants.

Acknowledgments

I wish to thank the Hawaii Natural Area Reserves System Commission, Division of Forestry and Wildlife of the Department of Land and Natural Resources for funding for this project.I also wish to acknowledge and thank J. W. Beardsley, F. G. Howarth, and J. Strazanac for their helpful comments and criticisms of this manuscript. This is Journal Series No. 3699 of the Hawaii Institute of Tropical Agriculture and Human Resources.

Resumen

La cadena de islas mas aislada en el mundo, son las islas Hawaiianas, las cuales tienen una fauna única comparada con el resto del mundo. La fauna hawaiana evolucionó sin la presencia de uno de los grupos de predadores mas eficientes, las hormigas. Durante los ultimos 100 años, se han establecido en Hawaii, 40 especies de hormigas causando consecuencias devastadoras a la fauna nativa. El impacto mas grande ha sido producido por *Linepithema humile [=Iridomyrmex humilis], Pheidole megacephala, Anoplolepis longipes, Solenopsis geminata, Solenopsis papuana, Paratrechina longicornis,* y *P. bourbonica,* siendo cada una de estas especies agresivas, unicolonialistas y polyginas. La devastación causada por estas hormigas está limitada mayormente a las areas bajas y a las areas montañosas disturbadas. La presente fauna de hormigas ha sido incapaz de penetrar el habitat de las areas virgenes localizadas 1,000 m sobre el nivel del mar.

La situación poco usual de una fauna de hormigas compuesta completamente de especies introducidas y una fauna nativa extremadamente diversa la cual evolucionó sin la presencia de hormigas, presenta una excelente oportunidad de investigación en invasiones biologicas. Estos resultados podrían proveer información importante aplicable a otras invasiones en el mundo, particularmente aquellas de hormigas.

References

Armstrong, R. W. 1973. *Atlas of Hawaii.* Univ. of Hawaii Press, Honolulu

Beardsley, J. W. 1979. Notes on *Pseudomyrmex gracilis mexicanus* (Roger). *Proc. Hawaii. Entomol. Soc.* 23:23.

Beardsley, J. W. 1980. Note of *Iridomyrmex glaber* (Mayr). *Proc. Hawaii. Entomol. Soc.* 23:186.

Britton, E. B. 1948. A revision of the Hawaiian species of *Mecyclothorax* (Coleoptera: Carabidae). *B. P. Bishop Museum Occ. Pap.* 19: 107-166.

Dalrymple, G. B., E. A. Silver, and E. D. Jackson. 1973. Origin of the Hawaiian Islands. *Am. Sci.* 61: 294-308.

Fellers, J. H., and G. M. Fellers. 1981. The status and distribution of ants in the crater district of Haleakala National Park. *Pacific Sci.* 36: 427-437.

Gagne, W. C. 1979. Canopy-associated arthropods in *Acacia koa* and *Metrosideros* tree communities along an altitudinal transect on Hawaii Island. *Pacific Insects* 21: 56-82.

Gillespie, R., and N. J. Reimer. 1993. The effect of alien predatory ants (Hymenoptera: Formicidae) on Hawaiian endemic spiders (Araneae: Tetragnathidae). *Pacific Sci.* (in press).

Hardy, D. E. 1979. An ecological survey of Puaaluu Stream. Part III. Report on a preliminary entomological survey of Puaaluu Stream, Maui. Coop. Nat. Park Resources Study Unit. Univ. of Hawaii, Manoa. Tech. Rept. 27: 34-39.

Hölldobler, B., and E. O. Wilson. 1990. The ants. The Belknap Press of Harvard Univ. Press, Cambridge, Massachusetts.

Holt, R. A., and B. Fox. 1985. Protection status of the native Hawaiian biota. Pp. 127-141. In: C. P. Stone and J. M. Scott [eds.]. *Hawaii's Terrestrial Ecosystems Preservation and Management.* Univ. of Hawaii Press, Honolulu.

Howarth, F. G. 1985. Impacts of alien land arthropods and mollusks on native plants and animals in Hawaii. Pp. 149-179. In: C. P. Stone and J. M. Scott [eds.]. *Hawaii's Terrestrial Ecosystems Preservation and Management.* Univ. of Hawaii Press, Honolulu.

_____. 1990. Hawaiian terrestrial arthropods: an overview. *B. P. Bishop Museum Occ. Pap.* 30: 4-26.

Huddleston, E. W., and S. S. Fluker. 1968. Distribution of ant species of Hawaii. *Proc. Hawaii. Entomol. Soc.* 20: 45-60.

Medeiros, A. C., L. L. Loope, and F. R. Cole. 1986. Distribution of ants and their effects on endemic biota of Haleakala and Hawaii Volcanoes National Park: a preliminary assessment. Pp. 39-52. Proc. 6th Conf. Nat. Sci., Hawaii Volcanoes National Park.

Moore, N. W., and W. C. Gagne. 1982. *Megalagrion pacificum* (McLachlan): A preliminary study of the conservation requirements of an endangered species. Report Odonata Specialists Group, International Union for Conservation of Nature and Natural Resources, No. 3. Ultrecht.

Mueller-Dombois, D., K. W. Bridges, and H. L. Carson. 1981. Island ecosystems: Biological organization in selected Hawaiian communities. Vol. 15. US/IBP Synthesis Series. Hutchinson Ross Pub. Co., Stroudsburg, PA.

Perkins, R. C. L. 1913. Introduction. Fauna Hawaiiensis 1(6): i-ccxxvii(xli-xlii).

Reimer, N. J., J. W. Beardsley, and G. Jahn. 1990. Pest ants in the Hawaiian Islands. Pp. 40-50. In: R. K. Vander Meer, K. Jaffe, and A. Cedeno [eds.]. *Applied Myrmecology: A world perspective.* Westview Press, Boulder, CO.

Simon, C. 1987. Hawaiian evolutionary biology: an introduction. Trends in Ecology and Evolution 2:175-178.

Smith, C. W. 1985. Impact of alien plants on Hawai'i's native biota, pp. 180-250. In: C. P. Stone and J. M. Scott [eds.]. *Hawaii's Terrestrial Ecosystems Preservation and Management.* Univ. of Hawaii Press, Honolulu.

Solem, A. 1976. Endodontid land snails from the Pacific islands (Mollusca: Pulmonata: Sigmurethra). Part I: Family Endodontidae. Field Mus. Nat. Hist. Chicago.

Wagner, W. L., Herbst, D. R., and R. S. N. Yee. 1985. Status of the native flowering plants of the Hawaiian Islands. Pp. 23-74. In: C. P. Stone and J. M. Scott [eds.]. *Hawaii's Terrestrial Ecosystems Preservation and Management.* Univ. of Hawaii Press, Honolulu.

Zimmerman, E. C. 1948. Insects of Hawaii. Vol 1. Introduction. xv. Univ. of Hawaii Press, Honolulu.

_____ . 1970. Adaptive radiation in Hawaii with special reference to insects. *Biotropica* 2(1): 32-38.

3

Characteristics of Tramp Species

Luc Passera

Introduction

There are a number of economically important ant species, the most significant of which are certainly the fire ants (Adams 1986, Lofgren 1986) and the leaf cutter ants (Cherrett 1986, Fowler et al. 1986), all limited to the American continents. Because of the damage they do to the environment and crop production, and the danger they pose to human health, these ants certainly warrant the term "pest ants". Among pest ants, attention in the last few years has focused on several species that exhibit a group of characters that has led to the term "tramp species". According to Hölldobler and Wilson (1990, p. 215), tramp species have the following characteristics: they are polygynous, unicolonial, they reproduce by budding, are largely dispersed throughout the world by human commerce and live in close association with humans. In this review the definition of tramp species is refined in light of recent work on three main species: the Pharaoh ant, *Monomorium pharaonis*, the Argentine ant, *Linepithema humile*, formerly *Iridomyrmex humilis*, and the little fire ant, *Wasmannia auropunctata*. Other less well-known species will be briefly discussed in order to better understand the strategies adopted by the principal pest species.

The Human Environment

Linepithema humile: since its description from the Buenos Aires area in 1868, this dolichoderine has spread throughout the world touching all the continents: the southern USA, starting with Louisiana, Mississippi, Alabama and California around 1891 (Newell and Barber 1913), South Africa in 1908 (Skaife 1955), Europe about 1904-1905 (Chopard 1921),

Australia in 1939, without sparing islands like Madeira or Hawaii (Wilson and Taylor 1967). Each time human commerce was the cause of introduction. In France it was apparently introduced with tropical plants (orchids and ferns) imported from South America to green houses in the Côte d'Azur. In South Africa it is suspected that trade with Argentina involving animal forage during the Boer War was responsible for the introduction. In Polynesia it followed the movement of troops during the second world war. Everywhere this species has invaded, it lives in close association with human habitations. For example, in the Languedoc-Roussillon region of France it is found in seaside resorts (Port-Leucate and La Grande Motte), but it is totally absent from the beaches that separate these resorts (Passera, unpublished). It appears to prefer areas modified or disturbed by human activity, such as cultivated fields, landscaped areas, garbage dumps, plaster rubble and dwellings.

Monomorium pharaonis: this cosmopolitan species probably originate from primary virgin forest of the Ethiopian zone (Bernard 1968). It is found in all tropical regions where it inhabits both natural outdoor situations and the inside of buildings (Wilson and Taylor 1967). This species has colonized temperate regions throughout the entire world, where it almost always lives inside houses. In Europe, for example, it is found especially in northern countries (Great Britain, Denmark, Belgium, Switzerland, Germany, Czechoslovakia, etc.) or in southern countries in which the altitude exceeds 1000 m, such as the principality of Andorra (Eichler 1978, Edwards 1986). Curiously, it seems less common in southern Europe because in 1976 it was not yet reported from Greece or Spain (Eichler 1976). Even though buildings are heated for a longer period each year in northern Europe and in countries with harsh climates like Andorra, perhaps *M. pharaonis* finds the short winter season in southern Europe unfavorable, because buildings are not heated as well as in northern Europe. It generally inhabits only the best-heated buildings: hospitals, houses with children, retirement homes, etc. Only exceptionally does it seem able to complete its lifecycle outdoors, e.g., in garbage dumps that generate heat (Kohn and Vlcek 1984). The Pharaoh ant is also found on the American continent (USA and Canada), Africa (Guinea), Asia (Japan) and in Australia. Like the Argentine ant, there is no doubt that it has spread through human commerce; we have found it in Toulouse in the packing carton of an electron microscope originating from Japan. At least two other species of *Monomorium, M. floricola* from Asia (Emery 1921) and *M. destructor* from Africa, which have pantropical distributions (Wilson and Taylor 1967), could also be considered tramp species.

Wasmannia auropunctata: this species originated from tropical America. It has colonized Central America, South America (Kempf 1972), North America (Florida and Canada) (Ayre 1977), Cameroon in Africa (Bruneau de Miré 1969), numerous islands in the Caribbean, New Caledonia, Wallis and Futuna Islands and the Galapagos Islands (Silberglied 1972; Ulloa-Chacon 1990). It prefers cultivated areas (citrus orchards, coffee fields, cacao plantations) and also buildings where it forages on various food sources (Smith 1965).

Other species: based only on the criterion of dispersal by human commerce, Wilson and Taylor (1967) counted 37 species of tramp ants in Polynesia. Several of these don't appear to merit the title of tramp species because they don't satisfy all of the conditions considered here (in particular, the absence of nuptial flight, intranidal mating and colony reproduction by budding). Others, *Tetramorium caespitum* or *Solenopsis geminata,* warrant special mention.

Cardiocondyla: *C. emeryi* and *C. nuda* originate from Africa and are found throughout the tropics (Wilson and Taylor 1967). *C. nuda* occurs in Malaysia, Oceana, India, Madagascar, northern Africa, Egypt, Cyprus, Australia and Polynesia (Bernard 1956). *C. emeryi* is known to occur in the Antilles, Madeira, Syria, Congo, Sudan and Madagascar (Bernard 1956), and South Africa where it causes damage in citrus orchards (Samways 1990). *C. wroughtoni* originates from tropical Asia and is known throughout the tropics (Wilson and Taylor 1967, Reimer et al. 1990) but is also found in Florida (Stuart et al. 1987), Japan (Kinomura and Yamauchi 1987) and Israel (Lupo and Galil 1985).

Pheidole megacephala: this species probably originates from Africa and has conquered almost all of the humid tropics (Wilson and Taylor 1967). It is an important pest in South America (Fowler et al. 1990). In South Africa it is a big problem in buildings (Prins et al. 1990). In Hawaii it tends scale insects on coffee (Reimer et al. 1990). In Morocco it inhabits the sidewalks of cities (Bernard 1968). In Europe it is found in heated greenhouses (Bernard 1968).

Paratrechina longicornis: this "crazy ant" is abundant in pantropical cities. It probably originates from the old world tropics (Wilson and Taylor 1967), but now occurs in Polynesia (Wilson and Taylor 1967, Reimer et al. 1990), the southern USA (Smith 1965, Trager 1984), in Cameroon (Dejean, personal communication), Brazil (Banks and Williams 1989) and South Africa (Prins et al. 1990). Depending on the climate, it lives outdoors or in heated apartments. In Europe, *P. longicornis:* is found in greenhouses (Bernard 1968).

Anoplolepis longipes: this "crazy ant" originating from Africa (Wilson and Taylor 1967) has invaded the old world tropics: Polynesia (Wilson and Taylor 1967), New Guinea (Baker 1976), several islands in the

Indian Ocean including Zanzibar, Mauritius, Reunion and the Seychelles (Haines and Haines, 1978a), Malaysia (Fiedler 1989), Sri Lanka, Myanmar (Burma), India (Veeresh 1990) and South Africa (Prins et al. 1990). *A. longipes* prefers cultivated fields (sugar cane, cacao, coffee), orchards (mangos, citrus) and dwellings.

Tapinoma: T. melanocephalum is a small species with a black head whose origin is not precisely known (Wilson and Taylor 1967). It is largely distributed in the tropics and subtropics of the world, where it occurs in association with humans (Smith 1965). It can be found in heated locations in northern regions such as Canada (Ayre 1977) and Germany (Steinbrink 1987). It is very common in the southern USA, Central America and South America (Kempf 1972; Nickerson and Bloomcamp 1988; Fowler et al. 1990; Harada 1990).

Lasius neglectus: this species is provisionally included in this list because it is perhaps the most recent member of the tramp ants. At present it has only been found in the city of Budapest. It nests in human structures and also in the soil (Van Loon et al. 1990). It is of unknown origin, but it is certain that it was accidentally imported by humans (Van Loon et al. 1990). Its status as a true tramp ant will be determined later by seeing if its distribution is linked to human activity.

Migration

The attraction of tramp ants to unstable and perturbed environments linked to human activity explains their strong tendency to move. As noted by Wilson (1971, p. 311), species observed in captivity exhibit a permanent nervousness. At the least disturbance (shock, vibration, manipulation of the nest, lighting, etc.) workers of *L. humile* pick up larvae and attempt to flee. In a natural setting, colonies of the Argentine ant relocate in response to changes in weather and/or human activity. This phenomenon was noted by Newell and Barber (1913) and Markin (1968) in the USA. In France, it has been well-studied by Benois (1973) on the Côte d'Azur, and we have verified this at Port-Leucate in the Roussillon near Perpignan. At the onset of winter, several colonies fuse to form large concentrations of ants that collect in places with southern exposure, including the base of walls, tree roots and the edges of sidewalks. In the spring the colonies split and seek out areas that are more shaded and humid. The spring and summer colonies are extremely mobile and react immediately to physical disturbance (trash disposal, gardening. etc.), to changes in weather conditions (increase or decrease in humidity) or to dietary modifications (exhaustion or discovery of a food source).

M. pharaonis, which habitually nests in buildings, also readily undergoes migrations in response to over-population or changes in nest site (Peacock et al. 1955a, Edwards 1986).

In Colombia and in the Galapagos, *W. auropunctata* prefers leaf litter, an unstable habitat that favors frequent migrations (Ulloa-Chacon 1990). In Israel *C. wroughtoni* was introduced inside figs of *Ficus sycomorus,* where it utilizes a cavity only 5 to 8 mm in diameter. It is also found in insect galls (Lupo and Galil 1985). Such habitats are very short lived requiring their occupants to change nest sites frequently.

The readiness with which *A. longipes* migrates has been noted by Baker (1976) who was able to trap entire colonies by placing bamboo stems in infested cacao plantations which were then rapidly colonized.

T. melanocephalum, which frequently nests in unstable and temporary habitats (plant stems, clumps of dried grass, debris), rapidly changes nest sites when conditions become unfavorable (Nickerson and Bloomcamp 1988).

Unicoloniality

According to Wilson (1971), unicolonial species are characterized by the absence of aggressive behavior between individuals from different nests occurring in one area. It is this absence of aggressivity that is responsible for the exchange of individuals between the different nests. These societies are called "open" in the sense of Le Masne (1952), as opposed to the "closed" societies that are typical of multicolonial species.

Unicoloniality is well-established in *L. humile.* As early as 1913, Newell and Barber remarked that even though colonies of this species are very intolerant toward other species, there is no antagonism between different colonies. In fact, in a heavily infested area it is impossible to distinguish among colonies, because all appear to belong to a single and enormous colony. This phenomenon was confirmed by Markin (1968) in California. It is identical in France: the site of Port--Leucate constitutes an immense society occupying some 700 by 300 meters. It is even possible to mix individuals from sites separated by distances of several hundred kilometers (Bonavita-Cougourdan 1988; Passera and Aron, unpublished data). This does not mean that there is an absence of colony recognition. Kaufmann and Passera (1991) showed that workers spend more time examining non-nestmate conspecifics coming from distant colonies. The absence of aggression applies equally well to mated queens. By contrast, males are attacked and killed when they are introduced into a queenright colony even though they are perfectly accepted in colonies that are deprived of a queen (Kaufmann and Passera 1991). This indicates that the openness of unicolonial societies has certain limits.

The different nests of the polydomous societies of the Pharaoh ant are open and individuals can be exchanged without aggression (Peacock et al. 1950, Edwards 1986, Berndt and Eichler 1987). Nevertheless, just as in the Argentine ant, certain individuals such as queens, can be attacked depending on their physiological condition (Petersen-Braun 1982).

W. auropunctata exhibits no aggressivity toward conspecifics: colonies collected in the field can be mixed without any problem in the laboratory (Clark et al. 1982, Ulloa-Chacon 1990). Observed seasonal variation in the composition of natural colonies, moreover, shows colony fusions alternate with colony divisions.

Less is known about other tramp species, but it seems they show the same tendency. For example, colonies of *T. melanocephalum* show no signs of aggression when mixed (Smith 1965) or when individuals are exchanged along foraging trails (Oster and Wilson 1978). In a same way in *A. longipes* individuals from different nests do not exhibit aggression towards each other (Haines and Haines 1978b)

Interspecific Aggression

The absence of intraspecific aggression seems to go along with strong interspecific aggressivity. Because tramp species are most often imported, they collide with native species that they must drive off. This is certainly the case with *P. megacephala* (Haskins and Haskins 1965, Fowler 1988). Undoubtedly the best known example is that of the Argentine ant. Several authors have studied the decline of native species in communities colonized by *L. humile* in Australia (Pasfield 1968) and in California (Erickson 1971, Tremper 1976). On the coastline of Languedoc-Roussillon in France, the Argentine ant has eliminated all the local species, in particular, *Tetramorium* spp. and *Tapinoma erraticum,* which are dominant in this area. Only a few colonies of *Plagiolepis pygmaea* and *Diplorhoptrum fugax* persist (Passera 1977, unpublished data). The more thorough study of Ward (1987) was conducted in different natural habitats in the Sacramento valley in California. This author found that even though *L. humile* alters the ant fauna in every locality, (16 of 27 species disappeared from the areas colonized by *L. humile),* there were differences among the species in the degree affected. Epigaeic species were more affected than hypogaeic species; less obvious species, like *Stenamma diecki* or *S. californicum* that forage in the soil and in the litter, are very resistant. Even among the epigaeic species, the extent of the effect depends on whether the species has a biology similar to that of the Argentine ant: *Liometopum occidentale* and *Tapinoma sessile,* which are dominant and opportunistic species that forage under the same conditions as *L. humile* are eliminated, whereas *Prenolepis imparis,* which forages at lower temperatures, remain present.

Strong competition can occur between two tramp species. This situation has been well-studied in Bermuda by Crowell (1968), Haskins and Haskins (1965), Fluker and Beardsley (1970) and Lieberburg et al. (1975). *P. megacephala* became established first around 1902 in displacing the other species present. At the time of the arrival of *L. humile* around 1953, it had conquered the whole island. The introduction of *L. humile* was followed by intense competition between these two tramp species, and about 1975 the Argentine ant seemed on track to eliminate its rival. However, the latest observation of Haskins and Haskins (1988) lead one to think that the situation is reaching an equilibrium, because *P. megacephala* had retaken several areas.

W. auropunctata also exhibits a remarkable level of aggression. Several studies have tracked its spread and its effects on the local ant fauna following its colonization of the Galapagos Islands (Silberglied 1972; Clark et al. 1982). Lubin (1984) estimates that only four of 29 species belonging to the genera *Solenopsis, Hypoponera* and *Strumigenys* (the last two genera are subterranean) are able to persist in its presence.

The recent introduction of *Lasius neglectus* into central Budapest has resulted in a reduction of the local ant fauna (Van Loon et al. 1990); in a period of 15 years, 17 local species disappeared or almost disappeared.

Polygyny

Tramp species are characterized by true polygyny, i.e., nestmate queens show no signs of hostility toward one another that could be construed as dominance. For example, in *L. humile,* although there is strong variation in individual oviposition rates, there is no clear dominance hierarchy (Keller 1988). The number of resident queens is high, although it is difficult to quantify due to the absence of clear boundaries between colonies. Another reason queen number is difficult to quantify is the fact that there is often seasonal variation linked to the production of new queens. In *L. humile,* queen number varies with season from 0.1 to 1.6 queens per 100 workers (Keller et al. 1989). Because the number of workers is often enormous, the number of queens can be immeasurable. Thus Horton (1918) estimated collecting 1,307,222 queens in one year in a 10-hectare orchard (6.3 queens per 100 workers). In *M. pharaonis,* Peacock et al. (1955b) studied colonies by

FIGURE 3.1. Average number of queens per 100 workers for several tramp species. The extreme values are represented by a bar. Data are from studies referenced in the text.

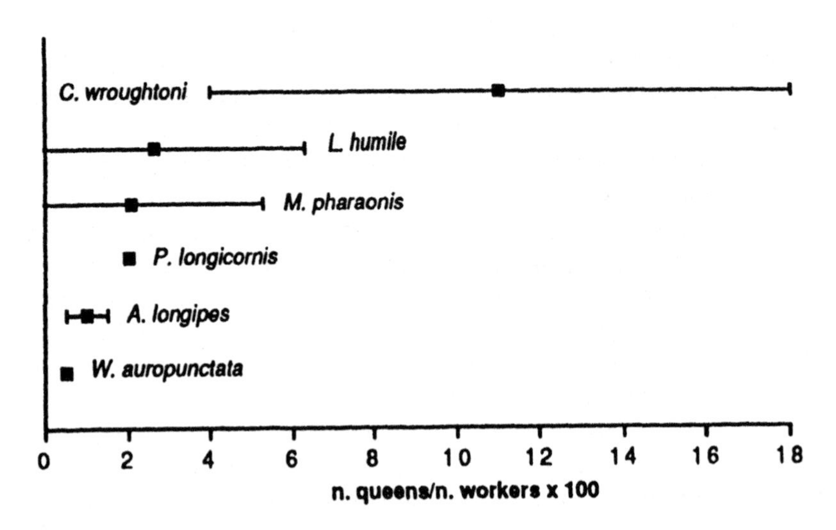

trapping in buildings and found from 0.8 to 5.3 queens per 100 workers, and they found a maximum of 110 queens in a single nest. Sequential trapping over a year by Bellevoye (1889) yielded 1,360,000 workers and 1,809 queens (or 0.13 queens per 100 workers). In *W. auropunctata,* Ulloa-Chacon (1990) found a mean of 1 to 13 queens per nest. The nests of this species range from small, containing only a few hundred workers, to very large with thousands of workers. Baker (1976) counted a few dozen queens and 10- to 20,000 workers in colonies of *A. longipes* in wet tropical forests and Haines and Haines (1978b) found up to 300 queens per nest in the Seychelles. *P. longicornis* nests contain about 2000 workers and 40 queens (Mallis 1982). The largest proportion of queens seems to be found in *C. wroughtoni,* which forms colonies containing up to 18 queens and a few dozen workers (Kinomura and Yamauchi 1987). In Figure 3.1 all these data have been converted to number of queens per 100 workers. The relationship is evidently highly variable, ranging from 0.1 to 11 but often surpasses 1.

Mating and Budding

These are undoubtedly the most important characters because, they are responsible for the unique dispersal of these species. A number of tramp ants, in which sexuals are normally winged, seem to have lost the capacity to undergo mating flights. Mating occurs inside the nest, and colony reproduction occurs via budding, i.e., departure on foot of workers and mated queens which will establish a new nest several meters away. This form of spread is slow and thorough and favors strong linkage among the different nests, but it does not allow colonization of distant areas. Long distance dispersal occurs only by passive transport, either natural (nests of the Argentine ant are transported along the Mississippi by pieces of driftwood; Newel and Barber 1913) or artificially by human commerce.

In this regard, the behavior of the Argentine ant is well-known: winged queens have not been observed to participate in nuptial flights by a number of authors (Skaife 1955, Markin 1970a, Benois 1973, Giraud 1982, Bartels 1983, Keller and Passera 1988) or only under exceptional conditions by others (Newell and Barber 1913; Passera and Vargo, unpublished data). By contrast, there are numerous accounts of males flying short distances, especially at dusk (Newell and Barber 1913; Skaife 1955; McCluskey 1963; Benois 1973; Giraud 1982; Bartels 1983; Keller, Passera and Vargo, unpublished data). Mating takes place in the nest, with the young virgin queens being inseminated either by foreign or nestmate males.

M. pharaonis behaves in a very similar manner; there is no nuptial flight and mating takes place in the nest (Peacock et al. 1950, Petersen and Buschinger 1971a, Edwards 1986, Berndt and Eichler 1987). Peacock et al. (1955a) studied colony reproduction under both natural and laboratory conditions. The smallest observed colony possessed one dealate queen, 35 workers and brood of all ages. Using traps these authors obtained colonies, the smallest of which contained three dealate queens, 150 workers and brood of all ages. Finally in the laboratory, these authors did not succeed in initiating new colonies by independent colony founding, not even by pleometrosis, although they succeeded in starting a new colony by isolating a single queen, several dozen workers and brood. However, Petersen and Buschinger (1971b) succeeded in starting new colonies beginning with a young queen in isolation with brood, and even partially succeeded in initiating independent colony founding by a single queen in isolation (a few workers were produced but the colony died a short time after). It is clear that throughout the world, new colonies of *M. pharaonis* originate from budding of larger colony fragments: Peacock et al. (1955b) demonstrated this experimentally by observing that a colony divided up into several fragments when small nests were connected to the principal nest.

In *W. auropunctata*, Ulloa-Chacon and Cherix (1989) observed mating inside laboratory nests, and, based on the behavioral repertoire of this species, it seems that nuptial flights are absent (Lubin 1984).

Independent colony founding by young queens is not possible, supporting the hypothesis that budding is the main mode of colony founding (Ulloa-Chacon 1990).

The observations of Van Loon et al. (1990) regarding *L. neglectus* are similar; no nuptial flight has been seen in the city of Budapest since the appearance of this new species and it seems that mating occurs in the nest followed by budding.

The other tramp ants are less well-studied. *M. floricola* must reproduce by budding with intranidal mating, because the queens are wingless (Smith 1965). The observations of Trager (1984) on *P. longicornis* are also worth mentioning. Males exit and assemble at nest entrances. Females then leave the nests and are undoubtedly inseminated on the ground near the nest entrance. In any case, Trager concludes that there is no nuptial flight. Given these conditions, colony reproduction can only occur via budding. In other species, the occurrence of a nuptial flight is possible, but intranidal mating has been observed, and it is not known what proportion of females fly or mate in the nest. This is the case with *C. wroughtoni*. Both winged and wingless males are found in this species, and both mate inside the nest with young winged females (Kinomura and Yamauchi 1987, Stuart 1990). Budding is undoubtedly the normal mode of colony reproduction, because attempts at haplometrosis fail (Stuart 1990). Nevertheless, it is possible in the laboratory to rear colonies successfully from groups of females (Stuart 1990). The biology of *A. longipes* is poorly understood; alates are able to fly but they were never definitely seen engaged in a nuptial flight so Haines and Haines (1978b) estimate that sexuals mate in the nest thus spread by budding. This is also the opinion of Veeresh (1990). Moreover, the number of dealate queens increases after sexuals eclose (Baker 1976).

Information is even more sketchy or non-existent for other species. It is not known whether mating flights occur in *T. melanocephalum* (Harada 1990), but it is known that new colonies are formed by budding (Smith 1965). Mating flights of *P. megacephala* have been seen in Hawaii (Wilson and Taylor 1967), but colony reproduction occurs via budding (Hölldobler and Wilson 1990). Trager (1984) observed sexuals of *P. bourbonica* flying, but the mode of colony foundation is not known.

Size and Monomorphism

Tramp ants are always very small in size; the smallest seems to be *T. melanocephalum* (worker length < 1.5 mm), *W. auropunctata* and *M. floricola* (worker length < 2 mm). All other species are smaller than 3.5 mm (Figure 3.2). The queens of course, are larger, but only rarely exceed

FIGURE 3.2. Mean body length of workers and queens of some tramp species. The extreme values are represented by a bar. Data are from studies referenced in the text and from personal measurements.

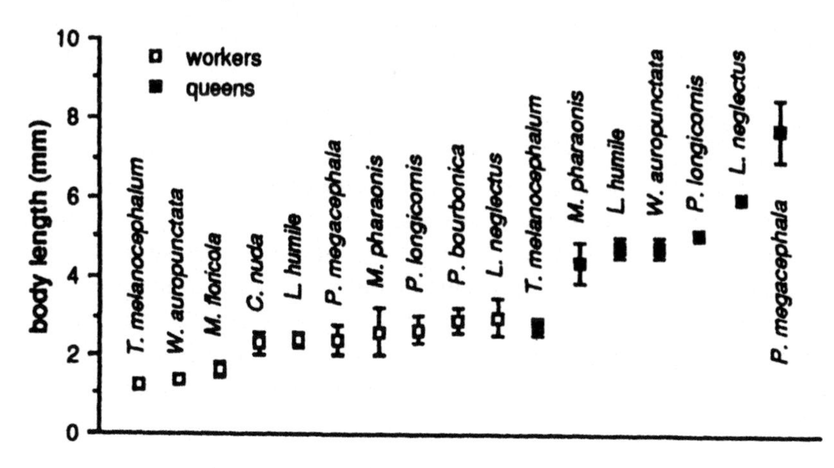

6 mm in length in *L. neglectus* and *P. megacephala.* All species, with the notable exception of *P. megacephala,* have monomorphic workers.

Lifespan of Queens

In general, queens of polygynous species are shorter lived than queens of monogyne species (Keller and Passera 1990). In the case of tramp ants, their lifespan can be extremely short. The record may belong to *T. melanocephalum,* in which, according to Harada (1990), queens live only a few weeks. The lifespan of queens of *M. pharaonis* has been well-studied; it does not surpass 29 weeks according to Petersen-Braun (1975), 38 weeks according to Edwards (1987), 39 weeks according to Peacock and Baxter (1950), 49 weeks according to Kretzschmar (1973). It is therefore considerably less than 12 months, but a few individuals may live a little longer than a year (Edwards 1986).

In *L. humile* the potential lifespan of queens is greater, but 90% of them are executed by their own workers in April or May. Since they eclose in June of the previous year, the effective lifespan is less than a year (Keller and Passera 1990).

The lifespan of queens of *W. auropunctata* is similar, because it barely exceeds a year: 11 to 14 months.(Ulloa-Chacon and Cherix 1989, Ulloa-Chacon 1990). Information on queen lifespan is lacking for other species.

A short queen life may seem disadvantageous, but it is compensated for by the large capacity of tramp species to produce and rear new queens. In the Pharaoh ant new queens can be produced all year long; it is only necessary to isolate a few dozen workers with a little brood in order to rear sexuals no matter what the season (Peacock et al. 1955a; Petersen-Braun 1975; Edwards 1986, 1987). The Argentine ant is almost as plastic. Although sexuals are naturally produced in springtime and it seems advantageous to produce sexuals seasonally, small queenless colony fragments in the laboratory almost always yield sexuals no matter what season the colonies are collected (Passera et al. 1988a, Vargo and Passera 1992). Ulloa-Chacon (1990) investigated for a period of 16 months the nest contents of *W. auropunctata* in Colombia. Except for December, winged female sexuals were present throughout the year. Trager (1984) reports that *P. longicornis* is capable of rearing sexuals throughout the year in warm regions, but it is limited to the period from May to September in cooler climates. In the Seychelles sexuals of *A. longipes* occurred throughout the year (Haines and Haines 1978b). It is evident that tramp species often succeed in adjusting to seasonal cycles.

Worker Sterility

Worker sterility may be a common character among tramp ants. This has been verified for *L. humile* by several authors (Markin 1970b, Benois 1973, Bartels 1983, Giraud 1983, Keller 1985, Passera et al. 1988b). This fact is equally well-established for *M. pharaonis;* Edwards (1986) and Berndt and Eichler (1987), among others, have established that the workers are sterile. Studies of *W. auropunctata* suggest a similar situation (Clark et al. 1982, Ulloa-Chacon and Cherix 1988); the latter authors reared queenless colonies for 80 days without observing oviposition. In *P. longicornis.* Aron and Passera (unpublished data) have held workers queenless for several weeks without seeing eggs appear.

Data are lacking for other species. One can speculate that workers of *P. megacephala* are sterile, because workers of this genus are reported not to lay eggs (Choe 1988). It will be interesting to study the situation in *L. neglectus,* because it belongs to a genus in which worker egg-laying is common (Choe 1988).

Discussion

Of the nine characteristics reviewed, the most exclusive may be the strong relationship tying tramp species to humans, whether it be for nest sites or long distance transport. A number of other ant species also exhibit an anthropophilic tendency (Bernard 1958, 1974; Smith 1965; Kondoh 1976; Pisarski 1982; Prins et al. 1990; Thompson 1990), but only tramp species are truly domestic since they follow humans in their movements. This anthropophilic tendency is accompanied by frequent nest changes. Nesting in the human environment implies utilization of very unstable habitats subject to frequent changes due to human activity, requiring tramp species to migrate often. Migration is by no means a unique attribute of tramp species. Nomadism is well-known in army ants in which it is linked to foraging for prey (Gotwald 1982). Changing of nests can also be provoked by intra- or interspecific competition, as in several species of *Pogonomyrmex* (Hölldobler 1976). Finally it may be linked, as in tramp species, to environmental instability. This is the case in *Leptothorax rugulatus* (Möglich 1978).

Small size and monomorphism (with the exception of *P. megacephala*) are characteristics that tramp species share with many other species. This goes for worker sterility, because it is known that the worker caste is sterile in entire genera, e.g., *Tetramorium, Pheidole* and *Solenopsis* (Fletcher and Ross 1985).

Three other characteristics that seem well-established for tramp species (unicoloniality, polygyny and colony reproduction by budding) are found to different degrees in other species and, in particular, in polydomous ants, which can form supercolonies containing many nests often occupying large areas.

Unicoloniality and its attendant properties, the absence of aggression and of clear nest boundaries, allow workers to be exchanged. This has been found in several species that form polycalic colonies, such as *Tapinoma sessile* (Smith 1928), or supercolonies, such as *Formica yessensis* (Ito and Imamura 1974), *F. lugubris* (Cherix 1980), *F. truncorum* (Rosengren et al. 1985), *Lasius sakagamii* (Yamauchi et al. 1981) and *Pseudomyrmex venefica* (Janzen 1973). The absence of aggression in these polycalic species is limited to members of the supercolony, whereas in tramp species (at least in *L. humile*) the tolerance seems to extend beyond the limits of a single habitat.

The degree of polygyny is often very great in tramp species, but it is not necessarily larger than in polydomous species; there may be more than 200 dealate queens in a single nest of *F. yessensis* (Ito 1973), 1,080,000 for the entire supercolony (Higashi et Yamauchi 1979), and up to 3000 to 5000 dealate queens per nest in *Formica polyctena* (Gösswald 1951, Lange 1956). On the contrary, it should be noted that there can be species that are both polydomous and monogynous, e.g., *Oecophylla longinoda* (Hölldobler 1979) and *Paratrechina flavipes* (Ichinose 1986), although tramp species are always polygynous.

Queen longevity seems to be characteristically short in tramp species. This particularity may be without equal in polydomous ants, but it needs to be confirmed by studying more species. It is the same for the readiness with which new queens are reared in tramp species. In polydomous species, like most ants, production of sexuals is seasonal and occurs in strict accordance with climatic and social factors (Brian 1980).

Colony founding by budding is a general tendency of polydomous species, e.g., *T. sessile* (Smith 1928, Kannowski 1959), *F. yessensis* (Ito 1973; Higashi 1976, 1983), *F. polyctena* (Mabelis 1979), *F. lugubris* (Cherix 1981), *L sakagamii* (Yamauchi et al. 1981) and *Polyrhachis dives* (Yamauchi et al. 1987). Initiation of new colonies by budding is facilitated by the existence of intranidal mating, which can occur together with nuptial flight, as in *T. sessile* (Smith 1928, Kannowski 1959), *F. yessensis* (Ito 1973, Higashi 1983) and *F. lugubris* (Fortelius et al. 1990) or without nuptial flight, as in *L. sakagamii* (Yamauchi et al. 1981).

Thus tramp species are very similar to polydomous ants with which they share in common certain attributes (unicoloniality, colony reproduction by budding, and, significantly, reduced nuptial flight), although these attributes are more strongly expressed in tramp species. To these characteristics of polydomous species, tramp species add their privileged relationship with the human environment assuring their ecological success.

In conclusion, the following definition of tramp ants is proposed: tramp ants are ant species with small, sterile workers, widely distributed throughout the world by human commerce, living in close association with humans. Tramp ants are polygynous, unicolonial and exhibit a reduction or the absence of nuptial flight leading to colony reproduction by budding.

Acknowledgments

This study was supported by a grant from the CNRS (URA 664) and the "European Network of Research Laboratories grant N° 418 210 57". I am very grateful to E. L. Vargo and J. P. Lachaud for reviewing the manuscript and respectively for assistance with the English version and the Spanish summary.

Resumen

En este trabajo, tratamos de esclarecer las caracteristicas comunes de las "hormigas vagabundas". Para llegar a este objetivo, utilizamos lo que se conoce de las tres especies vagabundas las mas estudiadas (*Linepithema humile* [=*Iridomyrmex humilis*], *Monomorium pharaonis* y *Wasmannia auropunctata*) en conjunto con los datos relacionados a otras especies vagabundas menos conocidas de los generos *Cardiocondyla*. *Paratrechina*, *Pheidole* y *Anoplolepis*. Asi, nueve caracteres son revisados.

Aparte de las fuertes relaciones trabadas con el Hombre para su diseminacion, los demas caracteres se encuentran compartidos con otras hormigas. Sin embargo, es con las especies polidomas que las semejanzas parecen las mas fuertes: unicolonialidad, poliginia, fundacion de las colonias por estacas y tendencia al apareamiento dentro del nido se encuentran en los dos grupos no obstante es en las especies vagabundas que estas estrategias estan mejor realizadas. A fin de cuentas, las especies vagabundas se pueden definir como especies de pequeno tamano con obreras esteriles, ampliamente distribuidas en el mundo por intermedio del comercio humano, y viviendo en asociacion estrecha con el Hombre; estas especies son poliginias, unicoloniales y presentan una ausencia o una reduccion del vuelo nupcial lo que conduce a una reproducion de las colonias por estacas.

References

Adams, C. T. 1986. Agricultural and medical impact of the imported fire ant. Pp. 48-57. In: C.S. Lofgren and R.K. Vander Meer [eds.]. *Fire ants and leaf-cutting ants: Biology & Management* Westview Press, Boulder, CO.

Ayre, G .L. 1977. Exotic ants in Winnipeg Canada. *Manitoba Entomol.* 11: 41-44.

Baker, G. L. 1976. The seasonal life cycle of *Anoplolepis longipes* (Jerdon) (Hymenoptera: Formicidae) in a cacao plantation and under brushed rain forest in the northern district of Paua New Guinea. *Insect Soc.* 23: 253-62.

Banks, W. A. and D. F. Williams. 1989. Competitive displacement of *Paratrechina longicornis* (Latreille) (Hymenoptera Formicidae) from baits by fire ants in Mato Grosso Brazil. *J.Entomol. Sci.* 24: 381-91.

Bartels, P. J. 1983. Polygyny and the reproductive biology of the Argentine ant. *Ph. D. Thesis.* Univ. of California at Santa Cruz.

Bellevoye, M. A. 1889. Observations on *Monomorium pharaonis* Latr. *domestica* Schrenck *Life* 2: 230-233.

Benois, A. 1973. Incidence des facteurs écologiques sur le cycle annuel et l'activité saisonniére de la fourmi d'Argentine *Iridomyrmex humilis* (Mayr) (Hymenoptera, Formicidae), dans la région d'Antibes. *Insect Soc.* 20: 267-295.

Bernard, F. 1956. Révision des Fourmis palearctiques du genre *Cardiocondyla* Emery. *Bull. Soc. Hist. Nat. Afr. Nord.* 47: 299-306.

_____ . 1958. Fourmis des villes et Fourmis du bled entre Rabat et Tanger. *Bull. Soc. Sc. Nat. et Phys. Maroc* 36: 131-142.

_____ . 1968. Les Fourmis (Hymenoptera: Formicidae) d'Europe occidentale et septentrionale. *Faune de l'Europe et du Bassin méditerraneen.* Masson, Paris.

_____ . 1974. Les Fourmis des rues de Kenitra (Maroc) (Hym.). *Bull. Soc. entomol. France* 79: 178-183.

Berndt, K. P., W. Eichler. 1987. Die Pharaoameise, *Monomorium pharaonis* (L.) (Hym., Myrmicidae). *Mitt. Zool. Mus. Berlin* 63: 3-186.

Bonavita-Cougourdan, A. 1988. Contribution à l'étude des communications et de leur rôle dans l'organisation sociale chez la fourmi *Camponotus vagus* Scop. *Thése*, Marseille.

Brian, M. V. 1980. Social control over sex and caste in bees, wasps and ants. *Biol. Rev.* 55: 379-415.

Bruneau de Miré, P. 1969. Une fourmi utilisée au Cameroun dans la lutte contre les Mirides du cacaoyer, *Wasmannia auropunctata* (Roger). *Café, Cacao, Thé* 13: 209-212.

Cherix, D. 1980. Note préliminaire sur la structure, la phénologie et le régime alimentaire d'une super-colonie de *Formica lugubris* Zett. *Insect Soc.* 27: 226-236.

____ . 1981. Contribution à la biologie et à l'écologie de *Formica lugubris* Zett. (Hymenoptera, Formicidae). Le probléme des super-colonies. *Thése*, Lausanne.

Cherrett, J. M. 1986. The economic importance and control of leaf-cutting ants. Pp. 165-192. In: S.B. Vinson [ed.]. *Economic Impact and Control of Social Insects*. Praeger, New York.

Choe, J. C. 1988. Worker reproduction and social evolution in ants (Hymenoptera: Formicidae). Pp. 163-187. In: J.C. Trager [ed.]. *Advances in Myrmecology*. E.J. Brill, Leiden.

Chopard, L. 1921. La fourmi d'Argentine *Iridomyrmex humilis* var. *arrogans* Santschi dans le Midi de la France. *Ann. Epiphyties* 7: 237-266.

Clark, D. B., C. Guayasamín, O. Pazmino, C. Donoso and Y. Paez de Villacís. 1982. The tramp ant *Wasmannia auropunctata*: Autecology and effects on ant diversity and distribution on Santa Cruz Island, Galapagos. *Biotropica* 14: 196-207.

Crowell, K. L. 1968. Rates of competition exclusion by the Argentine ant in Bermuda. *Ecology* 49: 551-555.

Edwards, J. P. 1986. The biology, economic importance and control of the Pharaoh's ant *Monomorium pharaonis* (L.). Pp. 257-271 In: S. B. Vinson [ed.]. *Economic Impact and Control of Social Insects*. Praeger, New York.

____ . 1987. Caste regulation in the pharaoh's ant *Monomorium pharaonis*: the influence of queens on the production of new sexual forms. *Physiol. Entomol.* 12: 31-39.

Eichler, W. 1976. Distribution, spread and hygienic importance of the Pharaoh's ants in Europe. *Proc. II Intern. Symp. IUSSI* (Varsovie), pp. 19-20.

____ . 1978. Die Verbreitung der Pharaoameise in Europa. *Memorabilia Zool.* 29: 31-40.

Emery, C. 1921. Hymenoptera: Formicidae: Myrmicinae. *Genera Insectorum* 174: 1-397. (cited by Wilson and Taylor, 1967).

Erickson, J. M. 1971. The displacement of native ant species by the introduced Argentine ant, *Iridomyrmex humilis* Mayr. *Psyche* 78: 257-266.

Fiedler, K. 1989. Russeltrillern: eine neue Form taktiler Kommunikation zwischen Blaulingen und Ameisen. Nachr. ent. Ver. Apollo NF. 10:125-132.

Fletcher, D. J. C., and K. G. Ross. 1985. Regulation of reproduction in eusocial Hymenoptera. *Ann. Rev. Entomol.* 30: 319-343.

Fluker, S. S., and J. W. Beardsley. 1970. Sympatric associations of three ants: *Iridomyrmex humilis, Pheidole megacephala* and *Anoplolepis longipes* in Hawaii. *Ann. Ent. Soc. Am.* 63: 1290-1296.

Fortelius, W., D.J.C. Fletcher, and D. Cherix. 1990. Queen recruitment in polygynous and polydomous *Formica* populations. Pp.243-244. In: G.K. Veeresh, B. Mallick and C.A. Viraktamath [eds.]. *Social Insects and the Environment*. Oxford and IBH Publishing Co. (PVT.) Ltd., New Delhi.

Fowler, H. G. 1988. Eradication of the native ant fauna by the introduction of an exotic ant in Itapirica, Bahia, Brazil during hydroelectric dam construction. *Environ. Conserv.* (in press) (cited by Fowler et al., 1990).

Fowler, H. G., J. V. Bernardi, J. C. Delabie, L. C. Forti, and V. Pereira-da-Silva. 1990. Major ant problems of South America. Pp. 3-14. In: R.K. Vander Meer, K. Jaffe and A. Cedeno [eds.]. *Applied Myrmecology: a World Perspective*. Westview Press, Boulder, CO.

Fowler, H. G., L. C. Forti, V. Pereira-da-Silva, and N.B. Saes. 1986. Economics of grass-cutting ants. Pp. 18-35. In: C. S. Lofgren and R. K. Van der Meer [eds.]. *Fire Ants & Leaf-Cutting Ants; Biology & Management*. Westview Press, Boulder.

Giraud, L. 1982. Contribution à l'étude de la biologie d'*Iridomyrmex humilis* (Hym. Dolichoderinae). Principe d'une lutte intégrée. *Thése 3eme cycle*. Université de Paris V.

_____ . 1983. Rôle inhibiteur du facteur "groupement de reines" sur l'apparition des mâles chez *Iridomyrmex humilis* (Hymenoptera: Formicidae). *C. R. Acad. Sc. Paris*. 296, s, III: 655-658.

Gösswald, K. 1951. *Die Rote Waldameise im Dienste der Waldhygiene: Forstwirtschaftliche Bedeutung, Natzung, Lebenweise. Zucht. Vermehrung und Schutz*. Kinau Verlag, Luneburg. (cited by Higashi, 1976).

Gotwald, W. H. Jr. 1982. Army Ants. Pp. 157-254. In: H.R. Hermann [ed.]. *Social insects* IV. Academic Press, New York.

Haines, I. H., and J. B. Haines. 1978a. Pest status of the crazy ant, *Anoplolepis longipes* (Jerdon) (Hymenoptera: Formicidae), in the Seychelles. *Bull. Entomol. Res.* 68: 627-638.

_____ . 1978b. Colony structure, seasonality and food requirements of the crazy ant, *Anoplolepis longipes* (Jerd.), in the Seychelles. *Ecol. Entomol.* 3: 109-118.

Harada, A. Y. 1990. Ant pests of the Tapinomini tribe. Pp. 298-315. In: R. K. Vander Meer, K. Jaffe and A. Cedeno [eds.]. *Applied Myrmecology: a World Perspective*. Westview Press, Boulder, CO.

Haskins, C. P., and E. F. Haskins. 1965. *Pheidole megacephala* and *Iridomyrmex humilis* in Bermuda--Equilibrium or slow replacement? *Ecology* 46: 736-740.

_____ . 1988. Final observations on *Pheidole megacephala* and *Iridomyrmex humilis* in Bermuda. *Psyche* 95: 177-184.

Higashi, S. 1976. Nest proliferation by budding and nest growth pattern in *Formica (Formica) yessensis* in Ishikari Shore. *J. Fac. Sci. Hokkaido Univ*. Ser. VI, Zool. 20: 359-389.

_____ . 1983. Polygyny and nuptial flight of *Formica (Formica) yessensis* Forel at Ishikari coast, Hokkaido, Japan. *Insect Soc.* 30: 287-297.

Higashi, S., and K. Yamauchi. 1979. Influence of a supercolonial ant *Formica (Formica) yessensis* Forel on the distribution of other ants in Ishikari coast. *Jap. J. Ecol.* 29: 257-264.

Hölldobler, B. 1976. Recruitment behavior, home range orientation and territoriality in harvester ants, *Pogonomyrmex*. *Behav. Ecol. Sociobiol.* 1: 3-44.

_____ . 1979. Territories of the african weaver ant (*Oecophylla longinoda* Latr.). A field study. *Z. Tierpsychol.* 51: 201-213.

Hölldobler, B. and E. O.Wilson. 1990. *The ants.* Springer-Verlag, Berlin.

Horton, J. R. 1918. The Argentine ant in relation to citrus groves. *USDA Bull.* 647: 1-73.

Ichinose, K. 1986. Occurence of polydomy in a monogynous ant, *Paratrechina flavipes* (Hymenoptera, Formicidae). *Kontyu* 54: 208-217.

Ito, M. 1973. Seasonal population trends and nest structure in a polydomous ant *Formica (Formica) yessensis* Forel. *J. Fac. Sci. Hokkaido Univ.*, Ser. VI, Zool. 19: 270-293.

Ito, M., and S. Imamura. 1974. Observations on the nuptial flight and internidal relationship in a polydomous ant: *Formica (Formica) yessensis* Forel. *J. Fac. Sci. Hokkaido Univ.*, Ser. *VI*, Zool. 19: 681-694.

Janzen, D. H. 1973. Evolution of polygynous obligate acacia-ants in Western Mexico. *J. Anim. Ecol.* 42: 727-750.

Kannowski, P. B. 1959. The flight activities and colony-founding behaviour of bog ants in southeastern Michigan. *Insect Soc.* 6: 115-162.

Kaufmann, B., and L. Passera. 1991. Premiére approche du probléme de la reconnaissance coloniale chez *Iridomyrmex humilis* (Formicidae, Dolichoderinae). *Actes Coll. Insectes Sociaux.* 7: 75-82.

Keller, L. 1985. Etude de la monogynie experimentale et de ses implications chez une espéce de fourmi polygyne (*Iridomyrmex humilis* (Mayr)). *Travail de diplôme.* Université de Lausanne.

_____ . 1988. Evolutionary implications of polygyny in the Argentine ant, *Iridomyrmex humilis* (Mayr) (Hymenoptera: Formicidae): an experimental study. *Anim. Behav.* 36: 159-165.

Keller, L., and L. Passera. 1988. Energy investment in gynes of the Argentine ant *Iridomyrmex humilis* (Mayr) in relation to the mode of colony founding in ants (Hymenoptera: Formicidae). *Int. J. Invert. Repr.* 13: 31-38.

_____ . 1990. Fecundity of ant queens in relation to their age and the mode of colony founding. *Insect Soc.* 37: 116-130.

Keller, L., L. Passera, and J. P. Suzzoni. 1989. Queen execution in the Argentine ant *Iridomyrmex humilis* (Mayr). *Physiol. Entomol.* 14: 157-163.

Kempf, W. W. 1972. Catálogo abreviado das formigas da Região Neotropical (Hym.: Formicidae). *Stud. Entomol.* 15: 3-344.

Kinomura, K., and K. Yamauchi. 1987. Fighting and mating behaviors of dimorphic males in the ant *Cardiocondyla wroughtoni. J. Ethol.* 5: 75-81.

Kohn, M., and M. Vlcek. 1984. The outdoor persistence of *Monomorium pharaonis* (Hymenoptera: Formicidae) colonies in Czechoslavakia. *Acta Ent. Bohemoslov.* 81: 186-189.

Kondoh, M. 1976. A comparison among ant communities under anthropogenic environment. *Proc. II Intern. Symp. IUSSI* (Varsovie), pp. 33-35.

Kretzschmar, Kh. 1973. Untersuchungen zur Biologie der Pharaoameise (*Monomorium pharaonis* L.) an Laborkolonien im Hinblick auf Bekampfungsmoglichkeiten. *Diss. A.*, Pad. Hochschule Potsdam. (cited by Berndt and Eichler, 1987).

Lange, R. 1956. Experimentelle Untersuchungen über die Variabilität bei Waldameisen (*Formica rufa* L.). *Z. Naturforschg.* llb: 538-543.

Le Masne, G. 1952. Classification et caracteristiques des principaux types de groupements sociaux réalisés chez les Invertébrés. Pp. 19-70. In: CNRS [ed.]. *34eme Coll. Intern. CNRS: structure et physiologie des sociétés animales.* Paris. 1950.

Lieberburg, I., P. M. Kranz, and A. Seip. 1975. Bermudian ants revisited: the status and interaction of *Pheidole megacephala* and *Iridomyrmex humilis. Ecology.* 56: 473-478.

Lofgren, C. S. 1986. The economic importance and control of imported fire ants in the United States. Pp. 227-256. In: S.B. Vinson [ed.]. *Economic Impact and Control of Social Insects.* Praeger, New York.

Lubin, Y. D. 1984. Changes in the native fauna of the Galapagos Islands following invasion by the little red fire ant, *Wasmannia auropunctata. Biol. J. Linn. Soc.* 21: 229-242.

Lupo, A., and J. Galil. 1985. Nesting habits of *Cardiocondyla wroughtoni* Forel (1890). *Israel J. Entomol.* 19: 119-125.

Mabelis, A. A. 1979. Nest splitting by the red wood ant (*Formica polyctena,* Foerster). *Netherlands. J. Zool.* 29: 109-125.

Mallis, A. 1982. *Handbook of pest control.* Franzak & Foster Co., Cleveland, Ohio. (cited by Thompson, 1990).

Markin, G. P. 1968. Nest relationship of the Argentine ant, *Iridomyrmex humilis,* (Hymenoptera: Formicidae). *J. Kansas Entomol. Soc.* 41: 511-516.

_____ . 1970a . The seasonal life cycle of the Argentine ant, *Iridomyrmex humilis,* (Hymenoptera: Formicidae) in southern California. *Ann. Entomol. Soc. Am.* 63: 1238-1242.

_____ . 1970b. Food distribution within laboratory colonies of the Argentine ant, *Iridomyrmex humilis* (Mayr). *Insect Soc.* 17: 127-157.

McCluskey, E. S. 1963. Rhythms and clocks in harvester and Argentine ants. *Physiol. Zool.* 36: 273-292.

Möglich, M. 1978. Social organization of nest emigration in *Leptothorax* (Hym. Form.). *Insect Soc.* 25: 205-225.

Newell, W., and T. C. Barber. 1913. The Argentine ant. *USDA Bur. Entomol. Bull.* 122: 1-98.

Nickerson, J. C., and C. L. Bloomcamp. 1988. *Tapinoma melanocephalum* (Fabricius) (Hymenoptera: Formicidae). *Florida Dept. of Agr. Cons. Ser., Div. Plant Ind. Entomol. Cir.* 307: 1-2.

Oster, G. and E. O. Wilson. 1978. *Caste ecology in the social insects.* Princeton Univ. Press, Princeton, N. J.

Pasfield, G. 1968. Argentine ants. *Aust. Nat. History.* 16: 12-15.

Passera, L. 1977. Peuplement myrmécologique du cordon littoral du Languedoc-Roussillon: Modifications anthropiques. *Vie et Milieu* 27: 249-265.

Passera, L., L. Keller, and J. P. Suzzoni. 1988a. Queen replacement in dequeened colonies of the Argentine ant *Iridomyrmex humilis* (Mayr). *Psyche* 95: 59-66.

_____ . 1988b. Control of brood male production in the Argentine ant *Iridomyrmex humilis* (Mayr). *Insect Soc.* 35: 19-33.

Peacock, A. D., and T. Baxter. 1950. Studies in Pharaoh's ant *Monomorium pharaonis* (L.). 3. Life history. *Entomol. Mon. Mag.* 86: 171-178.

Peacock, A. D., D. W. Hall, I. C. Smith, and A. Goodfellow. 1950. The biology and control of the ant pest *Monomorium pharaonis* (L.). *Misc. Publ. Depart. Agri. Scotland* 17: 1-51.

Peacock, A. D., J. H. Sudd, and T. Baxter. 1955a. Studies in Pharaoh's ant *Monomorium pharaonis* (L.). 11. Colony foundation. *Entomol. Mon. Mag.* 91: 125-129.

_____ . 1955b. Studies in Pharaoh's ant *Monomorium pharaonis* (L.). 12. Dissemination. *Entomol. Mon. Mag.* 91: 130-33.

Petersen-Braun, M. 1975. Untersuchungen zur sozialen Organisation der Pharaoameise *Monomorium pharaonis* (L.) (Hymenoptera, Formicidae). 1. Der Brutzyklus und seine Steuerung durch populationseigene Faktoren. *Insect Soc.* 22: 269-292.

_____ . 1982. Intraspezifisches Aggressionsverhalten bei der Pharaoameise *Monomorium pharaonis* L. (Hymenoptera. Formicidae). *Insect Soc.* 29: 25-33.

Petersen, M., and A. Buschinger. 1971a. Das Begattungsverhalten der Pharaoameise *Monomorium pharaonis* L. *Z. Ang. Ent.* 68: 168-175.

_____ . 1971b. Untersuchungen zur Koloniegründung der Pharaoameise *Monomorium pharaonis* L. *Sod. Anz. Schädl. Pflanz.* 44: 121-127.

Pisarski, B. 1982. Ants (Hymenoptera Formicoidea) of Warsaw and Mazovia. *Memorabilia Zool.* 36: 73-90.

Prins, A. J., H. G. Robertson, and A. Prins. 1990. Pest ants in urban and agricultural areas of Southern Africa. Pp. 25-33. In: R.K. Vander Meer, K. Jaffe and A. Cedeno [eds.]. *Applied Myrmecology: a World Perspective*. Westview Press, Boulder, CO.

Reimer, N., J. W. Beardsley, and G. Jahn. 1990. Pest ants in the Hawaiin islands. Pp. 40-50. In: R.K. Vander Meer, K. Jaffe and A. Cedeno [eds.]. *Applied Myrmecology: a World Perspective*. Westview Press, Boulder, CO.

Rosengren, R., D. Cherix, and P. Pamilo. 1985. Insular ecology of the red wood ant *Formica truncorum Fabr.*: I-Polydomous nesting, population size and foraging. *Mitt. Schweiz. Entomol. Ges.* 58: 147-176.

Samways, M. J. 1990. Ant assemblage structure and ecological management in citrus and subtropical fruit orchards in southern Africa. Pp. 570-587. In: R.K. Vander Meer, K. Jaffe and A. Cedeno [eds.]. *Applied Myrmecology: a World Perspective*. Westview Press, Boulder, CO.

Silberglied, R. 1972. The little fire ant, *Wasmannia auropunctata*, a serious pest in the Galapagos Islands. *Noticias de Galápagos*. 19: 13-15. (cited by Ulloa-Chacon, 1990).

Skaife, S. H. 1955. The Argentine ant, *Iridomyrmex humilis* (Mayr). *Trans. Roy. Soc. South Africa*. 34: 355-377.

Smith, M. R. 1928. The biology of *Tapinoma sessile* Say, an important house-infesting ant. *Ann. Ent, Soc. Am.* 21: 307-329, Plate 18.

_____ . 1965. House infesting ants of the eastern United States. Their recognition, biology and economic importance. USDA *Tech. Bull.* 1326.

Steinbrink, H. 1987. Ein weiterer Nachweis von *Tapinoma melanocephalum* (Hymenoptera, Formicidae) in der DDR. *Angew. Parasitol.* 28: 91-92.

Stuart, R. J. 1990. Experiments on colony foundation in the polygynous ant *Cardiocondyla wroughtoni*. Pp. 242. In: G. K. Veeresh, B. Mallick and C.A. Viraktamath [eds.]. *Social Insects and the Environment*. Oxford and IBH Publishing Co. (PVT.) Ltd., New Delhi.

Stuart, R. J., A. Francoeur, and R. Loiselle. 1987. Lethal fighting among dimorphic males of the ant *Cardiocondyla wroughtoni*. *Naturwissenschaften* 74: 548-549.

Thompson, C. R. 1990. Ants that have pest status in the United States. Pp.51-67. In: R.K. Vander Meer, K. Jaffe and A. Cedeno [eds.]. *Applied Myrmecology: a World Perspective*. Westview Press, Boulder, CO.

Trager, J. C. 1984. A revision of the genus *Paratrechina* (Hymenoptera: Formicidae) of the Continental United States. *Sociobiology* 9: 51-162.

Tremper, B. D. 1976. Distribution of the Argentine ant, *Iridomyrmex humilis* Mayr, in relation to certain native ants of California: ecological, physiological and behavioral aspects. *Ph. D. Dissert.* Univ. of California, Berkeley. (cited by Ward, 1987).

Ulloa-Chacon, P. 1990. Biologie de la reproduction chez la petite fourmi de feu *Wasmannia auropunctata* (R.) (Hymenoptera: Formicidae). *Thèse*. Lausanne.

Ulloa-Chacon, P. and D. Cherix. 1988. Quelques aspects de la biologie de *Wasmannia auropunctata* (Roger) (Hymenoptera, Formicidae). *Actes Coll. Insect Soc.* 4: 177-184.

_____ . 1989. Etude de quelques facteurs influençant la fécondite des reines de *Wasmannia auropunctata* R. (Hymenoptera, Formicidae). *Actes Coll. Insect. Soc.* 5: 121-129.

Van Loon, A. J., J. J. Boomsma, and A. Andrasfalvy. 1990. A new polygynous *Lasius* species (Hymenoptera: Formicidae) from central Europe. I. Description and general biology. *Insect Soc.* 37: 348-362.

Vargo, E. L. and Passera L. 1992. Gyne development in the Argentine ant *Iridomyrmex humilis* (Mayr): on the role of overwintering and queen control. *Physiol. Entomol.* 17:193-201.

Veeresh, G. K. 1990. Pest ants of India. Pp.15-24. In: R.K. Vander Meer, K. Jaffe and A. Cedeno [eds.]. *Applied Myrmecology: a World Perspective*. Westview Press, Boulder, CO.

Ward, P. S. 1987. Distribution of the introduced Argentine ant (*Iridomyrmex humilis*) in natural habitats of the lower Sacramento valley and its effects on the indigenous ant fauna. *Hilgardia* 55(2): 1-16.

Wilson, E. O. 1971. *The Insect Societies*. Belknap Press of Harvard Univ. Press, Cambridge.

Wilson, E. O. and R. W. Taylor. 1967. The ants of Polynesia (Hymenoptera: Formicidae). *Pacific Insects mono.* 14: 1-109.

Yamauchi, K., Y. Ito, K. Kinomura, and H. Takamine. 1987. Polycalic colonies of the weaver ant *Polyrhachis dives*. *Kontyu* 55: 410-420.

Yamauchi, K., K. Kinomura, and S. Miyake. 1981. Sociobiological studies of the polygynic ant *Lasius sakagamii*. 1. General features of its polydomous system. *Insect Soc.* 28: 279-296.

4

Coexisting Patterns and Foraging Behavior of Introduced and Native Ants *(Hymenoptera Formicidae)* in the Galapagos Islands (Ecuador)

Rolf E. Meier

Introduction

One of the major ecological problems of the Galapagos Islands (Figure 4.1) is the accidental introduction of exotic species by man. While the ecological impact and threat by introduced mammals (e.g. rats, dogs, cats and cattle) or many plants into the rich endemic fauna and flora is relatively well studied, little is known about the role of recently introduced insects. The little fire ant, *Wasmannia auropunctata*, probably reached the archipelago by human transport between 1920 and 1930 (Kastdalen 1982). Despite the small size of the worker ant (ca. 1.4 mm), this neotropical species causes considerable ecological damage by displacing native and endemic invertebrate fauna (Silberglied 1972; Clark et al. 1982; Lubin 1984, 1985; Meier 1985a).

Wasmannia auropunctata was initially located in inhabited areas, particularly on Santa Cruz island, where its burning sting is well known. From there, this unicolonial and polygynous species spread rapidly. Now it is not only found on four inhabited islands, Santa Cruz, San Cristobal, Isabela, and Floreana, and the previously populated Santiago (San Salvador), but also on the uninhabited island of Pinzon. On Santa Fe, *W. auropunctata* was eradicated recently after at least two earlier infestations. Prior studies showed the preferred habitat of *W. auropunctata* to be in the

FIGURE 4.1 Map of Galapagos Islands with study sites indicated by arrows.

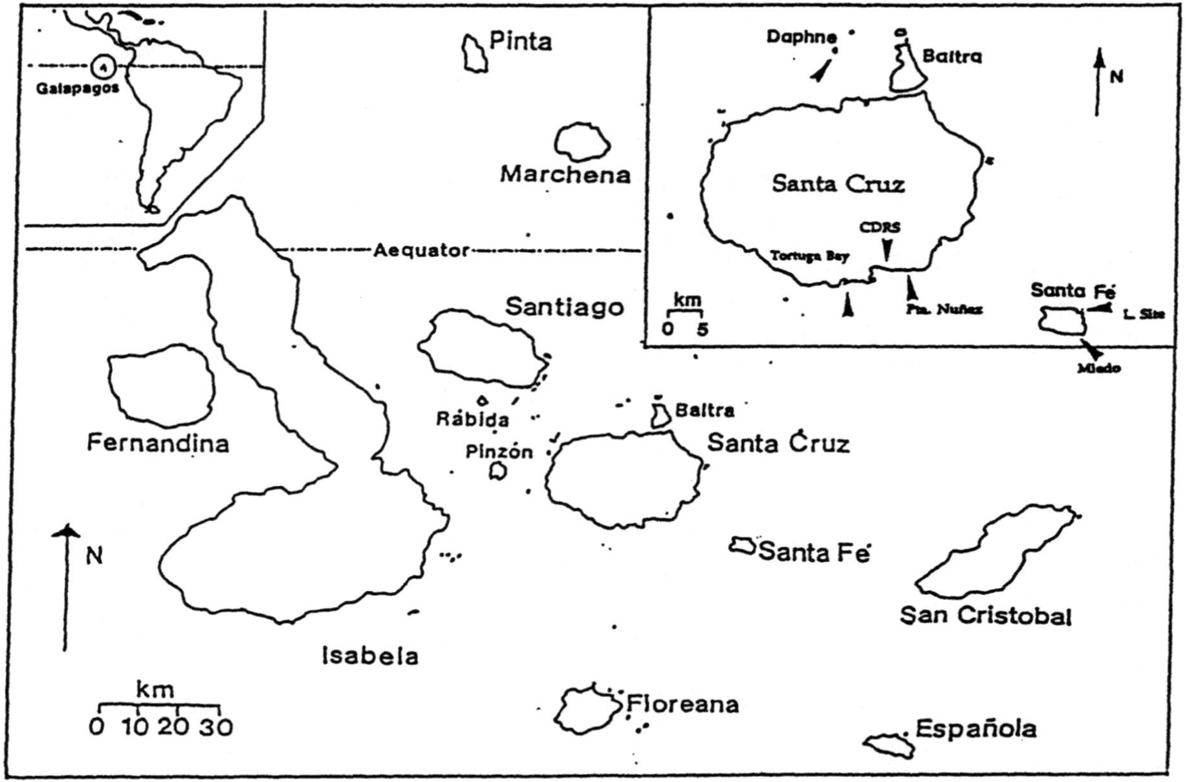

moist transition area and lower humid zones on Santa Cruz (Clark et al. 1982). This area is identical with the agricultural zone between 100m and 300m above sea level.

In 1982-83, an extraordinary El Nino Southern Oscillation (ENSO) raised the ocean and air temperatures in the Pacific region unexpectedly high, and induced 8 months of the heaviest rains within a century (Kiladis and Diaz 1984). This dramatic increase in precipitation caused the annual mean to rise from 374mm (1965-81) to more than 3500mm within eight months, strongly favoring the expansion of *W. auropunctata* into the arid zone. This arid habitat lies between the narrow coastal zone and the higher situated moist zone. Figure 4.2 is based on data from the Charles Darwin Research Station (CDRS) on Santa Cruz and illustrates the extraordinary distribution of rainfall during this 9-year study period (1981-1989) compared to the 25-year period lasting from 1965-90.

This study was conducted in 1983 and 1989 to determine the impact of ENSO 1982-83 on the coexistence patterns of native and introduced ants in the arid zone of 3 Galapagos Islands. The diversity and relative abundance of ants foraging on cacti and on the soil was monitored and compared with the data collected in 1981, 1983 and 1989. Thus, the long-term effect of the dispersal of moist-adapted ant species into formerly arid areas was studied and evaluated and compared to the dispersal of dry-adapted ant species after this event.

FIGURE 4.2 Annual rainfall (mm) (dark bar) at CDRS 1965-90 and mean of 25 years (clear bar).

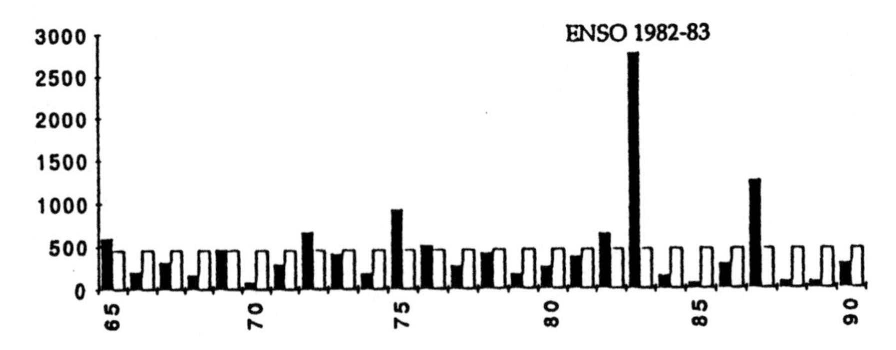

Sites and Methods of Investigation

Giant endemic cacti are the most conspicuous plants in the xeromorphic zone of many of the Galapagos Islands. Their numerous floral nectaries and permanent fruits represent an important and reliable food source for various birds (Darwin's finches and mockingbirds) and many insects. Ants feed mainly on the extrafloral nectaries at the base of young spines. On the island of Santa Cruz, the arborescent *Opuntia echios* (the prickly pear cactus) is found with the taller, but less abundant *Jasminocereus thouarsii* (the candelabra cactus) in the arid zone. This zone extends mainly between 6m and 120m above sea level (Wiggins and Porter 1971).

Extrafloral nectar is produced in abundance all year round (de la Vega 1988). In order to study the mode by which the ants partitioned this nectar, observations were started in 1981 on the centrally located island of Santa Cruz at the Charles Darwin Research Station (CDRS) and at Tortuga Bay. Then they were extended onto the neighboring islands of Santa Fe and Daphne Major in 1983 and 1989. As an added survey tool, tuna fish baits were placed in the same tests locations to attract soil-living ants, which feed only on fat, oil or proteins. The monitoring period was always between September and December, at the end of the dry season, and at the beginning of the wet, foggy garua season.

The study sites (Figure 4.1) and their length of transects were as follows:

1. Santa Cruz island: Charles Darwin Research Station (1650m), Tortuga Bay (850m), and Punta Nunez (605m);
2. Santa Fe island: Tourist Landing Site (570m), Miedo (850m);
3. Daphne Major island: Tourist trail and Crater crest (1840m).

All study sites looked rather similar in terms of vegetation. The transects contained 78-110 cacti and varied in length between 570m-1840m. Most transects were laid out at right angles to the coastline and reached altitudes of not more than 120m above sea level. A visual search (at least at 2 different times each day) was made for foraging worker ants on arborescent cacti. Their occurrence was monitored from the ground up to approximately 2m. When possible, the foraging and feeding behavior of individual as well as groups or masses of worker ants was studied during 24-hr observation periods.

The location of the *W. auropunctata* trails on the stem of the arborescent cacti was determined at the height of 1m by the aid of a compass. The number of outbound and inbound ants was registered independently with 2 counters during 60 sec periods.

The time required for workers to ascend or descend a vertical distance of 10cm-trail on the cactus stem was determined with a chronometer.

Voucher specimens have been deposited at the Museum of the CDRS, Santa Cruz, Galapagos, at the Museu da Zoologia, Universidade de Sao Paulo, Brasil, at the Los Angeles County Museum of Natural History, California, USA, and in the Museum Histoire Naturelle in Geneva, Switzerland.

The nomenclature of the taxa follows the classification of the neotropical Formicidae by Brandão (1991).

Results and Observations

Santa Cruz island

On Santa Cruz island, at the CDRS study site, *W. auropunctata* was the dominating ant species on cacti before and shortly after ENSO in 1982-83. During this time, it increased its foraging range to 17 additional cacti, but by October 1983, it abandoned one cactus it had been feeding on 2 years before. In contrast, the foraging range of 3 other ant species combined increased by only 4 cacti. The relative abundance of *W. auropunctata* increased from 88% to 91% between 1981-83, but decreased to 57% in 1989 (Figure 4.3). The total number of conspicuous ant trails on the cacti (*O. echios* var. *gigantea* and *J. thouarsii* var. *delicatus*) ranged from a peak of 124-131 trails in 1981-83, and to only 11 in 1989. In addition, the total mean number of outbound and inbound *W. auropunctata* worker ants on these trails decreased from 23.4 to 18.0 to 17.2 ants per minute respectively, for the years 1981, 1983 and 1989. An analysis with the non-parametric Kruskal-Wallis Test (a one-way analysis of variance), revealed these series of ant frequencies differed significantly from each other ($P<0.001$).

At CDRS, 3 species of ants coexisted on cacti nectaries in 1981 and 1989 compared to 6 species in 1983 after ENSO 82-83. In the daytime in 1983, only 1 endemic species was detected, *Solenopsis globularia,* with a relative abundance of 3%.

The true fire ant, *Solenopsis geminata,* was found on 29% of the cacti in 1989, but was not recorded before or shortly after ENSO 1982-83. While *S. geminata* foraged on giant cacti in 1989, *Paratrechina longicornis,* (the "*hormiga loca*" or crazy ant), vanished completely (7% to 0%). In October 1991, at the time this paper was presented at the Galapagos Ant Conference, a surprisingly different pattern of ant diversity was recorded (the data is not included in Figure 4.3). In 1989, *P. longicornis*

FIGURE 4.3 Relative abundance of ant species on cacti (A) and baits (B) on Santa Cruz island, Galapagos.

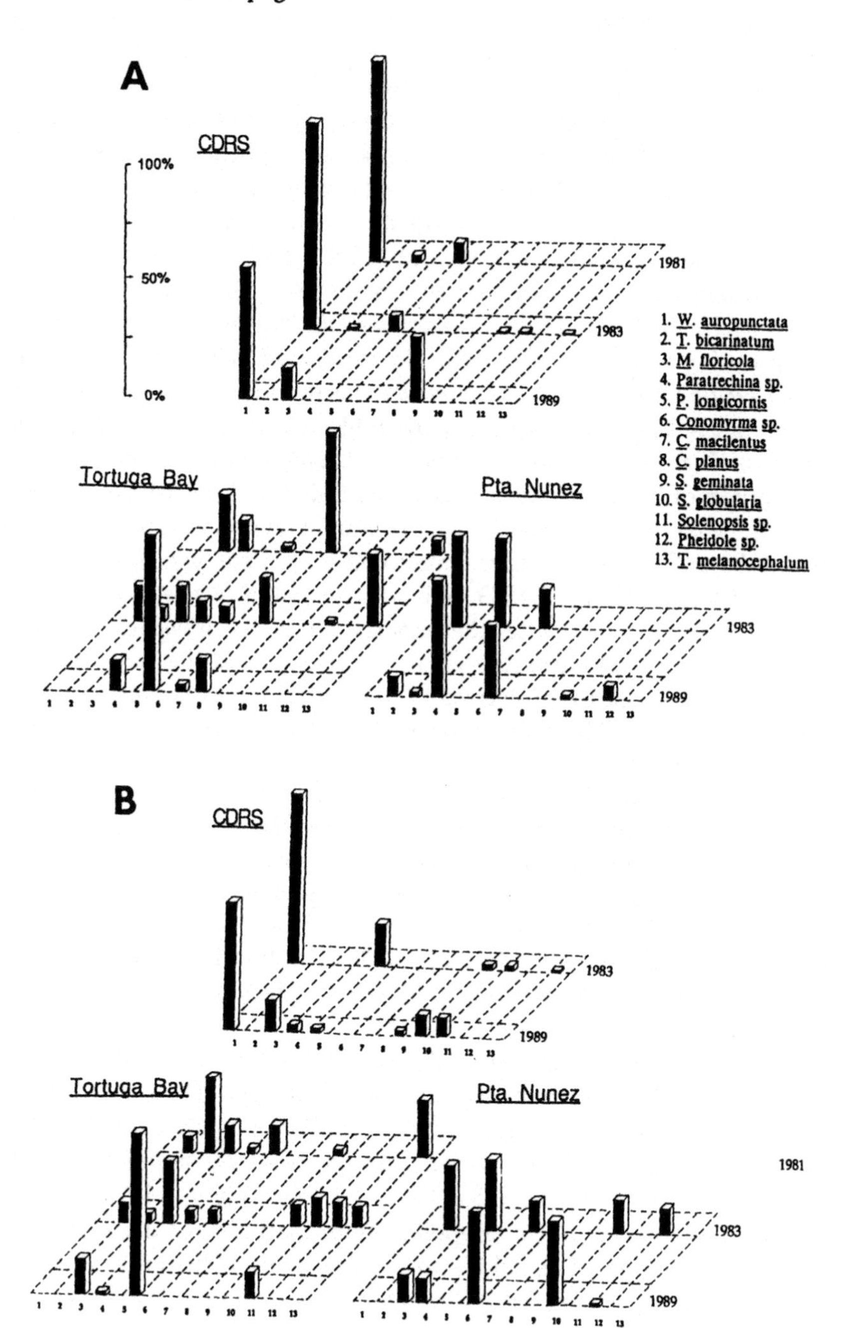

was detected on tuna fish baits on rare occasions (3%), but never on cacti. In 1991, *P. longicornis* exceeded *W. auropunctata* in abundance on cacti (42.3% to 28.8%). These fluctuating data corroborate that *P. longicornis* is one of the most opportunistic tropical ant species. It is timid in the presence of competitors and is displaced easily after perturbation, but quickly finds and settles in new sites or areas with low ant density. The tiny ant, *Monomorium floricola* increased slightly (4% to 10%) between 1981 and 1989. Thus, based on relative abundance on cacti at the end of 1991, *W. auropunctata* had lost its first rank, after being the most abundant ant species at CDRS, at least in 1981, 1983, and 1989.

Corresponding observations were made along the trail from Puerto Ayora to Tortuga Bay. Here, *W. auropunctata* had extended its range a distance of 3.4 km west from the Puerto Ayora-Bellavista Road after ENSO 1982-83. In 1981, *W. auropunctata* was not found 2.24 km from the beach. Since this trail became impenetrable and covered by 2-3 m high shrubs of *Parkinsonia, Scutia, Acacia,* and *Cryptocarpus,* the precise local distribution of *W. auropunctata* along this trail could not be determined anymore after ENSO 1982-83. A new trail, much closer to the beach (only 1150 m from it) was then monitored for *W. auropunctata.* Its rate of spread from 1982 until 1983, when extrapolated, was not more than 500 m per year. In 1983, after the rain, the sandy beach became covered by conspicuous vegetation (the grass, *Aristida sp.* and the vine, *Ipomoea pescaprae*). The moisture-preferring *W. auropunctata* was not encountered at Tortuga Bay in 1983, 1989, or in October 1991, nor along the new trail in 1989 and 1991.

The ENSO 1982-83 influenced the ant fauna in various ways. First, the intense rain blocked and drowned many of the soil-nesting ant colonies. Second, the rainfall, which lasted for a total of 8 months, soaked the roots and trunks of all the xeromorphic plants. The rate of feeding on cacti at CDRS, increased from 85% to 89.7% during the 1981-83 period. After a drastic reduction to 17.2% in 1989, it increased to 62.3% in 1991. These data may reflect variations in flux in the attraction of ants to the extrafloral nectaries.

From 1981-1989, *W. auropunctata* was invariably the most common ant species at CDRS, both on cacti and tuna fish baits. Thus, it was clear that *W. auropunctata* was competitive with other ants after ENSO 1982-83. But the census in 1989 revealed a decline in its density after several extremely dry years (1984, 1985, 1988 and 1989). Although *W. auropunctata* was able to hold and monopolize more than 57% of the surviving cacti, it had lost part of its habitat to *S. geminata* by 1989 and to *P. longicornis* by 1991. In October 1991, five ant species were monitored on cacti, two more than 1989.

The endemic, diurnally active ant species, *Camponotus planus*, which might have been expected on cacti nectaries at CDRS, was never observed probably because they were monopolized by *W. auropunctata* from 1981 to 1991. Only the endemic and nocturnally active *C. macilentus* was detected either alone or when one of the two tramp species *P. longicornis* or *M. floricola* were foraging on the cacti. While it was recorded on 38 of the 100 cacti (38%) in 1981, it was observed on only 2 of the surviving 86 cacti (2.3%) after ENSO 1982-83.

At the Tortuga Bay and Punta Nunez study sites, *W. auropunctata* was never observed. Tortuga Bay is situated 5 km west of CDRS, and Punta Nunez is located 5 km east of it. In both areas the diversity of ant species was greater than at CDRS. This was true for the tuna fish baits and the extrafloral nectaries. At Tortuga Bay, a total of 8 species were noted on cacti in 1983, the richest diversity so far, while at the same time at Punta Nunez, 6 species were found. Within period of 1983-1989, the most dominating ant species at Tortuga Bay was *P. longicornis* which was found on 8% to 66% of the observation periods; at Punta Nunez, *Paratrechina sp.* was observed on 37% to 49% of the observation periods. Also at Tortuga Bay, the relative abundance of *Tetramorium bicarinatum* reached 15% on cacti and 10% on tuna fish baits in 1983, but this species could not be detected by either observation method in 1989. *P. longicornis* was observed on 65.7% of the *Opuntias* and was found on 69% of the tuna fish baits during the same observation period. The relative abundance of the black and singly foraging *C. planus* decreased, ranging from 52%, 9% and 18%, respectively, in 1981, 1983, and 1989. Surprisingly, we also found the endemic and nocturnal *C. macilentus* feeding in the daytime on 3.2% of our observation times. Wheeler (1924) described it as a nocturnal species, since he never saw workers and soldiers during the daytime in his studies in April 1923.

The incidence of the *Paratrechina sp.* was relatively high (27%) on baits in 1983. The occurrence of *M. floricola* on baits decreased from 33% to 17.2% during 1981-1989. The endemic *S. globularia* was never observed on cacti but in rather low relative incidences on baits (10% to 12.1%) in 1983 and 1989, respectively.

At Punta Nunez, a steep drop in the relative abundance of *T. bicarinatum* on cacti (from 39% to 9%) and on baits (from 28% to 0%) was recorded in 1983 and 1989. The most common ants on cacti were *Paratrechina sp.* (37% to 49%) and a *Conomyrma sp.* (17% to 31%) which prefers dry conditions. On the baits, the *Conomyrma sp.* (14% to 38%) and the endemic *S. globularia* (16% to 36%) were most abundant.

Santa Fe island

Santa Fe island is located 30 km southeast of Santa Cruz. The Tourist Landing study site was located on the northeast corner of Santa Fe. *Wasmannia auropunctata* was detected on 60% of 115 tuna fish baits in November 1983. In the littoral zone, *W. auropunctata* nested mainly under shrubs of *Cryptocarpus pyriformis*. After this discovery, several eradication campaigns (Dec. 1985-89) were organized by the Galapagos National Park Service and the CDRS to reduce or eliminate *W. auropunctata* and prevent its spread to higher and wetter parts of the island. A check for this pest ant in an area of 10,000 m^2, was made by placing 419 baits in the treated areas at the end of the dry season in 1989. No *W. auropunctata* were found indicating that the measures taken were successful at that time. Whether or not the ants were completely eradicated can not be determined until repeated baiting programs during the wet garua season have provided negative results.

While *Pheidole sp.* was the second most common species (18% compared to *W. auropunctata* 60%) in 1983, this ant species apparently benefitted from the *W. auropunctata* eradication campaign since it occupied 62% of the baits in 1989 (Figure 4.4).

The second study site on Santa Fe island was Miedo, located on the southern coast of the island. The relative abundance of *Paratrechina sp.* clearly dropped both on cacti (75% to 25%) and on baits (48% to 18%) between 1983 and 1989. The most abundant ants on the cacti (*O. echios* var. *barringtonensis*) was *C. planus* (17% to 68%) while on the baits, *Conomyrma sp.* (6% to 46%) was most common. *Pheidole sp.* was detected exclusively on baits (8%) in 1989 (Figure 4.4).

FIGURE 4.4 Relative abundance of ant species on cacti or baits on Santa Fe and Daphne Major islands, Galapagos.

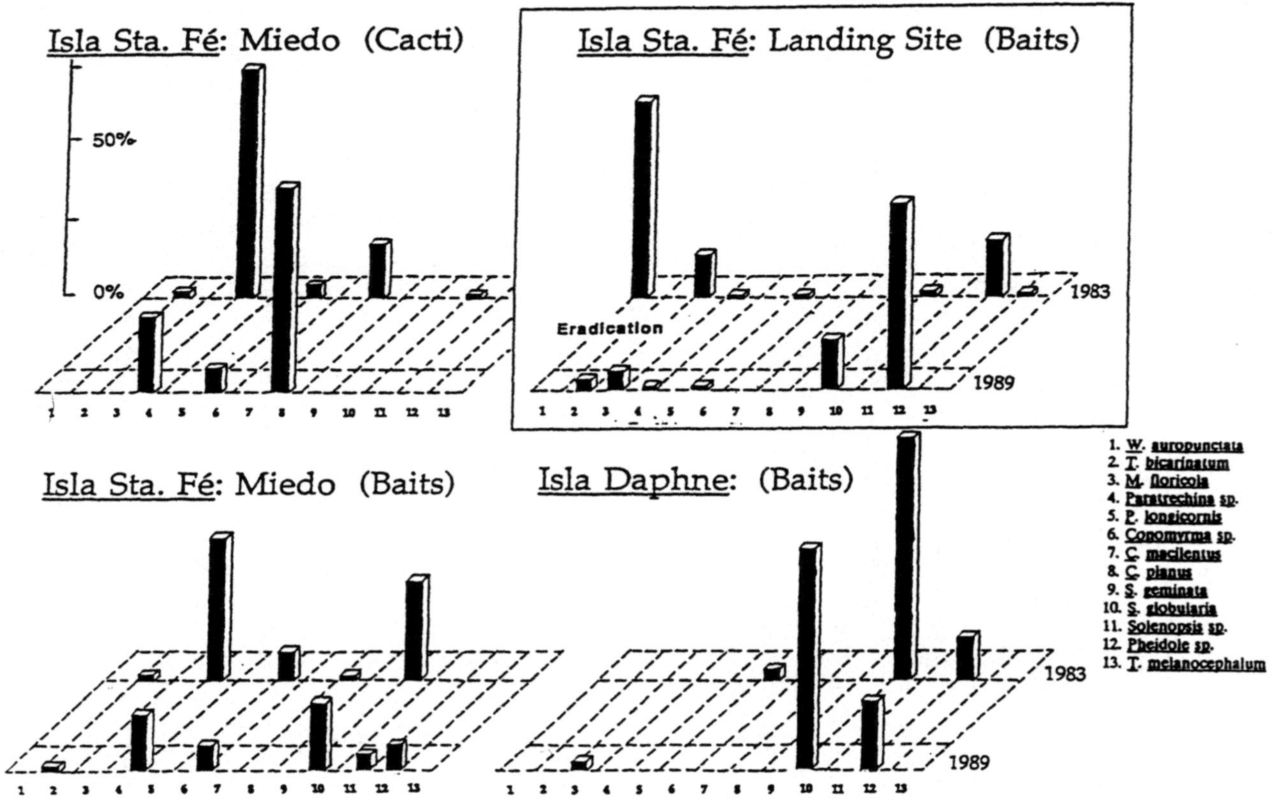

Isla Sta. Fé: Miedo (Cacti)

50%

0%

1983

1989

1 2 3 4 5 6 7 8 9 10 11 12 13

Isla Sta. Fé: Landing Site (Baits)

Eradication

1983

1989

1 2 3 4 5 6 7 8 9 10 11 12 13

Isla Sta. Fé: Miedo (Baits)

1 2 3 4 5 6 7 8 9 10 11 12 13

Isla Daphne: (Baits)

1983

1989

1 2 3 4 5 6 7 8 9 10 11 12 13

1. W. auropunctata
2. T. bicarinatum
3. M. floricola
4. Paratrechina sp.
5. P. longicornis
6. Conomyrma sp.
7. C. macilentus
8. C. planus
9. S. geminata
10. S. globularia
11. Solenopsis sp.
12. Pheidole sp.
13. T. melanocephalum

Daphne Major island

A few *Pheidole sp.* worker ants were detected on 5 out of 78 cacti (*O. echios* var. *echios*) in 1983 and on none of 62 cacti in 1989. Two factors may hinder or reduce the rate of foraging by ants above the soil on this small island; primarily, the permanent winds dry out the soil and plants, and secondarily, a lack of moisture which may be needed by the nectar-producing cacti. No foraging ants were detected in 1983 on 73 of 78 cacti (93.6%) on Daphne Major island. In comparison, Punta Nunez is situated on the southern and wetter slope of Santa Cruz. At this site, only 7 of 100 cacti (7%) were void of ants. The tuna fish baits attracted 5 ant species in 1983 and 3 ant species in 1989 (Figure 4.4). While ants were found on only 53.8% of the baits (77 of the 143) in 1983, 99.7% of the baits (107 of 108) were occupied by ants in 1989. In either census the relative abundance of *S. globularia* was highest and reached 81.2% in 1983 and 71.5% in 1989. *Pheidole sp.* was second in 1983 (15.3%) and 1989 (25.7%). In 1989, *M. floricola* (2.8%) was monitored exclusively at the lowest part of the Tourist trail. *C. macilentus* and *Paratrechina sp.*, which were present on cacti in 1983, could not be detected either on cacti or on tuna fish baits in 1989.

Foraging Activity and Interference Behavior

Location and use of ant trails on cacti

The most attractive nectaries for foraging ants were invariably those in the areoles of the developing stem tissue of young *Opuntia* pods. The location of the *W. auropunctata* trails on the cylindrical stem of the mature *Opuntia* cacti was non-random. The majority were found on the northern side of the trunks (94 of a total of 144; Sign-Test: $P<0.001$). During the observation period (September-December), this side was hidden from direct sunlight from 6 am till 3 pm. The trails were not absolutely fixed, but shifted back and forth within a range of 5-20 cm over several weeks. During the garua season, such a shift could be observed within 30 minutes towards the lee side. By following the descending worker ants from the *Opuntia* pods, it was determined that all of the nesting sites were under volcanic rocks or soil, but never above the soil surface.

Within a trail, the number of climbing (outbound) and descending (inbound) *W. auropunctata* per 60 sec was almost identical, indicating that these worker ants use the same trail for foraging to and from the food source to the nest. This is in contrast to the foraging behavior of *P.*

longicornis, which more often used the trail for returning to the nest only. Mass recruitment and trailing rate was detected on trails of both *W. auropunctata* and *S. geminata*. On small trails (<10 ants/min), the time for running a distance of 10 cm for *W. auropunctata* was 40.2+/-12.0 sec (N=92), while on mass foraging trails (10-30 ants/min), the speed increased and the time to cover 10 cm decreased to 33.4+/-10.1 sec (N=96) (t-Test, t=3.975; P<0.001). Twenty-four hour observations of foraging ant species revealed *W. auropunctata* to have the smallest range between maximum and minimum foraging speed (trailing rates) whether monitored during day and night or during any disturbances by rain, wind or direct sunlight. During identical stress situations *P. longicornis* and *S. geminata* foraging was hindered more than foraging by *W. auropunctata*.

Foraging time and habitat partitioning on cacti

Whereas *W. auropunctata* and *P. longicornis* foraged on cacti both day and night, *M. floricola* was predominantly active diurnally (Figure 4.5). However, between 1800 and 2400 hours a decline in activity was observed, and from midnight on, almost no foraging was seen. At CDRS, only in rare cases was more than one species found on the same cactus during daylight hours (6% in 1981; 7% in 1983; 0% in 1989; and 19.7% in 1991). During a 2-month observation period, *C. macilentus* and *M. floricola* were observed sharing a common food source by foraging on the same cactus at different time periods during the day. In contrast, *S. geminata*, *W. auropunctata* and *P. longicornis* in most cases, each monopolized a cactus 24 hours a day over the same 2-month period.

The interference behavior of *W. auropunctata* and *P. longicornis* was monitored in 12 series of studies. The larger and swifter *P. longicornis* detected the bait first. But in 10 experiments the slower but constantly recruiting *W. auropunctata* was able to displace *P. longicornis* after about 20 min and monopolized the bait for up to 45-60 min (Meier, unpublished data). These small "extirpators" (Holldobler and Wilson 1990) recruited effectively by using odor trails but seldom showed clear signs of aggressive behavior, e.g. by using their tiny stingers or their mandibles. The much faster (10X) *P. longicornis* (3.21+/-1.4 sec for 10cm; N=119) always tried to avoid contact with the smaller and slower *W. auropunctata* (38.8+/-11.1 sec for 10cm; N=91). This conspicuous behavior may be explained by their possession of a highly effective repellent venom or pheromone. Even a relative small number of *W. auropunctata* workers were able to keep a large number of *P. longicornis* at a distance of 2-3cm from the food source. Lethal effects on *P. longicornis* were never observed.

FIGURE 4.5 Daily activity pattern of foraging ants on cacti.

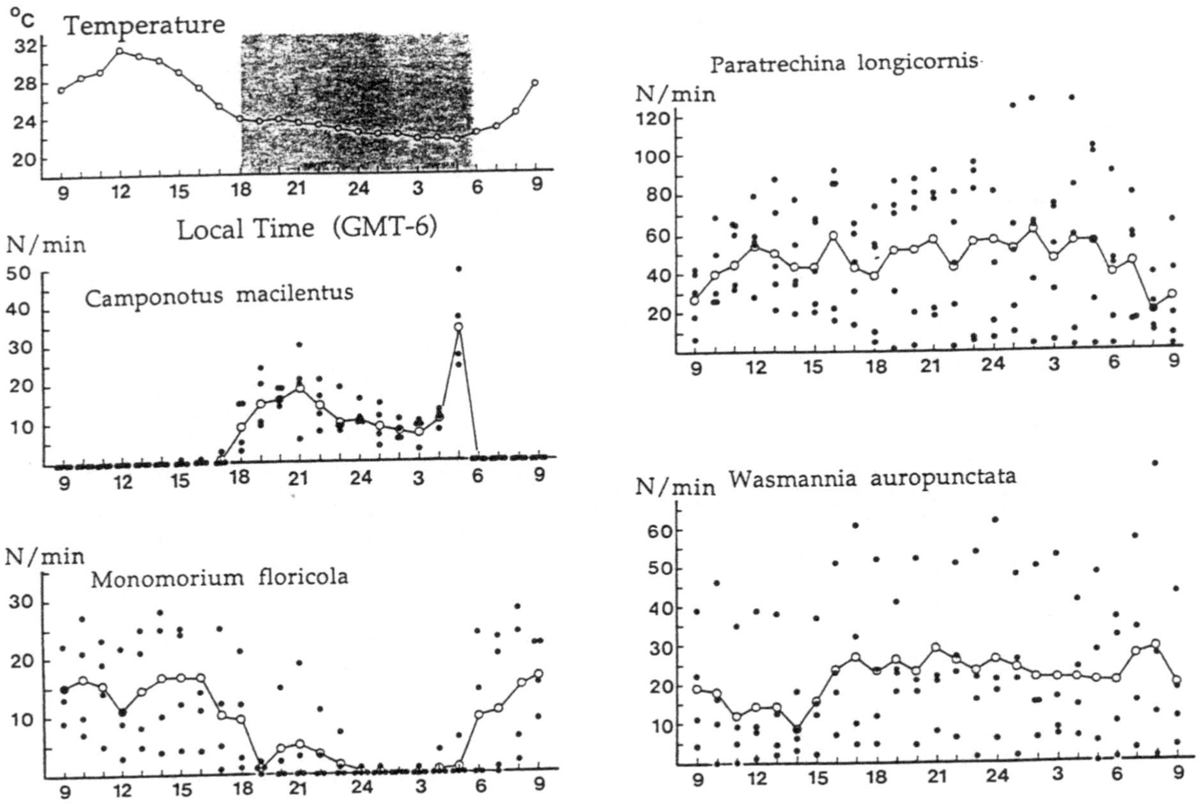
°C
Temperature
32
28
24
20
9 12 15 18 21 24 3 6 9
Local Time (GMT-6)
N/min
50
40
30
20
10
Camponotus macilentus
9 12 15 18 21 24 3 6 9
N/min
Monomorium floricola
30
20
10
9 12 15 18 21 24 3 6 9
Paratrechina longicornis
N/min
120
100
80
60
40
20
9 12 15 18 21 24 3 6 9
N/min
Wasmannia auropunctata
60
50
40
30
20
10
9 12 15 18 21 24 3 6 9

Discussion

Island biotas remote from continents usually contain relatively few species. Their fragile biological equilibrium is therefore highly sensitive to large oscillations or changes in climate and to newly introduced species. Such perturbations may limit the diversity and abundance of endemic and native species much more than in equivalent areas on the mainland. For example, a dramatic impact occurred on the Galapagos islands when the ENSO/El Nino in 1982-83, caused a sharp temperature increase followed by the heaviest rainfalls for the prior 100 years. ENSO 1982-83 had multiple ecological effects on the Galapagos (for review see Robinson and del Pino 1985). For instance, in the arid zone it impaired many cacti and favored the spread of humidity-preferring plants and animals, such as *W. auropunctata*. According to Lubin (1984) nearly half of the actual 29 ant taxa are tropical tramp species (Lubin 1984). Of all of them, *W. auropunctata* seems to have been the most successful ant species until now. During ENSO 1982-83 and for many of the following years, it expanded its range, mainly in the southern slope of Santa Cruz (Lubin 1985, Meier 1985a), on San Salvador (Lubin 1985), Santa Fe (Meier 1983), and Pinzon (L. Cayot, pers. comm.).

The present studies concentrated on arid zones, 3 of which were on Santa Cruz, 2 on Santa Fe, and 1 on Daphne Major. Besides the tiny *W. auropunctata* and *M. floricola*, the fire ant, *S. geminata* is expanding its range in the town of Puerto Ayora, on Santa Cruz (author's observations 1981, 1983 and 1989), and within the CDRS, mainly in the pens of the land iguanas and near the student dormitories (author's observations, 1989). *Solenopsis geminata* had already been described by early collectors on San Cristobal (Chatham) in the last century and interpreted as probably introduced (Emery 1893). Only recently has it become a pest in the village of Bellavista, located above Puerto Ayora (Williams and Whelan 1991). Within the arid coastal and the adjacent transition zone on several Galapagos islands, *T. bicarinatum, Tetramorium sp.* and *Tapinoma melanocephalum* have become rather abundant and at least, temporarily, appear to be expanding their range as well.

These studies indicate that the success of the *W. auropunctata* invasion into the lower and coastal arid zone of Santa Cruz, was much more modest than in the higher transition zone (Clark et al. 1982, Lubin 1984 1985). *W. auropunctata* had invaded CDRS already in the late 1970s (F. Koster, pers. comm.) and profited enormously from ENSO 1982-83. Their invasion of the lowlands was effective in only 1 (CDRS on the southern coast of Santa Cruz) of the 3 arid areas studied. They were not found at Tortuga Bay nor at Punta Nunez in 1983 nor 1989. Therefore, these 3 areas may have clear, but unknown differences in ecological habitat. Furthermore, it was noticed for the first time that the density of *W. auropunctata* has been shrinking at Puerto Ayora and at CDRS since 1989. Whether the extrafloral nectaries of *O. echios* also function in attracting protective ants, as was shown in *O. acanthocarpa* in Arizona (USA) (Pickett and Clark 1979), cannot be answered by this study.

On the neighboring island of Santa Fe, *W. auropunctata* was detected twice after infestations at the tourist landing site in 1975 (de Vries, unpublished data) and 1983 (Meier 1983). Various eradication attempts which used burning followed by the application of several insecticides had been partially successful during 1985-89. A recent check in 1991 did not detect this pest ant at all (S. Abedrabbo, pers. comm.).

On the continent in undisturbed rain forest habitat with a much richer ant diversity (e.g. Costa Rica), *W. auropunctata* has not been declared a pest because it is not a dominating and monopolizing species (see Tennant this volume). In cacao plantations in Cameroon, it has been reported that this ant species is appreciated because it protects against phytophagous insects (Bruneau de Miré 1969). In Brazil on the other hand, *W. auropunctata* is a serious pest, stinging pod pickers and tending mealybugs (see Delabie et al. this volume).

Why is *W. auropunctata* so successful on the Galapagos Islands and, therefore, a pest for endemic and native ants as well as for other arthropods and for humans? Various behavioral strategies of this tiny unicolonial Myrmicinae may lead to its superior ability of competitive displacement. First, *W. auropunctata* exploits habitat and food resources more efficiently than other native species. Due to the recruitment effect of its trail substance, this ant species can recruit a higher number of workers in a short time. It forages steadily day and night. Being a polyphagous generalist or opportunistic feeder (feeding on carbohydrates as well as on fat and proteins), it forages into more niches than food specialists. Thus, the energy consumption of their brood may be higher compared to sympatric ant species (for review of the biology see Ulloa-Chacon and Cherix 1990; or for a list of *W. auropunctata* literature see Ulloa-Chacon et al. 1991). Since it possesses a highly effective repellent venom or pheromone, its foragers often monopolize food sources in a short time, as has been demonstrated on tuna fish baits (Meier 1985b). In addition, *W. auropunctata* may use its sting in direct combat or in interspecific competition as well. Rapidly changing weather conditions, as for instance strong wind, heavy rain and full sunlight, affect the foraging behavior of *W. auropunctata* workers much less than that of sympatric ant species, which in the same situation may cease foraging completely.

Wheeler (1924) described the endemic C. *macilentus* and C. *planus* as the most abundant ant species in the xeromorphic parts of the islands. The yellow C. *macilentus,* which had already been collected by Charles Darwin at Floreana island in 1835, is still encountered on cacti or within houses at CDRS in 1989 and 1991. The black and smaller C. *planus* has never been recorded at CDRS on cacti, nor on tuna fish baits, nor within houses in 1981, 1983, or 1989, but exclusively on cacti at Tortuga Bay and at Punta Nunez. Maybe this day-active endemic ant species was displaced at the CDRS by the day-and-night active *W. auropunctata.* The poor ant species diversity on all the Galapagos Islands is only partially due to *W. auropunctata.* It may reflect unsaturated habitats after all. On all the islands, the number of ant species, currently 29, is strikingly low, when one compares it to the record of 43 on a single tree in the continental Peruvian Amazon forest (Wilson 1987).

Summary

The effects of the El Nino Southern Oscillation 1982-83, the heaviest since 1877 on ants was studied at the coastal arid zones and partially compared with the pre-El Nino situation at the CDRS in 1981 and 1989. Extrafloral nectaries on arborescent cacti (*O. echios* and *J. thouarsii*) represent a rich and reliable food source for many soil-nesting ants in the arid zone. During the garua season (September-December), foraging and feeding ants were observed on cacti on the islands of Santa Cruz, Santa Fe and Daphne Major. In 1983, the moisture-preferring ant species *W. auropunctata* and *T. bicarinatum* benefitted most within the xeromorphic sites. However, by 1989, after several very dry years, the opportunistic *P. longicornis* and *Pheidole sp.*, as well as the fire ant, *S. geminata* and the sand adapted *Conomyrma sp.* were dominant feeders on many cacti which *W. auropunctata* had been foraging before. The population of *W. auropunctata* is decreasing at this time.

After *W. auropunctata* was eradicated on Santa Fe in 1989, *Pheidole sp.* filled its ecological niche. The tramp species *W. auropunctata, P. longicornis,* and *S. geminata* which have numerous worker ants and are active 24 hrs, displayed mass foraging along large trails and thus were able to monopolize food sources. The endemic C. *planus* and C. *macilentus* have only a few foragers and are either active during the day (C. *planus*) or the night (C. *macilentus*). As a result, they can share cacti with various ant species. The interference of *W. auropunctata* with *P. longicornis* at protein baits revealed that the former is powerful in the sense that it can oust the bigger and much swifter *P. longicornis,* most probably due to a repellent pheromone rather than its tiny sting or small mandibles.

Acknowledgments

I wish to thank my field assistants in 1981: Ines de la Vega, Manuel Alvarez; in 1983: Elizabeth Potts and Simon Villamar; in 1989: Soraya Landetta, Tom Larson, Antonio Mendoza and Arnaldo Tupiza. I appreciate the help and logistic support by The Charles Darwin Research Station Directors Friedemann Koster and Daniel Evans and the staff of the Charles Darwin Research Station as well as the generous support by Miguel Cifuentes, Fausto Cepeda and Oswaldo Sarango, Directors of the Servicio Parque Nacional Galapagos. Andrew Laurie and Martin Wikelski allowed me to share their fascinating camp site at Miedo on Santa Fe island. Marco Robalino kindly provided the weather data. I received financial support by the Erziehungsdirektion des Kantons Zurich in 1989. Finally, I thank Yaël Lubin for introducing me to the ant problem. This study was supported by personal funds.

Resumen

En las zonas aridas del Archipielago de las Galápagos las nectarias extraflorales de los cactos gigantes (*Opuntia echios, Jasminocereus thouarsii*) son fuente de alimento estable para muchas especies de hormigas. Los aspectos ecologicos y de comportamiento de hormigas introducidas y nativas fueron estudiados durante la temporada de garua, entre septiembre y diciembre, en la isla de Santa Cruz, Santa Fe y Daphne Major. La influencia del Niño, el evento de los años 1982-83, que mantiene el record de lluvias desde 1877, fue investigado en la zona xermorfica en 1983 y 1989 y comparado en parte con los datos ganados ante el Niño en la Estacion Charles Darwin en 1981. La hormiga colorada *Wasmannia auropunctata* y la hormiga *Tetramorium bicarinatum* han disfrutado mucho de las lluvias de 1983. La hormiga loca *Paratrechina longicornis* y *Pheidole sp.*, y tambien la hormiga de fuego *Solenopsis geminata* y *Conomyrma sp.*, despues de algunos años muy secos, en 1989 han ocupado muchos cactos gigantes en que antes vivia la *W. auropunctata*. La abundancia de *W. auropunctata* esta disminuyendo desde aquel año.

Despues de la erradicacion de *W. auropunctata* en Santa Fe (1985-89) es la *Pheidole sp.* que se encuentra en este nicho ecologico. Las 3 especies de hormigas *W. auropunctata, P. longicornis*, y *S. geminata* con muchas obreras estan activas durante 24 horas, monopolizando alimentos y mostrando un efecto de grupo; ellas se mueven mas rapidamente en caminos con muchas obreras que en caminos con pocas obreras. En contro, las especies endemicas *Camponotus planus* y *C. macilentus* tienen pocas obreras, buscan solas y estan activas solo de dia (*C. planus*) o de noche (*C. macilentus*) compartiendo los cactos con otras especies. Con cebos de atun el comportamiento de competencia entre *W. auropunctata* y *P.*

longicornis ha revelado que la pequeña *W. auropunctata* ha hecho huir la mas grande y mas rapida *P. longicornis* probablemente mas con un "repellant pheromone" y menos con su pequeño agujon o con su debil mandibula.

References

Brandão, C. R. F. 1991. Adendos ao Catálogo Abreviado das Formigas da Região Neotropical. *Rev. Brasil. Entomol.* 35: 319-412.

Bruneau de Miré, P. 1969. Une fourmi utilisée au Cameroun dans la lutte contre les Mirides du cacaoyer, *Wasmannia auropunctata* (Roger). *Café Cacao Thé* 13: 209-212.

Clark, D. B., C. Guayasamín, O. Pazmiño, C. Donoso, and Y. Páez de Villacís. 1982. The tramp ant *Wasmannia auropunctata*: Autecology and effects on ant diversity and distribution on Santa Cruz Island, Galapagos. *Biotropica* 14: 196-207.

de la Vega, I. M. 1988. Comportamiento en busqueda de alimento de *Wasmannia auropunctata* y hormigas nativas, y competencia entre *Wasmannia auropunctata* y hormigas nativas en las islas Santa Cruz e Isabela, Galápagos. Tesis de la Universidad Central del Ecuador, Quito.

Emery, C. 1893. Notices sur qualques fourmis des Iles Galapagos. *Ann. Soc. Entomol. France* 63: 89-92.

Kastdalen, A. 1982. Changes in the biology of Santa Cruz Island between 1935 and 1965. *Not. de Galápagos* 35: 7-12.

Kiladis, G. and H. F. Diaz. 1984. A comparison of the 1982-83 and 1877-78 ENSO events. *Tropical Ocean-Atmosphere News* (Seattle, WA) 25: 7-8.

Hölldobler, B. and E. O. Wilson. 1990. *The Ants.* Springer-Verlag, Berlin.

Lubin, Y. D. 1984. Changes in the native fauna of the Galapagos Islands following invasion by the little red fire ant, *Wasmannia auropunctata. Biol. J. Linn. Soc.* 21: 229-242.

_____. 1985. Studies of the little red fire ant, *Wasmannia auropunctata,* in a Niño year. Pp. 473-493. In: G. Robinson and E. M. del Pino [eds.]. *El Niño en las Islas Galápagos: El evento de 1982-1983.* Charles Darwin Foundation, Quito, Ecuador.

Meier, R. E. 1982. Ecology and behavior of ants on giant cacti within the Darwin Station and Tortuga Bay (Island of Santa Cruz, Galapagos). Annual Report of the Charles Darwin Research Station 1982.

_____. 1983. Coexisting patterns and foraging behavior of ants within the arid zone of three Galapagos Islands. Annual Report of the Charles Darwin Research Station 1983.

_____. 1985a. Coexisting patterns and foraging behavior of ants on giant cacti on three Galapagos Islands, Ecuador. *Experientia* 41: 1228.

_____. 1985b. Interference behavior of two tramp ants (Hymenoptera: Formicidae) at protein baits on the Galapagos Islands, Ecuador. *Experientia* 41: 1229.

Pickett, C. H. and W. D. Clark. 1979. The function of extrafloral nectaries in *Opuntia acanthocarpa* (Cactaceae). *Am. J. Bot.* 66: 618-625.

Robinson, G. and E. M. del Pino [eds.]. 1985. *El Niño en las Islas Galápagos: El evento de 1982-1983.* Charles Darwin Foundation, Quito, Ecuador.

Silberglied, R. 1972. The little fire ant *Wasmannia auropunctata* a serious pest in the Galapagos Islands. *Not. Galápagos* 19/20: 13-15.

Ulloa-Chacon, P. and D. Cherix. 1990. Pp. 281-289. The Little Fire Ant, *Wasmannia auropunctata* (R.) (Hymenoptera: Formicidae). In: R.K. Vander Meer, K. Jaffe and A. Cedeno [eds.]. *Applied Myrmecology: a World Perspective.* Westview Press, Boulder, CO.

Ulloa-Chacon, P., D. Cherix, and R. Meier. 1991. Bibliografia de la hormiga colorada *Wasmannia auropunctata* (Roger) (Hymenoptera, Formicidae). *Not. Galápagos* 50: 8-12.

Wheeler, W. M. 1924. The Formicidae of the Harrison Williams Galapagos Expedition. *Zoologica* 5: 101-122.

Wiggins, I. L. and D. M. Porter. 1971. *Flora of the Galapagos Islands.* Stanford Univ. Press, Stanford, CA.

Williams, D. F. and P. Whelan. 1991. Polygynous colonies of *Solenopsis geminata* (Hymenoptera: Formicidae) in the Galapagos Islands. *Florida Entomol.* 74: 368-371.

Wilson, E. O. 1987. The arboreal ant fauna of Peruvian Amazon forests: A first assessment. *Biotropica* 19: 254-251.

5

Perspectives on Control of the Little Fire Ant (*Wasmannia Auropunctata*) on the Galapagos Islands

Patricia Ulloa-Chacon and Daniel Cherix

Introduction

The little fire ant, *Wasmannia auropunctata* (Roger), was discovered on the Galapagos Islands (Ecuador) by Silberglied (1972), but according to Clark et al. (1982) its arrival in the archipelago probably predates that time. In the humid conditions of Santa Cruz Island, it spread at a rate of about 170 m a year (Lubin 1984), but during "El Niño" years its expansion may reach 500 m a year. During these years the ants experience conditions of temperature and humidity that are optimal for their development and reproduction (Lubin 1985). Numerous ecological and behavioral studies have demonstrated the impact of *W. auropunctata* on native and endemic invertebrate fauna (ants, scorpions and spiders) of the Galapagos Islands (Clark et al. 1982). Moreover its presence is linked with outbreaks of other pest species (Homoptera, Coccidae) in cultivated areas of Santa Cruz Island (Lubin 1984, P. Ulloa-Chacon, pers. obs.).

On colonized Galapagos Islands, as well as other invaded areas around the world, control of this pest ant species was based on "classical" methods such as the use of highly poisonous and persistent chemicals. Tests were made with different organo-chlorinated substances, e.g. DDT (Fernald 1947, Osburn 1948) and phosphorylated chemicals such as malathion, parathion (Nickerson 1983), mirex (Sacoto 1981), endosulfan, fenthion and malathion (Delabie 1989). Mechanical control measures, i. e. the use of sticky strips (Spencer 1941, Delabie 1989), destruction of the nests (Spencer 1941) or the washing of infested fruits (Carjaval 1983) were also tested. In view of the special circumstances of the Galapagos Islands, Williams (1987) proposed that insect growth regulators (IGR) might be effective. Before starting a field assay we tested the effect of the juvenile hormone analog, methoprene, on laboratory colonies (Ulloa-Chacon and Cherix 1992). Our results showed that it had little effect on the adult worker population (Ulloa-Chacon and Cherix 1990a), but an important one on brood development and queen fecundity (Ulloa-Chacon and Cherix 1990a, 1990b, 1992; Ulloa-Chacon et al. 1991).

We present here the results of a field test using methoprene against *W. auropunctata* on Santa Cruz island.

Material and Methods

A site close to the village of Bellavista was selected. It was located within the transition zone (humid zone) in a cultivated area at about 194 m above sea level. The study was conducted from November 1989 to January 1990, which marked the end of the dry season. We used the same bait as used in our prior studies (Ulloa-Chacon and Cherix 1992) which contained 0.4% methoprene (Pharorid™). In the agricultural zone of Bellavista, the population density of *W. auropunctata* was very high (we found only one other species of Formicidae which belonged to the subfamily Ponerinae, but never on baits and very discrete). The nest density was about 9.0 $\pm$ 4.5 per square meter (Ulloa-Chacon 1990). Within this area we selected four stations (A, B, C and D), and within each of these we marked either three or four 5 m^2 plots. Two or three of these plots were treated (TR) and the other served as a control (T) (see Table 5.1). Each plot was subdivided into 1 m^2 quadrats and marked with colored sticks. At the base of each stick, a bait either with or without methoprene was placed depending on the area.

The ants consumed the baits in about 48 hours (1 cm^3). The bait was renewed in some plots at weekly intervals up to four weeks.

Twenty-four and 48 hours after distribution of the baits, we counted the number of baits with *W. auropunctata* workers and/or other invertebrates in all plots (T and TR). This was also performed at night in order to see if other animal species were attracted to the baits. One week after the beginning of the treatment, and for six weeks thereafter, a weekly evaluation of the ant population was done in randomly selected 1 m^2 subplots by scoring the number of nests, the number of workers, the

TABLE 5.1 Composition of experimental areas in the 4 stations at Bellavista, within the humid zone of Santa Cruz.

Station	Experimental plots Control	Treated	No. of applications
A	T 1	TR 1	4
		TR 2	2
B	T 2	TR 3	2
		TR 4	4
C	T 3	TR 5	3
		TR 6	3
		TR 7	1
D	T 4	TR 8	3
		TR 9	1
		TR 10	1

FIGURE 5.1. Occupation of baits by workers of *W. auropunctata* after 24 hours.

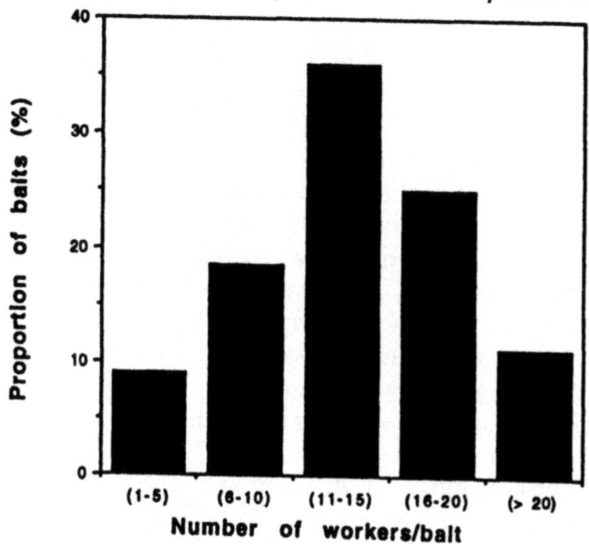

TABLE 5.2. General results of the survey of stations A, B, C and D from week 2 to week 6 after first bait application.

Time weeks no.	No. of nests / m² Control (n = 3)	Treated (n= 9)	No. of queens / m² Control (n= 3)	Treated (n = 9)	No. of workers / m² Control (n = 3)	Treated (n= 9)
2	6.3 ± 1.2	6.5 ± 1.1	11.3 ± 1.3	8.2 ± 1.9	1400 ± 208	1455 ± 304
3	8.3 ± 4.7	12.1 ± 1.5	7.5 ± 4.3	11.2 ± 2.1	1933 ± 1109	2844 ± 443
4	10.3 ± 2.7	10.8 ± 1.9	8.7 ± 5.2	12.4 ± 2.7	1866 ± 481	1988 ± 325
5	13.3 ± 1.4	9.9 ± 1.2	14.0 ± 5.6	8.8 ± 1.6	2367 ± 467	1644 ± 319
6	12.5 ± 3.5	9.0 ± 1.5	8.0 ± 3.0	11.3 ± 5.5	2300 ± 900	1600 ± 458
Mean/weeks	10.3 ± 1.3	9.8 ± 0.7	9.9 ± 1.8	10.2 ± 1.0	1950 ± 274	1954 ± 179

number of queens and the state of the brood. All queens were kept for further studies. Queens collected during each evaluation of the ant populations were taken to the laboratory for an oviposition test. Each queen was isolated in a small nest for 14 hours without workers and then the number of eggs laid during this period was counted. A sample of queens was dissected in order to estimate ovarian development (mean number of oocytes per queen).

Statistical analyses were performed on population density, egg laying and ovarian development. When the data were not normally distributed, we used the logarithm transformation (number of individuals) or the square root transformation (oviposition tests), followed analysis of variance (Anova) and t-tests.

Results

In Bellavista, *W. auropunctata* occupied 81% of all the baits (Figure 5.1). For the rest, a number of beetles (Nitidulidae) and cockroaches were seen, but constituted less than 10% of the total. Additionally 4.5% of baits were not visited or were lost.

The results for the entire study area are presented in Table 5.2 (mean calculated on 6 weeks). An Anova-test (log transformation) showed no significant differences between the treated and the untreated areas in relation to numbers of nests, ($F_{1,3}$=0.46; P=0.71), numbers of queens ($F_{1,3}$=1.59; P=0.20) and numbers of workers ($F_{1,3}$=1.93; P=0.14). Moreover, a comparison after 6 weeks showed there were still no significant differences between T and TR at that time.

A total of 127 oviposition tests were conducted. The mean number of eggs laid by queens from control areas was 9.7 $\pm$ 6.6 eggs per queen in 14 hours. This was significantly different (t-test: t = 6.11; df 126; P< 0.0001) from the queens of treated areas (4.1 $\pm$ 2.4 eggs). These results showed that the egg-laying rate was reduced by about 53% in treated queens. Table 5.3 shows the change in egg-laying from week three to week six. There was a progressive diminution from 7.4 $\pm$ 2.9 eggs per queen at the end of the third week to 2.7 $\pm$ 0.9 eggs by the end of the sixth week. Significant differences between treated and untreated queens first appear in week 4 (t-test, P< 0.01).

TABLE 5.3. Results of oviposition tests from week three to week six after first bait application (n = number of queens).

Week No	No of eggs /queens (mean ± standard error) control	treated	t-test (P<0.01)
3	9.2±4.1 (n=10)	7.4±2.9 (n=10)	NS
4	12.4±2.3 (n=10)	4.7±1.0 (n=35)	t=3.31; df 43
5	7.3±1.7 (n=10)	3.3±0.5 (n=40)	t=2.67; df 48
6	9.8±2.0 (n=10)	2.7±0.9 (n=10)	t=3.20; df 18

FIGURE 5.2. Influence of number of bait applications on queen fecundity; n = number of queens used for oviposition test, when the letters above the columns are different, there are significant differences (t-test; P<0.001).

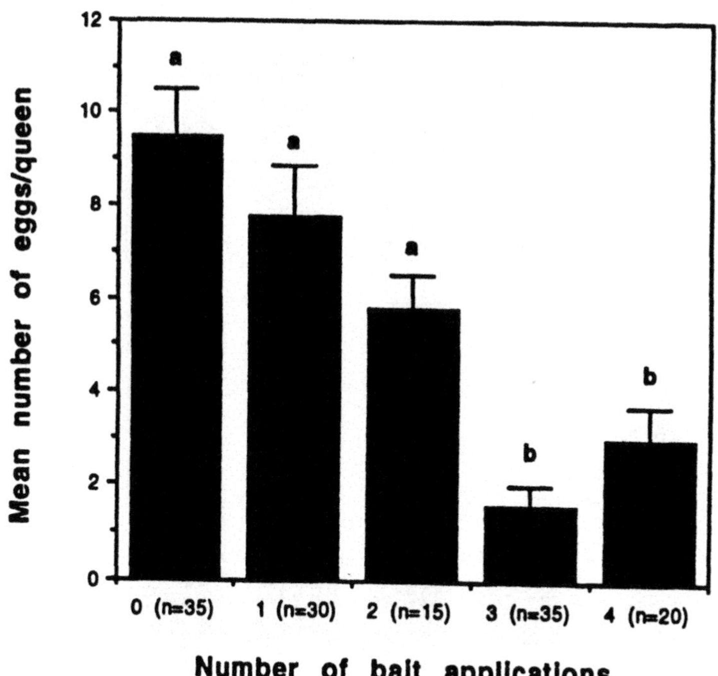

FIGURE 5.3. Comparison of ovarian development among queens from control areas (T) and treated areas TR) at stations A, B and C during five or six weeks after first bait application. When letters above columns are different, there are significant differences (P<0.01).

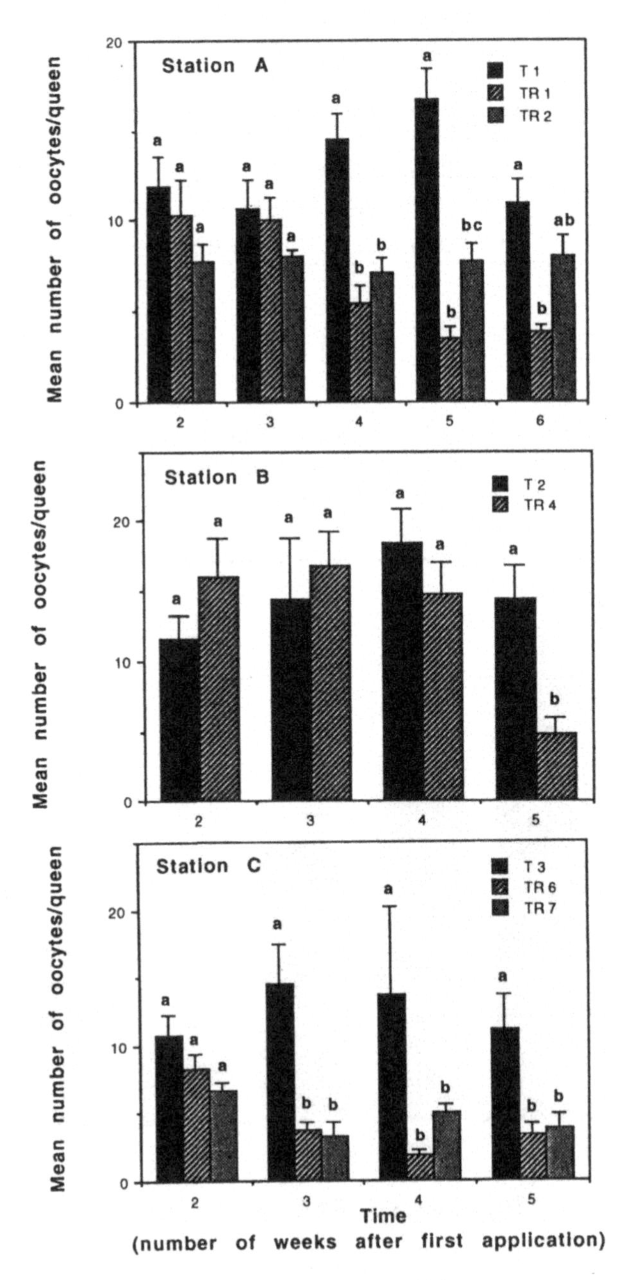

Moreover, we found that the number of bait applications was also of great importance. An Anova test showed significant differences between the egg-laying rates of queens in relation to the number of applications ($F_{4,130} = 16.14$; $P<0.01$). Comparisons between the number of applications and the control showed no statistically significant differences between the control and one or two applications, but high significance with three applications (t= 6.971; df 68; $P<0.001$) and four applications (t= 4.02; df 53; $P<0.001$) (Figure 5.2). We also compared the ovarian development of treated and untreated queens. A weekly estimation of the number of oocytes per queen is shown in Figure 5.3. These comparisons showed that there were small differences between each station (A, B and C). A significant reduction in oocytes occurred after three weeks at station C ($F_{3,36} = 14.8$; $P<0.0001$), after four weeks at station A ($F_{2,36} = 10.74$, $P<0.001$) and five weeks at station B ($F_{1,20} = 14.90$; $P<0.001$).

Discussion

The first important aspect when attempting to control a pest species by means of IGR baits is the attractiveness of the bait itself. In our case, more than 80% of the baits were visited and taken by the ants. This is comparable to the results obtained by Clark et al. (1982) who found 81.9% of sucrose baits visited by workers of *W. auropunctata*. Meier (1985) presented evidence that *W. auropunctata* dominates other species of ants on baits consisting of tuna fish or other high proteins. This is of great help, especially when effective control requires that the target species should feed for a protracted period on baits that might attract other species. In our case some baits disappeared, probably eaten by animals such as lava-lizards of the genus *Tropidurus*, rats, cats, or dogs. The low toxicity of methoprene to vertebrates (LD50 = 34,500 mg/kg, Siddall and Slade 1974) is a good argument to pursue these studies even if there are some negative affects on non-target invertebrate fauna (Retnakaran et al. 1985 and Pihan 1975).

Our field results confirmed those obtained under laboratory conditions (Ulloa-Chacon and Cherix 1992) insofar as the egg-laying rate of queens was reduced. Moreover, dissection of queens showed that there was also an important reduction in oocyte numbers after only three weeks. At the end of the first part of our investigation (six weeks) we could not see any diminution in the ant population, but our continuing observations demonstrated that populations were highly affected after three months without further treatment and have reached 50 to 75% reduction within the treated area (D. Cherix and P. Ulloa-Chacon, unpublished). It is important to note that effectiveness depends upon the number of treatments. One or two treatments were not enough to affect queens in our six week experiment. This substantiates results obtained with IGR's against other ant species, e.g. *Monomorium pharaonis* in which population reductions appeared eight to ten weeks following treatments (Edwards 1975, Edwards and Clarke 1978, Edwards et al. 1981, Rupes et al. 1988) or after six to ten weeks in *Pheidole megacephala* (Edwards et al. 1981, Horwood 1988). Field experiments conducted on *Solenopsis invicta* showed important population reduction occurred only after three to six months following treatment (Banks et al. 1983, 1986).

Acknowledgments

We would like to thank the people of the Charles Darwin Research Station at Puerto Ayora on Santa Cruz Island for their help during our stay and especially Maribel Martinez and Axa Luna who bravely bore the stings of the ants each day we worked in the field. Thanks are also due to Prof. Dave Fletcher for his thoughtful suggestions and improvements of the manuscript. Financial support by the Academic Society (Lausanne) to P. Ulloa-Chacon and the Museum of Zoology (Lausanne) is gratefully acknowledged.

Resumen

Cebos conteniendo metopreno a 0.4% fueron probados para el control de la pequeña hormiga de fuego u hormiga colorada, *Wasmannia auropunctata*, en Islas Galápagos (Santa Cruz). Seis semanas después de la primera aplicación de los cebos, no se observó reducción en el número de nidos, de reinas y de obreras, pero la fecundidad de las reinas se afectó significativamente. El número de huevos puestos por reina se redujo en aproximadamente un 53%, como tambien el número de oocitos por reina, en las áreas en las que se colocaron cebos 3 ó 4 veces.

References

Banks, W. A., L. R. Miles, and D. P.Harlan. 1983. The effects of insect growth regulators and their potential as control agents for imported fire ants (Hymenoptera: Formicidae). *Florida Entomol.* 66: 172-181.

Banks, W. A. 1986. Insect growth regulators for control of the imported fire ant. Pp. 387-398. In: C. S. Lofgren and R. K. Vander Meer [eds.]. *Fire ants and leaf-cutting ants* . Westview Press, Boulder, CO.

Carvajal, V. 1983. Mecanismo de difusion de la hormiga colorada *Wasmannia auropunctata* en la isla Santa Cruz. *Colegio Nacional Galápagos.*

Clark, D. B., C. Guayasamín, O. Pazmiño, C. Donoso, and Y. Páez de Villacís. 1982. The tramp ant *Wasmannia auropunctata*: Autecology and effects on ant diversity and distribution on Santa Cruz Island, Galapagos. *Biotropica* 14: 196-207.

Delabie, J. H. C. 1989. Avaliação preliminar de téchnica alternativa de controle da formiga "pixixica" *Wasmannia auropunctata* em cacauais. *Agrotrópica* 1: 75-78.

Edwards, J. P. 1975. The effects of a juvenile hormone analogue on laboratory colonies of Pharaoh's ant *Monomorium pharaonis* (L.). *Bull. Entomol. Res.* 65: 75-80.

Edwards, J. P. and B. Clarke. 1978. Eradication of Pharaoh's ants with baits containing the insect juvenile hormone analogue methoprene. *Pest Control* 20: 5-10.

Edwards, J. P., G. W. Pemberton, and P. J. Curran. 1981. The use of juvenile hormone analogues for control of *Pheidole megacephala* (F.) and other house-infesting ants. *Proc. Intern. Conf. Regulation of insect development and behaviour*. Pp. 769-779. Karpacz, Poland.

Fernald, H. T. 1947. The little fire ant as a house pest. *J. Econ. Entomol.* 40: 428.

Horwood, M. A. 1988. Control of *Pheidole megacephala* (F.) (Hymenoptera: Formicidae) using methoprene baits. *J. Aust. Entomol. Soc.* 27: 257-258.

Lubin, Y. D. 1984. Changes in the native fauna of the Galapagos Islands following invasion by the little red fire ant, *Wasmannia auropunctata. Biol. J. Linn. Soc.* 21: 229-242.

_______ . 1985. Studies of the little fire ant, *Wasmannia auropunctata* in a Nino year, Pp. 473-493. In: *El Niño en las Islas Galápagos El evento de 1982-1983.* Fundación Charles Darwin para las Islas Galápagos, Quito, Ecuador.

Meier, R. E. 1985. Interference behavior of two tramp ants at protein baits on the Galapagos Islands, Ecuador. *Experientia* 41: 1228-1229.

Nickerson, J. C. 1983. The little fire ant, *Ochetomyrmex auropunctata* (Roger) (Hymenoptera: Formicidae). *Florida Dep. Agr. Cons. Serv. Entomol. Circ.* 248.

Osburn, M. R. 1948. Comparison of DDT, chlordane and chlorinated camphene for control of the little fire ant. *Florida Entomol.* 31: 11-15.

Pihan, J. C. 1975. Utilisation des effets tératogenes des hormones juvéniles et de leurs mimiques dans la lutte contre les insectes. *Ann. Biol.* 14: 29-44.

Retnakaran, A., J. Granett, and T. Ennis. 1985. Insect growth regulators. Pp. 529-601. In: G. A. Kerkut and L. I. Gilbert [eds.]. *Comprehensive insect physiology biochemistry and pharmacology.* Pergamon Press, New York.

Rupes, V., Z. Vrba, L. Tomasek, J. Ryba, I. Hrdy, V. Rabas, M. Pokorny, J. Benesova, J. Brestovsky, J. Chmela, J. Cumpelik, J. Krecek, E. Kuklikova, V. Kobik., Z. Krejcikova., J. Ledvinka, D. Maslenova, V. Mazak, Z. Moric, J. Plichta, J. Teplan, F. Tondl, and E. Velebilova. 1988. The efficiency of two variants of bait with juvenoid methoprene (Lafarex N and Lafarex N'86) on the Pharaoh's ant, *Monomorium pharaonis,* in domestic locations. *Acta Entomol. Bohemoslov.* 85: 418-427.

Sacoto, X. L. 1981. Pruebas en laboratorio para el control de la hormiga *Wasmannia auropunctata* mediante la utilizacion de cuatro cebos. Universidad Central del Ecuador, Quito.

Siddall, J. B. and M. Slade. 1974. Tests for toxicity of juvenile hormone and analogues in mammalian systems. Pp. 345-348. In: W. J. Burdette [ed.]. *Invertebrate endocrinology and hormonal heterophylly*. Springer-Verlag, New York.

Silberglied, R. 1972. The little fire ant, *Wasmannia auropunctata* a serious pest in the Galapagos Islands. *Noticias de Galápagos* 19/20: 13-15.

Spencer, H. 1941. The small fire ant *Wasmannia* in citrus groves -- a preliminary report. *Florida Entomol.* 24: 6-14.

Ulloa-Chacon, P. 1990. Biologie de la reproduction chez la petite fourmi de feu *Wasmannia auropunctata* (Roger) (Hymenoptera, Formicidae). Ph. D. thesis, Université de Lausanne, Faculté des Sciences.

Ulloa-Chacon, P. and Cherix, D. 1990a. Perspectives de contrôle chimique de la petite fourmi de feu *Wasmannia auropunctata* au moyen d'analogues de l'hormone juvénile. *Actes Coll. Insectes Sociaux.* 6:187-194.

_____ . 1990b. The little fire ant, *Wasmannia auropunctata* (R.) (Hymenoptera, Formicidae). Pp. 281-289. In: R. K. Vander Meer, K. Jaffe and A. Cedeno [eds]. *Applied Myrmecology: A World Perspective.* Westview Press, Boulder, CO.

_____ . 1992. Effect of the insect juvenile hormone analogue methoprene on the little fire ant, *Wasmannia auropunctata* (Hymenoptera, Formicidae) *J. Econ. Entomol.* (in press).

Ulloa-Chacon, P., D. Cherix, and R. Meier. 1991. Bibliografia de la hormiga colorada *Wasmannia auropunctata* (Roger) (Hymenoptera, Formicidae). *Noticias de Galápagos* 50: 8-12.

Williams, D. F. 1987. Foreign travel report in Galapagos Islands. *Attini* (An International Newsletter on Pest Ants) 18: 16-17.

6

Food Searching Behavior and Competition Between *Wasmannia Auropunctata* and Native Ants on Santa Cruz and Isabela, Galapagos Islands

Ines de la Vega

Introduction

The little fire ant, *Wasmannia auropunctata* (Roger), is widely distributed and is found on the inhabited Galapagos Islands (Santa Cruz, Floreana, San Cristobal, Isabela), as well as uninhabited San Salvador (Lubin 1984). It is considered a menace to many of the invertebrate and some vertebrate fauna of the Islands due to its aggressiveness.

This ant was probably introduced accidentally to the Galapagos, on the island of Santa Cruz between 1930-1940 (Kastdalen, pers. comm.). It has been determined that, where *W. auropunctata* exists in large densities, no other ant species exist (Clark et al. 1982, Lubin 1984). For this reason studies were initiated on (1) the feeding behavior and daily activity of *W. auropunctata* and other native ant species and (2) the extent of competition between *W. auropunctata* and native species.

Materials and Methods

Studies of *W. auropunctata* behavior and competition with other ants were conducted during two climatic periods: the hot season and the cool or "garua" season. The test sites were located primarily on Santa Cruz and included 3 locations in the dry zone (Charles Darwin Research

Station (CDRS), Barranco, Punta Nuñez) and one each in the transition zone (Tortoise Reserve [La Caseta]), the humid zone (Cerro Maternidad [*Scalesia*]), and the *Miconia* zone (Media Luna). One site was located on Isabela and one on Santo Tomas (Corazon Verde, Los Tintos).

Ant behavior and activity in the dry zone at the CDRS was measured by the number and speed of foraging ants on trees along a 100m section of the tourist trail. A total of 66 trees of *Opuntia echios* and *Jasminocereus thouarsii* on both sides of the trail were numbered and used for observation sites. The search for ants on these trees was made from the soil upwards. The number of trails of *W. auropunctata* and/or other species were noted for each tree. Observations were made monthly and the percentage of trees occupied by *W. auropunctata* and the other ants species was determined each month. "T" tests were calculated to determine the highest number of infested trees and the largest number *W. auropunctata* trails per tree for the two seasons.

The daily activity of *W. auropunctata* was determined at 3 separate times (0700, 1300, and 1900) on 10 of the 66 trees in the prior study. The 10 trees were chosen at random. On each of the 10 trees, an area of 10x10 cm was marked off at a height of 1m above the soil in a section of the tree with the most trails. The following data were noted: (1) number of ants ascending and descending in 60 seconds, and (2) foraging speed of the ants in both directions (sec/10cm). The temperature in the shade was noted for each observation. Observations were made in triplicate at the times indicated above. These data were collected over a 12-month period. The total numbers of descending and ascending ants in each tree were summed and the mean for the 10 trees was calculated. Using these means, graphs of the activities were drawn. Similiarly, means of the ants foraging speed were used to construct graphs of how fast the ants move up and down the trees.

Correlations were calculated for *W. auropunctata* foraging activity on cactus trunks between the number of ants:

1. ascending in 60 seconds and the speed of the ascending ants
2. descending and the speed of descending ants
3. ascending and descending and
4. ascending and descending and the speed of the descent

In order to compare the activities of *W. auropunctata* with other ant species, we counted the ants on trees in each of the study areas over the two seasons. A search was made for ants on each tree and those trees with ants were marked. Observations of activity were carried out for 24 hours in the first season and for 72 hours in the second season. The same activities were observed as in the transect, except that the observations were carried out three times every three hours. Ant activity was graphed using the log of the mean of each three observations. For the results from the upper part of the Tortoise Reserve (La Caseta), Wilcoxon's Test was used to verify significant differences.

Observations of competition and displacement of *W. auropunctata* and other species of ants were made during experiments in which foods were offered to each ant species over 30 minutes in each zone except for Media Luna and La Caseta (N=1536). The total number of species attacked and the total number of species attacking while feeding on the baits was noted. All of the results of the observations on food, for all the sites and in the two seasons were totalled, including the number of attacks of each species and the number of species attacked.

TABLE 6.1. Number of trees with *W. auropunctata* and other ant species.

Month	*No.trees. with Wa*	*% of 66 Trees*	*Mf*	*Pl*	*Wa+other species*
Dec. 1981	45	68.00			
Jan. 1982	37	56.00	1		2 Mf
Feb.	40	60.61	4		1 Mf
Mar.	40	60.61	1		1 Mf
Apr.	33	50.00	3		2 Pf
May	27	40.90	7	1	2M & Pl
Jun.	40	60.61	4	1	1 Pl
Jul.	23	34.85	4		1 Pl
Aug.	21	31.80	2		
Sep.	33	50.00	5		
Oct.	27	40.90	2		1 Pl
Nov. 1982	22	33.30	4	1	

Wa=Wasmannia, Mf=Monomorium floricola, Pl=Paratrechina longicornis
X=37.79 Y=25.200 t=3.797 P<0.01

TABLE 6.2. Number of trails on trees with *W. auropunctata*.

Months	*No. trails with W. auropunctata*	*No. trails per occupied tree*
Dec.	104	44
Jan.	59	31
Feb.	73	41
Mar.	63	38
Apr.	63	37
May	53	35
Jun.	93	44
Jul.	43	27
Aug.	28	23
Sep.	57	38
Oct.	51	32
Nov.	44	26

X=72.42 Y=44.60 t=2.92 P<0.02

Results

The seasonal study of *W. auropunctata* behavior established that more trees were infested with *W. auropunctata* during the months of December to June (hot season) than for the months of July to November (garua season) ($P<0.01$, Table 6.1). The number of trails of *W. auropunctata* was also higher from December to June than from July to November ($P< 0.02$, Table 6.2).

It was determined that at 0700, when feeding conditions were optimal (in February and May), there was a marked increase in *W. auropunctata* activity.

The speed of *W. auropunctata* in descending the cactus trunk was lowest in October.

The observations at 1300 for *W. auropunctata* showed no variation in the activity of descending or ascending. The speed of ascending and descending was constant for all of the months.

However, the activity of *W. auropunctata* at 1900 showed most descending activity in May, which is the month when there were fewer extrafloral nectaries.

The speed from December to May is slightly less than from June to November.

The correlation between the number of ants and the speed over 10cm was not significant in the 12 months, except for December, which was significant at 1% level, and for April and August at 5%. From these results, the speed of *W. auropunctata* does not depend on the density of the ants. It may depend on other conditions such as available food requirements.

The study comparing the daily activity of *W. auropunctata* with other ants using the census taken in the two seasons resulted in 15 different analyses, only one of which will be treated here.

Humid zone: Cerro Maternidad. The activity curves in the humid zone at Cerro Maternidad *(Scalesia)* for *W. auropunctata* and *Pheidole* sp. were very similar. *Camponotus planus* was not active at 0700 or at 0400. *Paratrechina nesiotis* had a similar activity pattern to *Tetramorium bicarinatum,* except for a decrease in *T. bicarinatum*'s activity at 2200, which was possibly due to low temperature.

The activity of *W. auropunctata* in the two *Scalesia* trees was almost equal except that in one of the *Scalesia* trees the ant activity was slightly higher at 1300, possibly due to wind variation. There was no *Pheidole* sp. activity at 0400 because of a strong drizzle and low temperature. It was most active at nights and in the morning with less activity occuring in the afternoon. *T. bicarinatum* was not active at 0700 on the first day of observations. There was a strong drizzle at 1300 and *T. bicarinatum* were active in the soil but not on the trees. The same occurred on the 2nd and 3rd days at 1000, while at 1600 they were not active due to excessive drizzle.

In December to June *W. auropunctata, Pheidole* sp., *T. bicarinatum,* and *P. nesiotis* were active over the 24-hour period. *Camponotus planus* is not active in the morning.

Changes in the activity patterns of *T. bicarinatum, W. auropunctata* and *C. planus* were observed in June to November due to increased rain, strong winds, variation in temperature and changes in vegetation. *T. bicarinatum* was more active in the soil. *Paratrechina nesiotis* had no changes in its behavior between the seasons. *Pheidole* sp. was observed near *W. auropunctata* in the sample tree in the first season, while in the second season it was found away from the tree, in the soil.

Of all the different attacks recorded between the different ant species, most of the attacks were carried out by *T. bicarinatum* and *S. geminata.* The species most attacked were *P. nesiotis, Pheidole* sp. and *W. auropunctata* (Table 6.3).

W. auropunctata received 11 attacks and attacked other ants 3 times (Table 6.3). This may be interpreted as an effect of strong repellant pheromones which the *W. auropunctata* workers emit which repel other species so as to gain access to the source of food artificially presented (Meier 1983). It was observed that with naturally available food, ants such as *T. bicarinatum* and *S. geminata* attack weaker species and as such also cause damage to the Galapagos ecosystem.

TABLE 6.3. Competitive interactions between ants.

	Attacking Ants:												
Ants attacked:	*W. a*	*P. n*	*Ph.*	*T. b*	*T. m*	*S. g*	*O. b*	*C. a*	*Ca.*	*C. m*	*Cy.*	*M. f*	*Total Attacked*
Wasmannia auropunctata	-	8	0	1	0	1	1	0	0	0	0	0	11
Paratrèchina nesiotis	-	1	9	18	0	20	1	0	1	0	-	0	50
Pheidole sp.	1	13	0	0	0	0	0	0	0	0	-	0	14
Tetramorium bicarinatum	0	2	2	0	0	0	0	0	0	0	2	0	6
Tapinoma melanocephalum	0	1	0	1	0	0	0	0	0	0	0	0	2
Solenopsis geminata	2	1	0	2	0	0	0	0	0	0	0	0	5
Odontomachus bauri	0	0	0	0	0	0	0	0	0	0	0	0	0
Conomyrma albemarlensis	0	5	0	0	0	0	0	0	0	0	0	0	5
Cardicondyla sp.	0	0	0	0	0	0	2	0	0	0	0	0	2
Camponotus macilentis	0	0	0	4	0	0	0	0	0	0	0	0	4
Cylindromirex sp.	0	4	0	2	0	0	0	0	0	0	0	0	6
Monomorium floricola	0	0	0	0	2	0	0	0	0	0	0	0	2
Total attacking	3	35	11	28	2	21	2	2	1	0	2	0	107

Resumen

Se llevó a cabo un estudio de la hormiga introducida *Wasmannia auropunctata* desde diciembre 1981 hasta noviembre 1982 en cuatro zonas en la Isla Santa Cruz y una sola zona en la Isla Isabela en el Archipiélago de Galápagos. Se observó durante la época de calor y de frío (Garúa) la actividad de *W. auropunctata* y otras hormigas y competencia entre especies.

Observaciones de actividad de hormigas en cactus indica que *W. auropunctata* fue más abundante en la época de calor que en época de garúa. Se observó también que el patrón de actividad cambio entre épocas y lugar.

Se observo competencia entre hormigas para alimento con *Solenopsis geminata, Paratrechina nesiotis,* y *Tetramorium bicarinatum* atacando otras hormigas con más frecuencia. *W. auropunctata* fue una de las tres especies de hormigas más atacadas.

References

Clark, D. B., C. Guayasamín, O. Pazmiño, C. Donoso, and Y. Páez de Villacís. 1982. The tramp ant *Wasmannia auropunctata*: Autecology and effects on ant diversity and distribution on Santa Cruz Island, Galapagos. *Biotropica* 14: 196-207.

Lubin Y. D. 1984. Changes in the native fauna of the Galapagos Islands following invasion by the little red fire ant, *Wasmannia auropunctata*. *Biol. J. Linn. Soc.* 21: 229-242.

Meier, R. E. 1983. Patrones de coexistencia y comportamiento alimenticio de hormigas dentro de la zona árida en tres islas de Galápagos. Informe Anual de la Estación Científica Charles Darwin, 1983: 26-28.

7

The Ecology of *Wasmannia Auropunctata* in Primary Tropical Rainforest in Costa Rica and Panama

Leeanne E. Tennant

Introduction

The little fire ant, *Wasmannia auropunctata,* is a widespread ant species native to the New World tropics (Kusnezov 1951, Fabres and Brown 1978). In areas where it has been introduced outside of its natural range, it has become a problem both economically, in agriculture, and ecologically, in its effects on native organisms, particularly local ant communities. In areas of high *W. auropunctata* densities, few if any ant species are usually found (Clark et al. 1982, Lubin 1984). However, in its natural habitat of primary lowland rainforest, *W. auropunctata* does not dominate as it does in disturbed areas (Levings and Franks 1982, S. Cover unpub. obs., author's pers. obs.). Ecological and behavioral explanations for this difference in the effects of *W. auropunctata* on the local ant fauna are being investigated as part of research on the interactions between ants (including *W. auropunctata*) and the ant-plant *Conostegia setosa* in primary rainforest in Costa Rica and Panama. This paper presents preliminary data from a long-term study of this ant-plant system.

My research concentrates on how ecological factors such as ant species life histories and competitive interactions between ant species determine the occupants of *C. setosa.* From these studies, preliminary data on *W. auropunctata* have been obtained regarding frequency of occupation of *C. setosa,* nesting habits, foraging strategies, and interactions with other ant species. Studies of *W. auropunctata* in its native habitats can provide insight into the basic biology of the species and perhaps also into possible methods of control.

Study Sites and Organisms

Studies of *C. setosa* (Melastomataceae) and its ant occupants began in 1989 at the La Selva Biological Station, Heredia Province, Costa Rica (lowland tropical forest, 50-100 meters above sea level) and in 1991 at the Nusagandi Pemasky Station, Comarca de Kuna Yala (San Blas), Panama (tropical wet forest, approximately 350 meters above sea level). *C. setosa* is an understory plant found almost exclusively as large clumps of plants (ranging from 1-110 stems). The formicaria ("ant-houses") of *C. setosa* consist of bi-lobed pouches located at the base of each leaf.

Where *C. setosa* occurs, only a small subset of the local ant fauna interacts with this plant because the small size of the domatia allows a limited size range of ants to occupy the plants. At La Selva, *C. setosa* is most often occupied by two dominant, generalist ant species, *W. auropunctata* and *Pheidole annectans,* and is occasionally occupied by small colonies of two species of *Solenopsis* (subgenus Diplorhoptrum). The colonies of the dominant ant species are large and usually occupy many stems of a clump as well as nest sites in the soil and other vegetation around the plants. The *Solenopsis* spp. colonies generally occupy only one pouch per plant. Although one plant only houses one ant species, more than one ant species is often found in different plants within a single clump. At Nusagandi, the plants are most commonly occupied by a small red *Pheidole* species (*Pheidole* sp.1, sp. nov., S. Cover, pers. comm.). This *Pheidole* species is probably an obligate ant-plant nester since it has only been found in ant-plants and never forages off the plant, even when baited with tuna. More rarely, *C. setosa* is occupied by *W. auropunctata,* another *Pheidole* species (*Pheidole* sp. 2), and *Brachymyrmex* sp. 1.

Methods

Occupation of* C. setosa *plants. Fifteen clumps of *C. setosa* plants were chosen along a 3200 meter trail loop through primary forest at the La Selva Biological Station in March 1991. Eleven clumps were selected along two trails through primary forest at Nusagandi in September 1991. In each clump, all plants were tagged and numbered and the ant occupants of each plant were identified.

Interactions between species at baits. Interactions between ant species at baits were observed in the five *C. setosa* plots containing *W. auropunctata* at La Selva during June and July 1991. Fifteen baits were placed in each plot in the same arrangement as in the bait surveys described above. Each bait consisted of approximately four grams of tuna and a Kimwipe® soaked in sucrose water placed on a (8 cm^{2}) white plastic tray. Observations of interactions between ant species, two baits at a time, were conducted immediately after placement and continued for 15 minutes. Each bait was then checked after 30, 45, and 60 minutes (from placement) and the number of ants of each species present at the bait was recorded. Ants of each species were collected and preserved in 70% ethanol for later identification.

Bait surveys of ant communities. A $4m^2$ plot was established around each of the *C. setosa* clumps and the plot was baited with tuna to identify the species of ants present in the area. These bait surveys were conducted during the wet season at both sites: June and July 1991 at La Selva and October 1991 at Nusagandi. Baits of approximately four grams of tuna were placed directly on the ground around the perimeter of each plot (two baits on each side) and two baits were placed in the middle of the plot (10 ground baits per plot). Five additional baits were placed in low vegetation on $8cm^2$ white plastic trays. Baits were put out between 0700 and 1200. After one hour, a $30cm^2$ area of leaf litter and surface soil around each ground bait was collected and placed in a Ziploc® bag. In addition, each vegetation bait was examined and all ants were collected. The litter samples were later sorted in a tray either in the field or in the laboratory, and a representative number of each ant species found was collected from each sample and placed in 70% ethanol for later identification.

Tuna baits do not sample the entire ant community but do attract the widely-foraging ant species that are most likely to interact with the ants in the *C. setosa* plants and with *W. auropunctata*. Furthermore, by collecting leaf-litter samples, a haphazard sample of leaf-litter and soil-dwelling ant species that are not attracted to the baits was also collected. The bait method used here proved to be the most effective for collecting a large number of ant species in a reasonable amount of time in April 1991 trials at La Selva.

In addition, 80 plots without *C. setosa* were surveyed for the presence of *W. auropunctata* and *P. annectans* using the same tuna bait survey technique described above for the *C. setosa* plots. These plots were established every 50 meters along the same trail loop as the *C. setosa* plots. In general, these areas were fairly open and were located from one to 30 meters from the trail. Ant specimens from all baits were examined for the presence of *W. auropunctata* and *P. annectans.*

The identity and number of ant species in each plot were recorded. Species diversities for each plot were not calculated because estimates of diversity require not only the number of species, but also the abundance of each species, which for ants requires knowledge of the number of colonies present. Ant species have very different modes of foraging (i.e. some recruit heavily while others do not recruit at all) so that the number of foragers at baits is not representative of the number of colonies nearby.

Results

Occupation of C. setosa plants. At the La Selva Biological Station, *W. auropunctata* occupied plants in five of the 15 plant clumps surveyed (Table 7.1). *W. auropunctata* was the sole occupant of three of these clumps and was present along with *P. annectans* in the other two clumps. *W. auropunctata* was found living in 15.6% of the occupied plants surveyed while *P. annectans* was found in 50% of the plants (n=128 plants). Thirty-three percent of the clumps contained only one ant species, 27% had two ants species and 40% had three ant species (Table 7.1). All the ant species occupying *C. setosa* at La Selva foraged off the plants and were also found living in other nest sites such as organic matter caught in palms and in rotting wood.

At Nusagandi, the dominant ant occupant was a small red *Pheidole* (sp. 1) which did not forage off the plant and is most likely an obligate ant-plant occupant. *Pheidole* sp. 1 occupied plants in nine of the 11 *C. setosa* clumps surveyed while *W. auropunctata* was found in only four clumps (Table 7.1). At Nusagandi, *W. auropunctata* occupied only 10.7% of the occupied plants surveyed while *Pheidole* sp. 1 occupied 69.3% of the plants. Forty-five percent of the clumps had one species of ant occupant, 36% had two ant species, and 18% had three ant species (Table 7.1). *W. auropunctata* was the sole occupant of two of its four clumps (both of which only contained one plant), and co-occurred with *Pheidole* sp. 1 and with both *Pheidole* sp. 1 and *Pheidole* sp. 2 in the other two plots.

Bait surveys of ant communities. Surveys of all 11 *C. setosa* plots at Nusagandi revealed an average of 13.5 $\pm$ 2.1 ant species per plot at ground baits (Table 7.2) and an average of 2.6 $\pm$ 1.3 ant species per bait (range=0-8 species, n=110 baits). Vegetation baits have not yet been analyzed. Although *W. auropunctata* occupied plants in only four of the eleven plots, it was present at baits in eight plots. The number of ant species per plot in the eight plots with *W. auropunctata* was significantly greater than the number of ant species in plots without *W. auropunctata* (Students' t-test, $t=4.7$; $df=1,9$; $p<0.01$; Table 7.2). This difference may be

TABLE 7.1. The total number of *C. setosa* plants occupied by each ant species in 15 clumps at La Selva, Costa Rica and 11 clumps at Nusagandi, Panama during 1991 and the number of *C. setosa* clumps in which each ant species was present.

Ant Species	*La Selva*		*Nusagandi*	
(or No. of ant species)	*No. of clumps (n=15)*	*No. of plants (n=128)*	*No. of clumps (n=11)*	*No. of plants (n=75)*
W. auropunctata	5	20	4	8
P. annectans	12	64	-	-
Solenopsis sp. 1	7	25	-	-
Solenopsis sp. 2	6	19	-	-
Pheidole sp. 1	-	-	9	52
Pheidole sp. 2	-	-	5	10
Brachymyrmex sp. 1	-	-	1	5
one ant species	5		5	
two ant species	4		4	
three ant species	6		2	
W. auropunctata only	3		2	

TABLE 7.2. The average number of ant species (± standard deviation) collected from ground baits in each plot at La Selva, C. R. and at Nusagandi, Panama.

Location, Plot type	*Avg. No. of Ant Species (± standard deviation)*	*No. of Plots Analyzed*
La Selva, Costa Rica		
C. setosa plots with *W. auropunctata*	14.4 ± 1.1	5
Transplant plots	14.0 ± 1.1	39
Plots with *W. auropunctata*	14.3 ± 1.8	18
Plots without *W. auropunctata*	13.6 ± 1.7	21
Nusagandi, Panama		
C. setosa plots	13.5 ± 2.1	11
Plots with *W. auropunctata*	15.3 ± 1.7	8
Plots without *W. auropunctata*	8.7 ± 2.2	3

due to the fact that the areas without *W. auropunctata* were extremely wet (one was beside a stream) and most ant species, including *W. auropunctata,* were unable to nest in the soil there. Most of the ant species collected in these plots were nesting arboreally. *W. auropunctata* was attracted to a greater number of ground baits per plot (average=5.6 ± 1.6) at Nusagandi than any other ant species collected at baits. *Hypoponera* sp. 1 was the next most common ant species but was collected from an average of only 2.8 ± 1.5 baits per plot.

In the five *C. setosa* clumps containing *W. auropunctata* at La Selva, there was an average of 14.4 ± 1.1 ant species per plot (Table 7.2) and 3.0 ± 1.2 ant species per bait (n=50 baits). This was very similar to the number of ant species found in the *C. setosa* clumps at Nusagandi. The overall average number of ant species in 39 of the 60 transplant plots was 14.0 ± 1.1 ant species. Plots in which *W. auropunctata* was present had a significantly higher number of ant species per plot than did plots from which it was absent (Students' t-test, $t=2.82$; $df=1,37$; $p<0.01$; Table 7.2).

Interactions between ant species at baits. In the bait observations of the five *C. setosa* clumps containing *W. auropunctata* at La Selva, *W. auropunctata* was found at 27 of the 75 baits (36%) and was the sole occupant of only 6 baits (8%). An average of 2.8 ± 1.2 ant species were present per bait (n=75 baits). Where present, *W. auropunctata* was always one of the first ant species to arrive at a bait and usually recruited at least ten other workers and often many more (>40 workers). At the end of 60 minutes, *W. auropunctata* was the sole occupant of 48% of the baits it visited, was replaced by another species at only 11% of baits (by *Solenopsis* sp., *Pheidole* sp. 3, and *Atta cephalotes*), and co-occurred with one or two other ant species at 41% of baits. *W. auropunctata* tended to share baits with two types of ant species: (1) species with large solitary foragers such as *Aphaenogaster* sp., *Ectatomma* sp., *Pachycondyla* sp., and *Odontomachus* sp. and (2) heavy-recruiting species including *Solenopsis (Diplorhoptrum)* sp., *Paratrechina* sp., and *Pheidole* sp 3.

Few encounters were observed between *W. auropunctata* and other ant species at baits. Aggressive interactions were occasionally documented between *W. auropunctata* and several *Solenopsis (Diplorhoptrum)* species. In most encounters, *W. auropunctata* won the bait by recruiting enough foragers to chase other species off. However, one large black *Pheidole* species (*Pheidole* sp. 3, possibly undescribed, E. O. Wilson (pers. comm.) consistently won encounters against *W. auropunctata* at baits. *Pheidole* sp. 3 workers were first attacked and repelled by *W. auropunctata* workers but then soldiers were recruited which were able to snip the *W. auropunctata* workers in two and drive them off the baits.

Discussion

These preliminary results indicate that *W. auropunctata* is quite common in at least some areas of primary lowland rainforest in the New World Tropics (Kusnezov 1951, Fabres and Brown 1978, Levings and

Franks 1982) but does not have the devastating effects on the local ant community that it has in areas where it has been introduced.

Wasmannia auropunctata occupied 14% of the *C. setosa* plants in Costa Rica and 10% in Panama, while two species of *Pheidole* were much more frequent occupants. *Wasmannia auropunctata* may eventually be outcompeted by *P. annectans* at La Selva since it initially occupies many plants (unpub. data), but *P. annectans* is the most frequent occupant overall. The mode of colonization of these plants by *W. auropunctata* is most likely by expansion of a colony or budding off of a part of the colony, often without queens, since dissection of plants occupied by *W. auropunctata* usually revealed no queen.

In contrast to the potentially competitive system at La Selva, colonization of *C. setosa* at Nusagandi is dominated by *Pheidole* sp. 1, which may be chemically attracted to *C. setosa* and most likely is the first to find and colonize the plants. An additional survey of the ant occupants of 119 *C. setosa* plants along two trails at Nusagandi during October 1991 revealed that 65% of the plants were occupied by *Pheidole* sp. 1 and 11% by *W. auropunctata*. *Wasmannia auropunctata* most often occupied plants near gaps or in disturbed areas of forest and occasionally appeared to be occupying plants abandoned by *Pheidole* sp. 1. *Wasmannia auropunctata* did not dominate occupation of clumps of plants. In both locations *W. auropunctata* co-occured in clumps with *Pheidole* spp.

Interactions between ant species. When present, *W. auropunctata* was attracted to more baits per plot than any other ant species. This suggests possible superior exploitative abilities regarding food sources, especially since *W. auropunctata* was always one of the first ant species to discover a bait in the La Selva *C. setosa* plots and tended to persist at the baits. Lubin (1984) and Clark et al. (1982) found that *W. auropunctata* usually dominated baits and was the sole occupant of greater than 80% of their baits. In the La Selva *C. setosa* plots, *W. auropunctata* occurred alone at only 48% of baits where it was present and coexisted with one or two other ant species at 44% of baits. Most often these other ant species were large solitary foragers which were ignored by *W. auropunctata* and were able to get away with large pieces of tuna. However, *W. auropunctata* also occasionally occurred at a bait with another heavily-recruiting ant species, usually keeping spatially separated at the bait. Similar to observations in the Galapagos Islands by many researchers, including Clark et al. (1982) and Lubin (1984), I observed few interactions between *W. auropunctata* and other ant species at baits. Most often this was because *W. auropunctata* was the first to find and dominate a bait. *W. auropunctata*'s success at discovering baits and nest sites before other species is likely due to its polydomous nesting habits, with nests spread out in parts over a large area, and its mass recruitment foraging system. Both of these characteristics may allow *W. auropunctata* to encounter and exploit a greater number of food items and nest sites than other species.

W. auropunctata's ability to persist at baits is likely due to its defensive abilities, which include a powerful sting and a very well-defended rugose exoskeleton, and to its ability to recruit large numbers of workers very quickly. Clark et al. (1982) suggested that interference competition may be involved in *W. auropunctata*'s success in the Galapagos Islands since *W. auropunctata* lost at baits only 25% of the time. Interference competition may also play a role in primary rainforest communities in this study. *W. auropunctata* was replaced by another ant species at only 11% of the baits. However, eventual replacement of *W. auropunctata* by another ant species after the 60 minute observation period is possible. Meier (1984) suggested that *W. auropunctata* emits a repellent pheromone which hinders the access of other ant species to the baits. This may occur to some extent at my sites but my results do not support this as a general rule because many other species were able to co-occur with *W. auropunctata* at the baits.

Ant communities. All plots surveyed in both Costa Rica and Panama had a relatively high number of ant species per plot, averaging aproximately 14 species (Table 7.2). Plots with *W. auropunctata* present even had a higher number of ant species than plots without *W. auropunctata*. These results clearly underestimate the actual number of ant species present per plot because mainly general-foraging ant species were attracted to the tuna baits and only a small number of soil-dwelling ant species were collected. However, this indicates even more strongly that *W. auropunctata* is able to coexist at these sites with ant species that have similar dietary requirements and foraging behaviors.

The presence of *W. auropunctata* does not lower the local number of ant species in primary rainforest, as it does in areas where it has been introduced such as the Galapagos Islands (Clark et al. 1982, Lubin 1984, Meier 1984). Levings (1983) also found that many ant species were able to coexist with *W. auropunctata* in primary deciduous tropical forest on Barro Colorado Island, Panama. However, nest densities of *W. auropunctata* on Barro Colorado Island were low (0.05-0.13 nests/m^2, Levings and Franks 1982) compared to many areas where *W. auropunctata* has been introduced (0.75-1.75 aggregations/m^2 in the Galapagos Islands, Lubin 1985). When *W. auropunctata* is present in such high densities, it is difficult or impossible for other ant species to coexist (Clark et al. 1982, Lubin 1984). Nest densities appear to be equally low at La Selva (M. Kaspari, pers. comm.) and at Nusagandi (pers. obs.). Similarly, Levings (1983) found *W. auropunctata* present but not always locally abundant in all six of her study sites on Barro Colorado Island. Horvitz and Schemske (1990) found *W. auropunctata* in low abundance at each of their six study sites in Mexico, and S. Cover found low *W. auropunctata* densities in Cuzco Amazonico, Madre de Dios, Peru, though distinctly higher near small gaps in the canopy (pers. comm.).

The rare cases of high densities of *W. auropunctata* coexisting with other ant species in introduced areas have been explained by the low overlap in nesting and dietary requirements between *W. auropunctata* and the sympatric species (Lubin 1985). These other ant species are usually specialist feeders or live in areas unsuitable for *W. auropunctata*. This explanation may also apply to some extent in primary rainforest. However, it cannot explain the coexistence of *W. auropunctata* with many other heavy-recruiting generalist species such as *P. annectans* which have nesting and foraging behaviors similar to *W. auropunctata*. As Levings (1983) observed, many ant species in tropical ground nesting communities (and in lower arboreal niches as well) that have similar nesting and dietary requirements often nest in close proximity and their foraging ranges may overlap considerably. Levings (1983) suggested that each species in a tropical ant community uses a slightly different mode of foraging and the first one to discover a food item usually is the most successful. Results from this study indicate that *W. auropunctata* is particularly successful at discovering and exploiting baits. In a tropical forest with its diverse availability of nesting sites and food sources, *W. auropunctata* may be able to exploit many of these sources while still leaving enough food and nest sites for other ant species in the area. Whether most tropical ant species are food or nest site limited is not known.

Several rainforest ant species such as *P. annectans* and *Pheidole* sp. 3 appear to be able to compete successfully against *W. auropunctata* and may help keep population densities low. Diffuse competitive interactions from a large number of neighboring ant species may also limit the success of *W. auropunctata* in primary forest. In new areas, *W. auropunctata* may be free of such competitors as well as from natural predators and parasites. The army ant, *Neivamyrmex pilosus,* has been observed carrying *W. auropunctata* brood on Barro Colorado Island, Panama (M. Kaspari, pers. comm.) and might play a large role in controlling *W. auropunctata* densities. Other aggressive, predaceous ant species may also limit *W. auropunctata* populations. Furthermore, parasitic flies in the family Phoridae have been found to have a large impact on the behavior and ecology of many ant species (Feener 1981), as well as on interactions between ant species (Feener 1988). Studies of predators and parasites are needed in both *W. auropunctata*'s native habitats and in areas where it has been introduced. Continued studies of *W. auropunctata* in its native habitat will provide useful information on how *W. auropunctata* interacts with other members of the native ant community and how it may be held in check by these and other biotic factors.

Acknowledgments

Support for research at La Selva, Costa Rica was provided by The Organization for Tropical Studies and the Pew Charitable Trusts. Research in Panama was supported by a Short-term Fellowship from The Smithsonian Tropical Research Institute. Dr. E. O. Wilson generously provided the financial assistance which allowed me to attend this conference. I would like to thank S. Cover and M. Kaspari for helpful comments and discussion of this work, B. Traw for field assistance, and E. O. Wilson for his continued support. I also wish to thank the staff and researchers at La Selva and STRI for their company and encouragement.

Resumen

Estoy estudiando la ecologia y los comportamientos de *Wasmannia auropunctata* en bosque tropical primario en Costa Rica y Panama. Este estudio es parte de mis investigaciones de la influencia mutua entre hormigas y la planta *Conostegia setosa (Melastomataceae)*. Aunque *W. auropunctata* es una plaga ecologica en muchas regiones donde ha sido introducida, no es dominante en las comunidades de hormigas en sus lugares nativos en bosque tropical. *W. auropunctata* es muy commun en mis sitios en Costa Rica y Panama. La encontre en plantas en cinco de 15 sitios en Costa Rica y en cinco de 11 sitios en Panama. Colecte un promedio de 14 especies de hormigas de cebos de atun en areas donde hay *W. auropunctata* en cada lugar. Esto es diferente de los efectos de *W. auropunctata* en areas donde ha sido introducida donde no hay muchas otras especies de hormigas. *W. auropunctata* se encontro en mas cebos de atun en cada sitio que todas las otras especies pero no se influye mucho con las otras hormigas. En los cebos y en las plantas, *W. auropunctata* fue a veces apropriada de antemano o sacada de su sitio por otras especies de hormigas. Es posible que otras especies competitivas limiten las densidades de *W. auropunctata* en bosque tropical.

References

Clark, D. B., C. Guayasamín, O. Pazmiño, C. Donoso, and Y. Páez de Villacís. 1982. The tramp ant *Wasmannia auropunctata*: Autoecology and effects on ant diversity and distribution on Santa Cruz Island, Galapagos. *Biotropica* 14: 196-207.

Fabres, G. and W. L. Brown, Jr. 1978. The recent introduction of the pest ant *Wasmannia auropunctata* into New Caledonia. *J. Aust. Entomol. Soc.* 17: 139-142.

Feener, D. H. 1981. Competition between ant species: outcome controlled by parasitic flies. *Science* 214: 815-817.

_____ . 1988. Effects of parasites on foraging and defensive behavior of a termitophagous ant, *Pheidole titanis* Wheeler (Hymenoptera: Formicidae). *Behav. Ecol. Sociobiol.* 22: 421-427.

Horvitz, C. C. and D. W. Schemske. 1990. Spatiotemporal variation in insect mutualists of a neotropical herb. *Ecology* 71: 1085-1097.

Kusnezov, N. 1951. El género *Wasmannia* en la Argentina (Hymenoptera: Formicidae). *Acta Zool. Lilloana* 10: 173-182.

Levings, S. C. 1983. Seasonal, annual, and among-site variation in the ground ant community of a deciduous tropical forest: some causes of patchy species distributions. *Ecol. Monog.* 53: 435-455.

Levings, S. C. and N. R. Franks. 1982. Patterns of nest dispersion in a tropical ground ant community. *Ecology* 63: 338-344.

Lubin, Y. D. 1984. Changes in the native fauna of the Galápagos Islands following invasion by the little red fire ant, *Wasmannia auropunctata*. *Biol. J. Linn. Soc.* 21: 229-242.

_____ . 1985. Studies of the little fire ant, *Wasmannia auropunctata*, in a Niño year, 473-493. In: El Niño en las Islas Galápagos: El evento de 1982-1983. Fundación Charles Darwin para las Islas Galápagos, Quito, Ecuador.

Meier, R. E. 1984. Coexisting patterns and foraging behaviour of ants within the arid zone of three Galapagos Islands. Charles Darwin Research Station Annual Report.

8

Relations Between the Little Fire Ant, *Wasmannia Auropunctata*, and Its Associated Mealybug, *Planococcus Citri*, in Brazilian Cocoa Farms

J. H. C. Delabie, A. M. V. da Encarnação and I. M. Cazorla

Introduction

The little fire ant, *Wasmannia auropunctata* (Roger), is a serious problem in Brazilian cocoa plantations. Not only is its sting highly feared by pod pickers, but the ant also tends homopterans, such as *Planococcus citri* Risso *(Coccidea:Pseudococcidae)* (Delabie 1988). This mealybug is one of the most damaging sap-sucking insects in the world, particularly in citrus and mango orchards (Salazar 1972). It also can live on 150-200 plant species according to Kirkpatrick (1953). It is one of the most common mealybug species in cocoa plantations in the world, though it is generally considered a minor pest (Strickland 1951, 1952; Kirkpatrick 1953; Hanna et al. 1956). It is, however, the main vector of a cocoa virus disease in Trinidad (Kirkpatrick 1953, Entwistle 1972).

In West Africa, it is one of the most important vectors of the Swollen-Shoot virus disease (Strickland 1951, 1952; Hanna et al. 1956; Entwistle 1972). *Planococcus citri* is also the most common mealybug on Bahia and is almost always associated with *W. auropunctata* (erroneously identified as *Solenopis* spp in some papers) (Silva 1944, 1950; Delabie 1988). It reduces bean production up to 10% by sap sucking (Delabie and Cazorla 1991). There is no evidence of the existence of males in the population studied here, even though males were generally described in *P. citri* populations in other countries (James 1937, Kirkpatrick 1953, Salazar 1972). Thus it is suspected of being parthenogenetic in this region. Valuable information on the biology of *P. citri* and other mealybug species, regarding their association with ants, can be found in Nixon (1951), Kirkpatrick (1953), Entwistle (1972), and Salazar (1972).

Aside from its association with *P. citri, W. auropunctata* is known for its ability to associate with a variety of coccids on a range of plants, including many with economic importance (Spender 1941, Smith 1942, Salazar 1972). *W. auropunctata* has also achieved renown for a number of different reasons (Delabie 1988, Ulloa-Chacon and Cherix 1990), the most important of which is its very competitive habits and its efficacy as a colonizer (Fabres and Brown 1978; Lubin 1984; Delabie 1988, 1990).

Our study of the association between *W. auropunctata* and *P. citri*, two species interacting with tropical crops, is particularly important from both the agronomic and ecological points of view. Our objective was to gain a better understanding of how both partners benefitted by this association. The association here is not obligatory, and is characterized as a nonsymbiotic mutualism (Cushman and Beattie 1991).

Study Site and Methods

The relationship between the two species of insects was studied on cherelles and pods originated by the simultaneous hand pollination of 400 flowers on producing cocoa trees. The experimental area was located at the Unitaria Farm, belonging to the "Centro de Pesquisas do Cacau", Ilheus, State of Bahia, Brazil. The experiments lasted 32 weeks, from April to November 1990. Other observations were casually carried out in the same area until June 1991.

Observations were made weekly on pod growth (length in mm), pod maturation, and the number and distribution of mealybugs. Pod pathology and mortality were also observed, including the effects of a virulent epidemic of Black Pod Rot, *Phytophthora* spp. Variations in mealybug populations were studied pod by pod, and the mean number of insects for all the observations was determined. Also, occurrences of the first and subsequent colonizations of mealybugs on pods were analyzed separately. The success of mealybug colonization by individuals and by groups was also observed. Furthermore, sap-sucking and mobile mealybugs as well as individuals carried by ants were collected and later measured at the laboratory.

Other observations made at the tree sites (pods) were the distance between each fruit and the tree base, and the location of ant nests. All the pods were observed and the presence or absence of an ant-mealybug association noted.

Observations

Location of ant-mealybug colonies on trees.

The colonies formed by the association between *W. auropunctata* and *P. citri* were located mainly on the pods (Silva 1944, Delabie 1988). In the absence of fruit, the ants may tend the mealybugs on young shoots of the same plant or on shade trees, such as some Myrtaceae and *Gallesia gorazema* (Silva 1944).

All our observations were made during the fruiting period and consequently, only on pods. The colonies were dependent on the distance of the pod from the tree base where the ant nest was located. Almost all the fruits were infested from the base to up to two meters high. The number of infested pods decreased as the distance from the nest increased (Figure 8.1). Almost all the pods stayed free of mealybugs from a distance of 5.5 m or greater.

FIGURE 8.1 Percent of pods with mealybugs and ants, as a function of the distance of the pods from the tree base (n=459 pods on 31 trees).

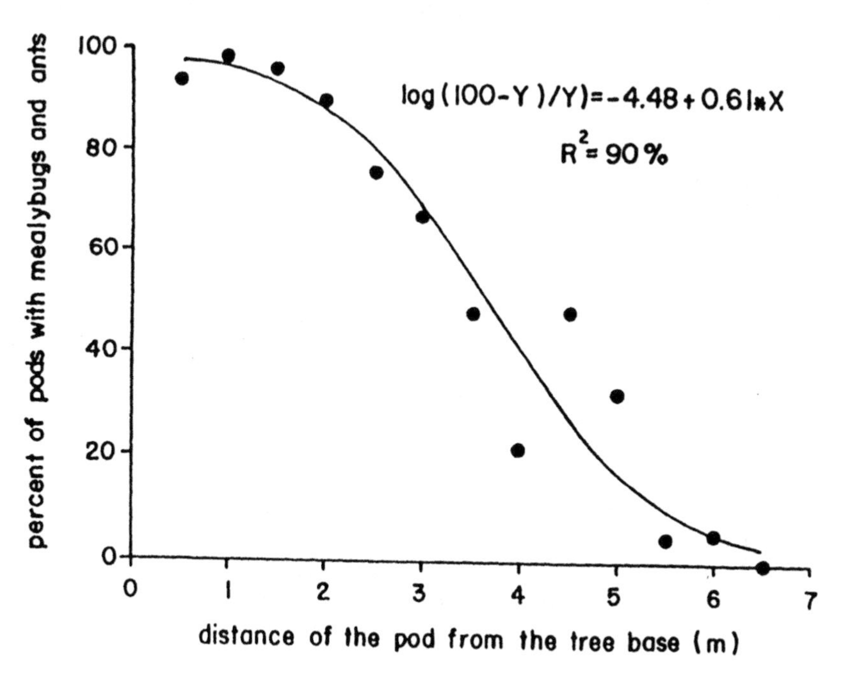

FIGURE 8.2 Variation of mealybug infestation and survival of pods.

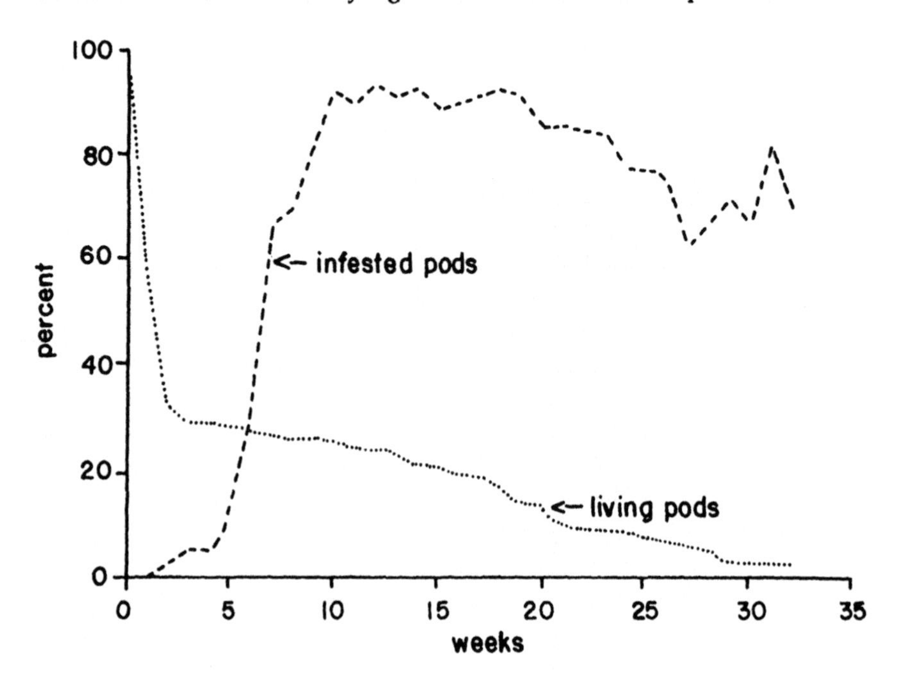

FIGURE 8.3 Fluctuation of mealybugs per pod during the maturation period.

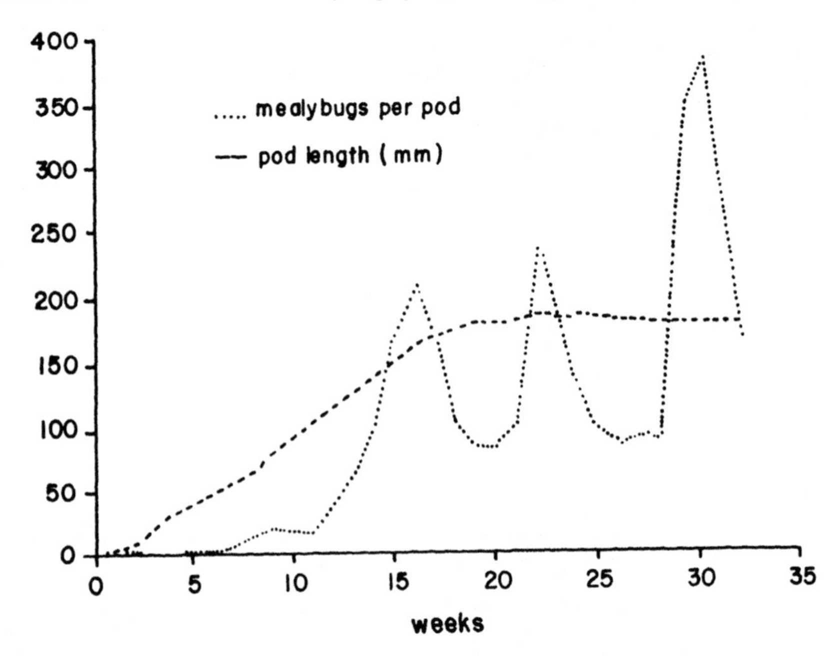

Evolution of mealybug populations during pod growth

Only 118 (29.5%) of the 400 simultaneously hand-pollinated flowers had formed cherelles three weeks later (Figure 8.2). Of this number, 38 developed into pods after 22 weeks at which time their average length was 185 mm (Figure 8.3). This high mortality occurred during the last months due to Black Pod Rot. The number decreased to 24 fruits at harvest time (27-28 weeks), when maturation was achieved and the pods were characterized by a yellow golden color. Observations were carried out until the 32nd week (Figures 8.2 and 8.3).

Cherelles with only a few mealybugs (1-5) were observed two weeks after pollination. The quantity of feeding mealybugs grew quickly, especially from the 5th to the 9th weeks. After this, 85-90% of pods remained infested with ant-mealybug associations (Figure 8.2). The mealybug populations on each pod were evaluated weekly and some examples are reported in Figure 8.4. The populations of mealybugs were highly variable on the pod itself, and also from one pod to another. Due to this variability, the mean of the *P. citri* population of all the pods was determined each week. This data are shown graphically in Figure 8.3, so that natural variations of the populations are more easily viewed. Four peaks are evident. The low peak at 8 to 9 weeks after pollination corresponds to the first *Pseudococcidae* generation of the pod colonizers, which generally established on the cherelle peduncle. The three subsequent peaks are separated by successive intervals of 7 weeks, each corresponding to one generation. This represents twice the time determined by Kirkpatrick (1953) for the life cycle of *P. citri* in Trinidad. These three generations show a mean of up to 200 individuals per pod, after 15-17, 22-23 and 29-30 weeks. These intervals correspond to the maximal growth of the second, third and fourth generations, respectively (Figure 8.3). The location of the mealybug populations was not dependent on pod shape and length; generally, they established all around the pod with some preference for the top.

FIGURE 8.4 Fluctuation of mealybug populations on four pods during the experiment.

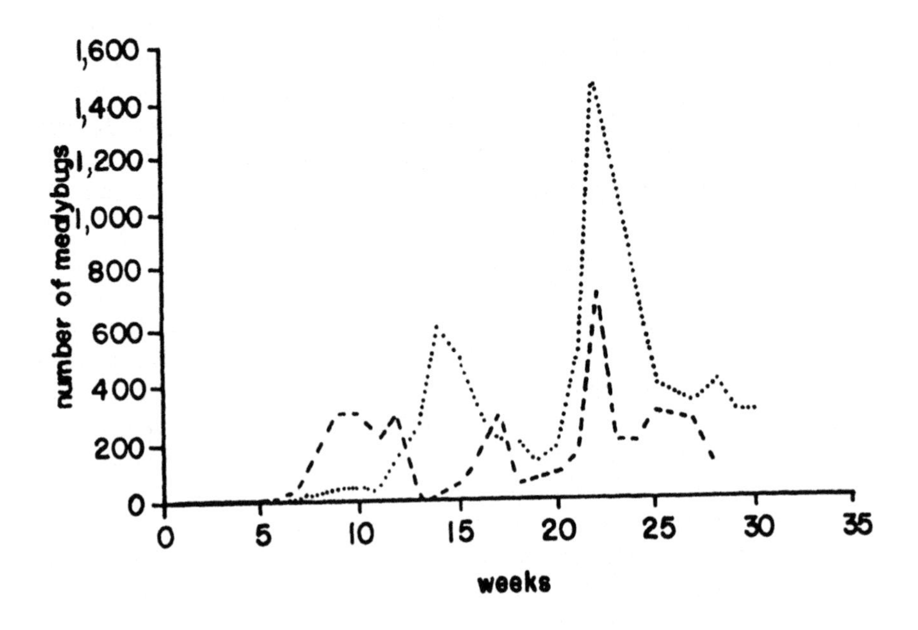

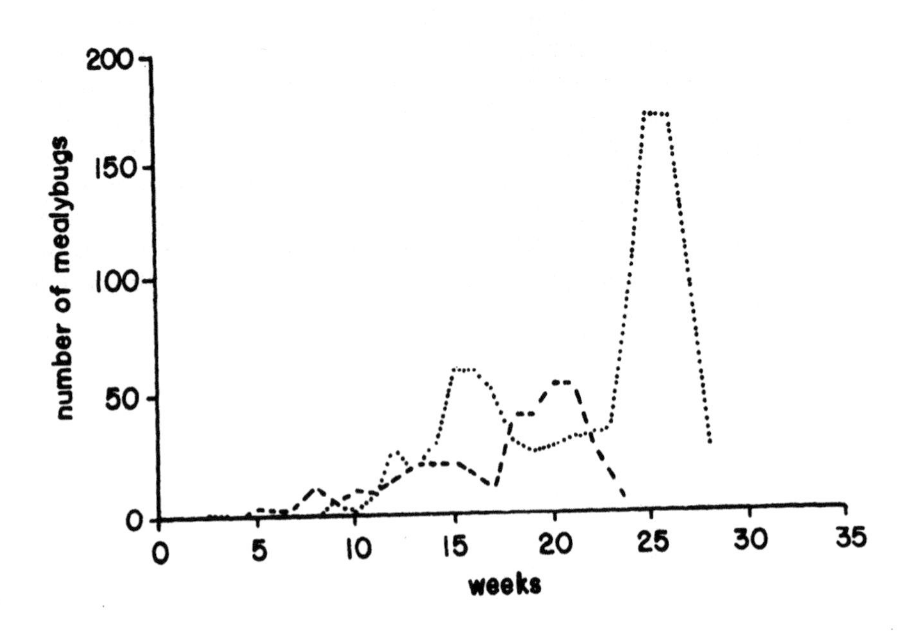

FIGURE 8.5 First-order and second-order pod mealybug colonizations and colonies aborted.

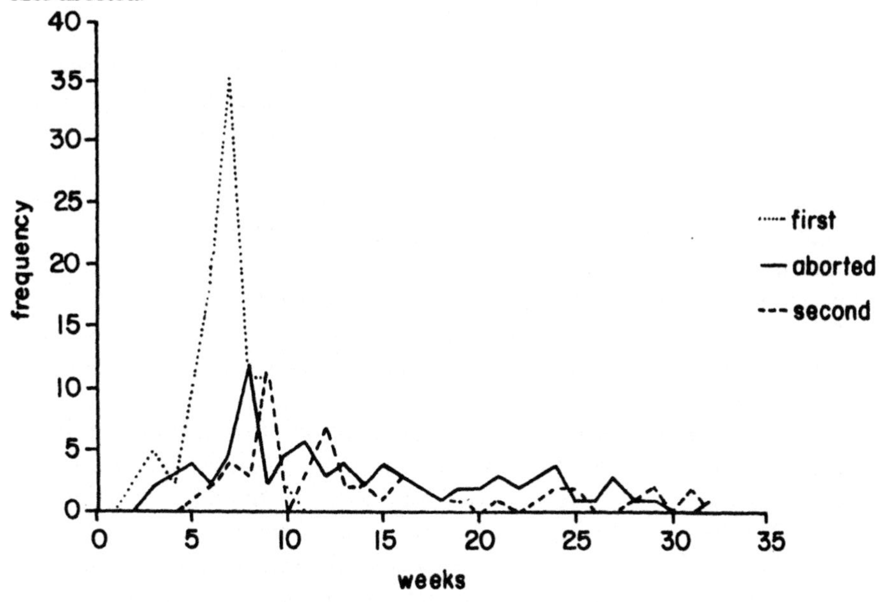

FIGURE 8.6 Number of colonies as a function of the initial number of mealybugs in the colonization of pods.

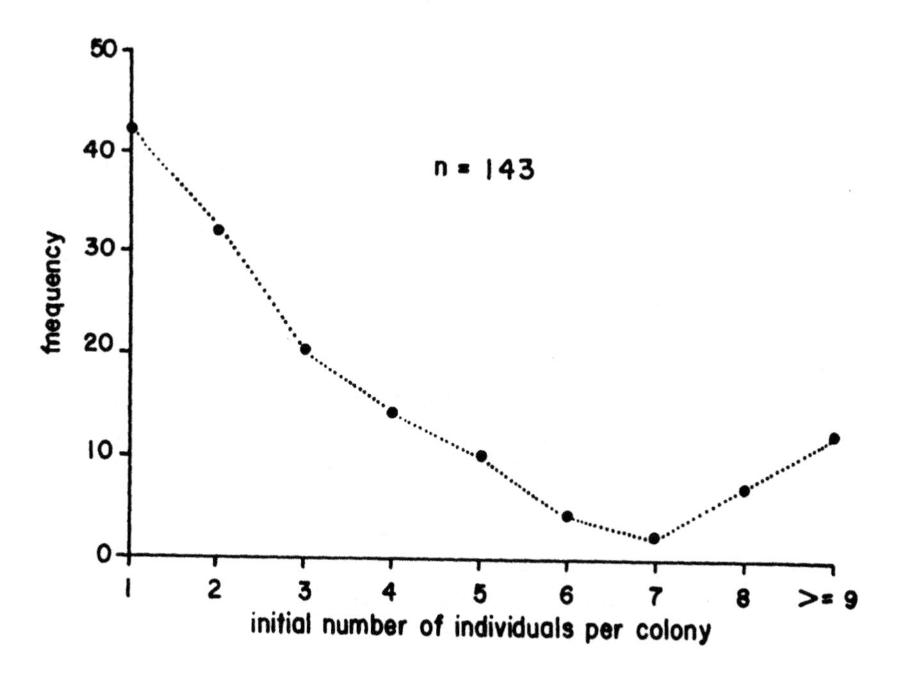

Colonization of cherelles by mealybugs

The occurrences of first-order and second-order colonizations, as well as aborted colonizations, are reported in Figure 8.5. The principal peak of colonization (36% of the total) is observed on 7-week- old pods. The development of second-order colonizations occurs in a sequence almost identical to that of the aborted colonizations only one or two weeks later. This shows that the occupation of sites is continuous, but that the most favorable period is at the beginning of cherelle development. The relative synchrony between the different populations which, for example, resulted in the population trends shown in Figure 8.3, was due mainly to the development of populations from first-order pod colonizations.

If all the initial data on mealybug formation of colonies on pods are joined (including second-order colonizations), the majority of colonies are founded by only a few individuals (42% by one individual, 38% by two, 20% by 3, etc.) (Figure 8.6). The descending part of the curve, around 85% of the observations, corresponds to colonizations made by mobile single mealybugs or mealybugs carried by ants. The second part of the curve in Figure 8.6 shows the initial colonies made by 8 to 80 individuals. They represent 15% of the observations. It is highly improbable that single mealybugs were able to form colonies of 40 or 80 individuals by themselves in the period of one week. A detailed series of observations showed that the ants were found carrying two types of mealybugs: (a) female nymphs transported one by one from one pod to another; and (b) ovisacs laid by mature mealybug females, transported in the same manner (Figure 8.7). Each of them contained from 12 to 295 eggs (mean=120 eggs per ovisac), or even newly-hatched nymphs. The number of eggs laid by a single female given in the literature is very variable (50-600) and depends on the host-plant nature and its location (Kirkpatrick 1953, Entwistle 1972, Salazar 1972). Thus, the ants are able to implant future well-populated colonies of mealybugs on pods, carrying only an ovisac full of eggs.

As shown in Figure 8.8, the success of pod colonization depends on the number of mealybugs in the initial phase. After two weeks, only 38% of the colonies formed initially by a single individual were alive, compared to the 94% of those formed with an initial number greater than 5. After one generation (7 weeks, approximately 45 days), 28% of colonies formed by a single individual were still active compared to 77% of the second category. Thus, the greater the initial population on a pod, the more chance it has to survive and reproduce.

FIGURE 8.7 Percent of (A) single mealybugs (n=133) and (B) mealybugs carried by ants (n=76) in each class (length).

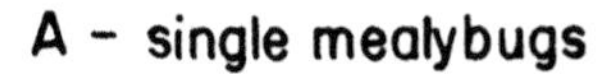

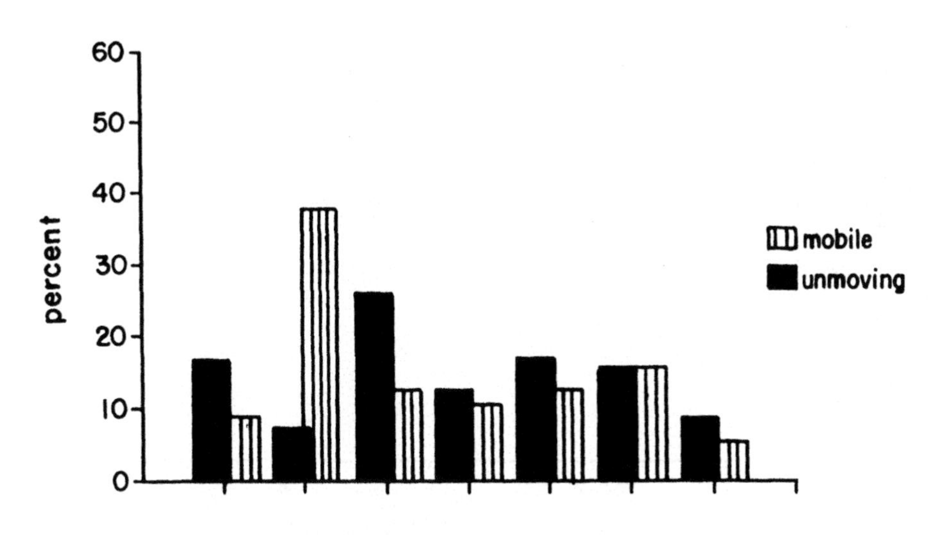

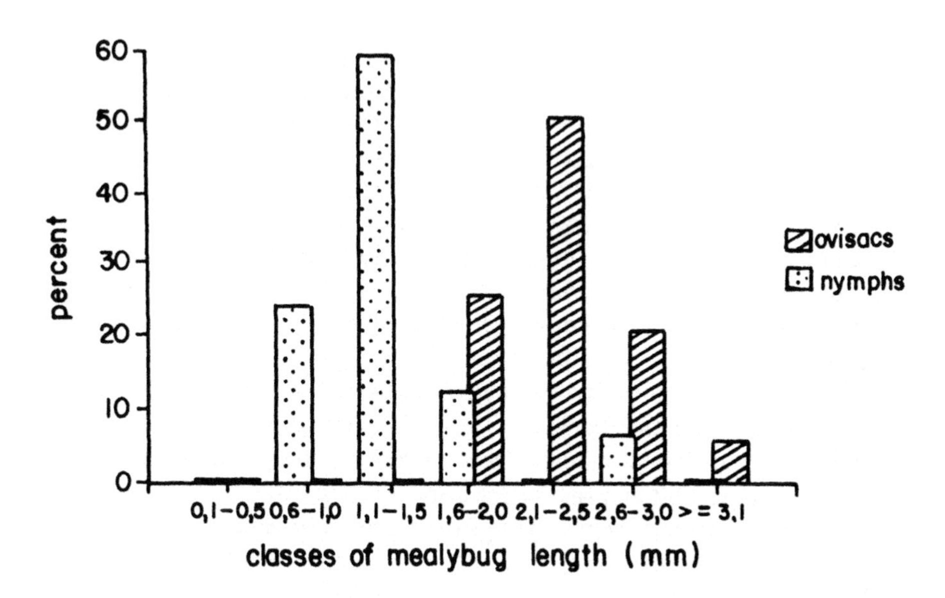

FIGURE 8.8 Variations in the survival of mealybug colonies on pods as a function of the initial number of individuals.

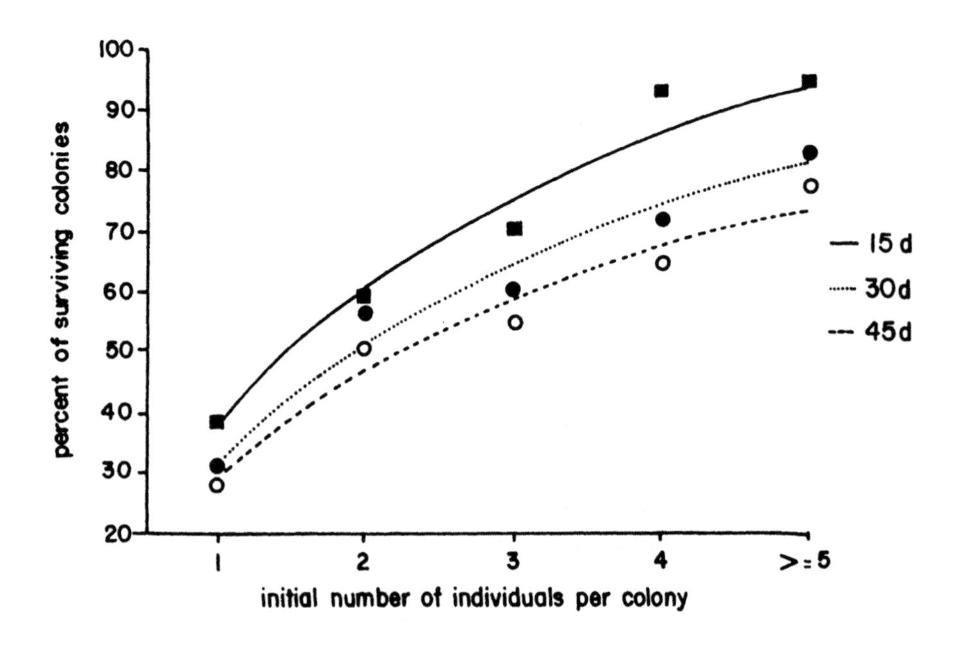

Discussion and Conclusions

The distribution of mealybug colonies on the cocoa trees is a direct consequence of the energetic necessities of ants located in nests at the tree base. It may be disadvantageous for the ants, from an energetic point of view, to maintain mealybugs too distant from the nests. In other words, the gain from honeydew is less than the rising energy costs with distance and time. This pattern is in agreement with the general lines of the central place foraging theory (Holldobler and Wilson 1990). On the other hand, ant numbers may decrease when distance from the nest increases, and protection against mealybug and ant predators or parasites is probably less efficacious.

Our observations on the mutualism between *W. auropunctata* and *P. citri* show that the ants benefitted from the honeydew produced by the mealybugs and, less obviously, from catching some predators attracted to *P. citri*, (regarded as prey by the ants). On the other hand, predation by the ants protected the mealybugs from predators and parasites. Also, ingestion of honeydew protected against opportunist fungi, that could grow on excess honeydew or wax. The ants keep the mealybugs clean, concentrating on the juvenile forms. Our conclusions are generally similar to those formulated in other papers relative to the relations between ants and homopterans (Nixon 1951, Wood 1982, Buckley 1987, Sudd 1987).

In conclusion, we made the following observations regard the mutualistic relationship of *W. auropunctata* and *P. citri*. The ants carry single nymphs and ovisacs of mealybugs and distribute them on available and favorable sites (pods) in their foraging area. If the number of homopterans on a pod justifies continuous searching by the ants for honeydew, the newly formed mealybug colony has a good chance to survive and reproduce, and will form successive generations so long as plant sap is available. This is true because the number of ants on a honeydew source is proportional to its productivity (Dreisig 1988). Small numbers of mealybugs have little chance to attract foraging ants, because the small amount of honeydew they produce does not compensate, from an energetic standpoint, the activity of the ants. Aggregations like these would receive only casual visits by ants and would never induce recruitment. This is the main reason that such colonies will suffer relatively high and rapid mortality, while larger populations are generally more successful. On the other hand, it is almost impossible to find, in the same experimental area, *P. citri* not attended by *W. auropunctata*. The reason is that, under our field conditions, the mealybugs are not only dependent on the ants for their survival but are also very dependent on favorable sap-sucking sites, the best sites being in the foraging territory of the ants, where aggregations of mealybugs are more likely to be formed.

The transport of mealybug nymphs or ovisacs by the ants to favorable sites for sap-sucking has to be regarded as a strategy for maintaining the ants' foraging territory. An adequate distribution of mealybug colonies will justify constant activity by the ants from the nest at the tree base up to the pods.

Acknowledgments

The authors are grateful to Bernardo dos Santos, J. Crispim S. do Carmo, J. Raimundo M. dos Santos, for technical assistance, Antonio Carlos Moreira for drawing the figures, Luis C. E. Milde for helpful support, Dr. Daniele Matile-Ferrero for the confirmation of P. citri *identification and Dr. Keith Alger for revision of the English manuscript. This project was partially supported by a grant from CNPq.*

Resumen

Se estudió la asociación entre la hormiga *Wasmannia auropunctata* y el piojo harinoso, *Planococcus citri* en frutos de cacao. Durante 6-7 meses de desarrollo de los frutos se observaron 4 generaciones sucesivas de *P. citri*, hasta que 94% de los frutos fueron colonizados. La primera generación, la cual corresponde a los colonizadores se estableció frequentemente en el pedunculo del fruto. Las siguientes generaciones se presentaron a continuación en intervalos de 7 semanas, independientemente del crecimiento del fruto. Aunque las ninfas son móbiles, frecuentemente las hormigas son las que transportan el ovisaco o las ninfas para que continuen la colonización. La colonización es continua durante el desarrollo del fruto, pero es mas importante en frutos de 5 semanas. Si la primera colonización no es exitosa, los frutos son reinfestados varias veces.

Colonizaciones con un gran número de piojos harinosos tienen mas chance de sobrevivir que aquellos con pocos individuos porque las poblaciones pequeñas de piojos son menos atractivas para las hormigas que las poblaciones grandes, en la forma en que las hormigas proporcionan limpieza y defenza de los piojos. Las hormigas distribuyen huevos o ninfas de piojos en sitios disponibles o favorables (frutos) y la presencia de los piojos contribuye a fortalecer el territorio de las hormigas.

References

Buckley, R. C. 1987. Interactions involving plants, Homoptera, and ants. *Ann. Rev. Ecol. Syst.* 18: 111-135.

Cushman, J. H. and A. J. Beattie. 1991. Mutualism: assessing the benefits to hosts and visitors. *TREE* 6: 193-195.

Delabie, J. H. C. 1988. Ocorrência de *Wasmannia auropunctata* (Roger, 1863) (Hymenoptera, Formicidae, Myrmicinae) em cacauais na Bahia, Brasil. *Revista Theobroma (Brazil)* 18: 29-37.

_____ . 1990. The ant problems of cocoa farms in Brazil. Pp. 555-569. In: R.K. Vander Meer, K. Jaffe and A. Cedeno [eds.]. *Applied Myrmecology: a World Perspective.* Westview Press, Boulder, CO.

Delabie, J. H. C. and M. I. Cazorla. 1991. Danos de Planococcus citri (Homoptera, Pseudococcidae) na produção do cacaueiro. *Agrotropica (Brazil)* 3: 53-57.

Dreisig, H. 1988. Foraging rate of ants collecting honeydew or extrafloral nectar and some possible constraints. *Ecol. Entomol.* 13: 143-154.

Entwistle, P. F. 1972. *Pests of cocoa.* Tropical Sciences Series, Longman, London.

Fabres, G. and W. L. Brown, Jr. 1978. The recent introduction of the pest ant *Wasmannia auropunctata* into New Caledonia. *J. Aust. Entomol. Soc.* 17: 139-142.

Hanna, A. D., E. Judenko and W. Heatherington. 1956. The control of *Crematogaster* ants as a means of controlling the mealybugs transmitting the Swollen-Shoot virus disease of cacao in the Gold Coast. *Bull. Entomol. Res.* 47: 219-226.

Holldobler, B. and E. O. Wilson. 1990. *The ants.* Springer-Verlag, Berlin.

James, H. C. 1937. Sex ratios and the status of the male in Pseudococcinae. *Bull. Entomol. Res.* 28: 429-461.

Kirkpatrick, T. W. 1953. Notes on minor insect pests of cacao in Trinidad. Pp. 62-71. In: A Report on Cacao Research. The Imperial College of Tropical Agriculture, St. Augustine, Trinidad.

Lubin Y. D. 1984. Changes in the native fauna of the Galapagos Islands following invasion by the little red fire ant, *Wasmannia auropunctata. Biol. J. Linn. Soc.* 21: 229-242.

Nixon, G. E. 1951. *The Association of Ants with Aphis and Coccids.* Commonwealth Inst. Entomol., London.

Salazar, T. J. 1972. Contribucion al conocimiento de los Pseudococcidae del Peru. *Rev. Per. Entomol.* 15: 277-303.

Silva, P. 1944. Insect pests of cacao in the state of Bahia, Brazil. *Tropical Agriculture* 21:8-14.

_____ . 1950. The coccids of cacao in Bahia, Brazil. *Bull. Entomol. Res.* 41: 119-120.

Smith, M. R. 1942. The relationship of ants and other organisms to certain scale insects on coffee in Puerto Rico. *J. Agric. Univ. Puerto Rico* 26: 21-27.

Spender, H. 1941. The small fire ant *Wasmannia* in citrus groves--a preliminary report. *Florida Entomol.* 24: 6-14.

Strickland, A. H. 1951. The entomology of Swollen-Shoot of cacao. I-The insect species involved, with notes on their biology. *Bull. Entomol. Res.* 41: 725-748.

_____ . 1952. The entomology of Swollen- Shoot of cacao. II-the bionomics and ecology of the species involved. *Bull. Entomol. Res.* 42: 65-103.

Sudd, J. H. 1987. Ant aphid mutualism. Pp. 355-365. In: A. K. Minks and P. Harrewijn [eds.]. *Aphids: their biology, natural enemies and control.* Elsevier, Amsterdam.

Ulloa-Chacon, P. and D. Cherix. 1990. The little fire ant *Wasmannia auropunctata* (R.) (Hymenoptera : Formicidae). In: R.K. Vander Meer, K. Jaffe and A. Cedeno [eds.]. *Applied Myrmecology: a World Perspective.* Westview Press, Boulder, CO.

Wood, T. K. 1982. Selective factors associated with the evolution of Membracid sociality. Pp.175-179. In: M. D. Breed, C. D. Michener and H. E. Evans [eds.]. *The biology of social insects.* Westview Press, Boulder, CO.

9

Biology and Importance of Two Eucharitid Parasites of *Wasmannia* and *Solenopsis*

John M. Heraty

Introduction

The biological control of pestiferous ants by means of self-sustaining and host-specific insect parasites is obviously desirable over almost any other control alternative. Chemical control of *Wasmannia auropunctata* (Roger) in the Galapagos Islands is not a desirable alternative because of the unique and fragile nature of the ecosystem. Pesticides can be effective in controlling *Solenopsis invicta* Buren and *S. richteri* Forel in the United States, but they have not been able to eradicate fire ants. Although chemicals can be useful for localized treatment, they cannot provide long term control and should not be used in non-agricultural habitats in which the diversity of other insect groups may be already under severe stress.

Unfortunately, ants present formidable defenses to most insect parasitoids and few groups have been able to adapt to ants as hosts. Except for Kistner's (1982) review of myrmecophilous *Diapriidae,* there has been little attention paid to the Hymenoptera that are true parasitoids of ants. Among such Hymenoptera, the *Eucharitidae* is by far the largest and most diverse group, with 42 genera and over 400 species distributed world-wide. The majority of species are included in two subfamilies, the *Oraseminae* and *Eucharitinae,* all of whose members are parasitoids of ant pupae.

Can *Eucharitidae* be considered for biological control? Johnson (1988) tried to address their potential for controlling ants but ended up raising more questions than answers. His major complaint was the "scarcity of biological and systematic information for the eucharitids". In general, I must agree. Biological information exists for only 19 of the 42 recognized genera of *Eucharitidae*. However, through reclassification and grouping of species into monophyletic groups based on morphological analyses, there is enough information available to form predictions and identify certain taxa for more intensive study. Earlier lists of eucharitid genera and their ant hosts show little congruence, for example, species of *Myrmecia* (Myrmeciinae) were listed as hosts for at least three genera of eucharitids (Wheeler and Wheeler 1937, Johnson 1988). However, recent taxonomic changes in *Eucharitidae* (Boucek 1988, Heraty 1990) have resulted in higher levels of congruence and the hosts of most eucharitid genera are clearly restricted to only one or a few closely related genera of ants (Heraty 1990); *Myrmecia* are now regarded as the host of only one genus, *Austeucharis*. Several reviews of the general biology of *Eucharitidae* have been published to which I would refer the interested reader (Clausen 1940a,b, 1941; Heraty and Darling 1984; Heraty 1990; Johnson 1988). In this chapter, I will focus on the biology of the genus *Orasema*, and, in particular, two species that are parasitoids of *W. auropunctata* and *S. invicta*.

Background

The genus *Orasema* is distributed worldwide throughout the tropical regions. In the Old World, 17 species are distributed among five species groups (Heraty 1990). In the New World, 80-100 species can be recognized, which extend north to southern Canada and south to central Argentina; these can be allocated to several species groups, none of which are shared between the Old and New World. Only 38 of the New World species have been described, and few of these can be recognized based on the original descriptions. I am currently revising these species and the predictions presented here are based on my initial efforts to group species into monophyletic groups.

The two species of *Orasema* known to attack *W. auropunctata* and *S. invicta* occur in two species groups (or possibly as distantly related species within one larger group). The first group (dealt with in this paper) includes the following species: *O. costaricensis* Wheeler and Wheeler, *O. minutissima* Howard, *O. smithi* Howard, and an undescribed species (*O.* sp. C1, nr *O. costaricensis*) from the southeastern United States, with *O. minutissima* and *O. smithi* considered as closely related, derived members; the second group includes *O. worcesteri* (Girault) and *O. xanthopus* (Cameron). *O. minutissima* is a parasite of *Wasmannia. auropunctata* and *W. sigmoides* Mayr (Table 9.1). *O. xanthopus* is a parasite of species in the

Solenopsis saevissima-subcomplex of species (following Trager 1991), which includes *S. invicta*. Information on recognition, description of immature stages, and the taxonomic history of *O. xanthopus* is being prepared for separate publication.

TABLE 9.1 Host associations for species of *Orasema*. Names for ant species follow recent combinations in Kempf (1972), Smith (1979), and Trager (1991).

Parasite	*Host Ant*	*Country*
O. argentina	*Pheidole nitidula strobeli* [3]	Argentina
O. assectator	*Pheidole* sp. [1,7]	India
O. costaricensis	*Pheidole flavens* [12]	Costa Rica
O. coloradensis	*Solenopsis molesta* [11]	United States (CO)
	Pheidole bicarinata [11]	United States (CO)
	Formica subnitens [6]	United States (ID)
	Formica oreas comptula [6]	United States (ID)
O. fraudulenta	*Pheidole megacephala* [9]	Ethiopia
O. minuta	*Pheidole* nr *tetra* [5] [W]	United States (FL)
	Tetramorium sp. [5]	United States (FL)
O. minutissima	*Wasmannia auropunctata* [2,15]	Cuba, Puerto Rico
	Wasmannia sigmoides [15] [S]	Puerto Rico
O. rapo [?]	*Eciton quadriglume* [5]	Brazil
O. sp. [C1 nr costaricensis]	*Pheidole dentata* [5,10]	United States (FL)
O. sixaolae	*Solenopsis tenuis* [12]	Costa Rica
O. sp.	*Pheidole* sp.[5] [W]	Solomon Islands
O. sp. [B1 nr *bakeri*]	*Solenopsis geminata* X *xyloni* [5] [T]	United States (TX)
O. sp. [B2 nr *bakeri*]	*Pheidole* nr *clementensis* [5] [W]	Mexico
	Pheidole nr *californica* [5] [W]	Mexico
	Pheidole sp. [5] [W]	Mexico
	Tetramorium sp. [5] [W]	Mexico
O. susanae	*Pheidole* nr *tetra* [5] [W]	Argentina
O. tolteca	*Pheidole hirtula* [8] [W]	Mexico
O. valgius	*Pheidole* sp. [4]	Australia
O. wheeleri	*Pheidole ceres* [11]	United States (TX)
	Pheidole sciophila [11]	United States (TX)
	Pheidole tepicana [11]	United States (TX)
O. worcesteri	*Pheidole radoszkowskii* [5]	Argentina
O. xanthopus	*Solenopsis invicta* [5,13,14]	Brazil
	Solenopsis saevissima-complex[13]	Brazil, Uruguay

[1]*Das (1963);* [2]*Gahan (1940);* [3]*Gemignani (1933);* [4]*Girault (1913);* [5]*Heraty (1990);* [6]*Johnson et al. (1986);* [7]*Kerrich (1963);* [8]*Mann (1914);* [9]*Reichensperger (1913);* [10]*Van Pelt (1950);* [11]*Wheeler (1907);* [12]*Wheeler and Wheeler (1937);* [13]*Williams and Whitcomb (1973);* [14]*Wojcik et al. (1987);* [15]*New Record. Single letters in brackets refer to identifier: S=R. Snelling; T=J. Trager; W=E. O. Wilson; parentheses refer to U.S. state where reared from host.* Orasema *species C1 and B1-2 are undescribed species.*

Biology of Orasema

Adults of *Eucharitidae* do not parasitize the host directly but instead deposit their eggs on or into plant tissue. Females of *Orasema* use their ovipositor to form a chamber in the plant tissue in which a single egg is deposited (Clausen 1940a, Johnson et al. 1986). Eggs are deposited into a wide variety of plants including leaves of tea *(Theaceae),* mango *(Anacardiaceae),* oak *(Fagaceae)* and olive *(Oleaceae),* involucral bracts of flower heads (of several families) and even young banana fingers *(Musaceae)* (Nicolini 1950, Tocchetto 1942, Roberts 1958, Johnson et. al. 1986, Heraty 1990). Species are consistent in their choice of plant structure for oviposition, and within species groups trends in choice of plant structure are apparent (Heraty 1990).

The first-instar larva, termed a planidium (Figures 9.1, 9.2), must gain transport to the host brood. Larvae of *Eucharitinae* enter the colony by various means of phoretic attachment to foraging workers of the host ants (Clausen 1941). Planidia of *Orasema* are often associated with Thysanoptera or Homoptera as intermediate hosts (Das 1963, Clausen 1941, Beshear 1974, Wilson and Cooley 1972, Johnson et al. 1986, Heraty 1990). It is unknown whether these intermediate hosts serve as a temporary food source, or as an obligatory association in which ants collect the host as prey items and inadvertently transport planidia to the brood (Johnson et al. 1986, Heraty 1990).

Initially, the planidium parasitizes the ant larva. In *Orasema,* the parasite larva burrows just under the host cuticle and begins feeding. Upon pupation of the host, the first-instar larva becomes external and resumes feeding in the ventral region of the thorax. First-instar larvae can undergo almost a 1000-fold increase in size before molting to the next instar (Figure 9.3). Further development takes place on the host pupa. *Orasema* larvae often do not consume all of the host and the remains of these deformed host pupae have been termed phthisergates, phthisaners or phthisogynes based on sex and caste of the host (Wheeler 1907). Separation of these three forms is difficult and usually unneccesary, and application of the single term phthisergates to all pupae deformed in this manner is appropriate. Pupation and emergence of adults takes place within the nest. Mature larvae and pupae of *Orasema* have a very characteristic shape, and possess series of swollen pustules or warts (Figures 9.4, 9.5). Cuticular hydrocarbons of immature stages, and probably adults within the nest, mimic those of the host brood (Vander Meer et al. 1989), so that the parasitoids are generally well treated by the ants. Ants have been known to protect immature stages, assist in eclosion, and feed adults within the nest (Wheeler 1907, Williams 1980, Wojcik 1989). These "factors" appear to break down

FIGURES 9.1-5. *Orasema xanthopus.* 1-3, first-instar larva: 1. habitus of unfed larva; 2. mouthparts; 3. fully distended larva in external feeding phase, tergites dorsal and separated; 4. anterior region of third-instar larva; 5. worker of *Solenopsis* sp. clasping pupa. Scale in mm.

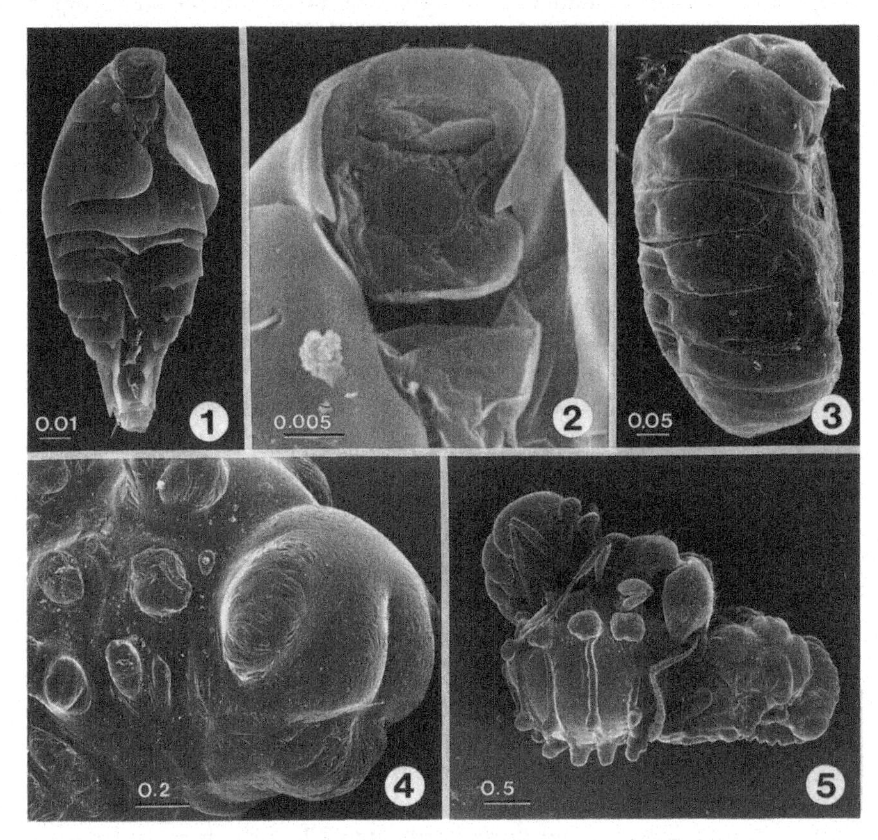

within a few days after eclosion and adults are recognized by their host as either refuse to be ejected from the colony or as an invader and attacked (personal observations). Adults must leave the nest to mate and deposit eggs (Heraty 1990).

Are Orasema *host specific?*

Recent taxonomic changes (Boucek 1988, Heraty 1990) have shown that associations between *Eucharitidae* and their ant hosts are relatively congruent. Species of *Oraseminae* are almost exclusively parasitoids of *Myrmicinae,* and *Eucharitinae* are parasitoids of *Formicinae, Myrmeciinae* and *Ponerinae* (Heraty 1990). Among New World species of *Orasema,* the majority of records are from various species of *Pheidole, Solenopsis, Wasmannia*, and *Tetramorium* (all *Myrmicinae*), with uncommon records from *Formica (Formicinae)* and *Eciton (Ecitoninae)* (Table 9.1). All of the Old World host records for *Orasema* are from species of *Pheidole*. Because species of *Pheidole* are the only known host for species of the closely related genus *Orasemorpha, Pheidole* is the probable ancestral host for *Orasema*.

Accurate host information exists for 20 species of *Orasema* (Table 9.1). Four of the species have been reared from more than one species of ant host. Of these four species, two have been reared from different ant genera and one has been reared from two different subfamilies (Table 9.1). In some cases, the actual number of host associations may be fewer. *O. coloradensis* Wheeler was reared from species of *Pheidole* and *Solenopsis* (both *Myrmicinae*), and *Formica (Formicinae)* (Table 9.2). Wheeler (1907) acknowledged that the *Solenopsis molesta* (Say), which were the host of *O. coloradensis*, were living in cleptobiosis with *Formica ciliata* Mayr, and the parasitoids may have originally come from the *Formica* colony. Even so, *O. coloradensis* belongs to a distinct species group that is distantly related to the species groups being discussed here. Of the two species groups in question, species of *Orasema* have been reared from Myrmicinae only.

Because the number of accurate host associations are few, it is still difficult to assess the degree of host specificity. In south Texas, *Orasema* sp. [B1, near *bakeri* Ashmead] was reared from colonies of *S. geminata* X *xyloni* (hybrid species, cf. Trager 1991). Females deposited their eggs in large numbers on *Acacia* shrubs *(Leguminosae)* that were less than four

TABLE 9.2 Rates of parasitism calculated for a single colony of *P. dentata* attacked by *Orasema* sp. [C1] in Huntsville State Park, Texas.

Unparasitized *Pheidole*		Parasitized *Pheidole*	
Larvae	32	*Orasema* planidia	16
Pupae	204	*Orasema* larvae	5
		Orasema pupae	17
		Phthisergates	234
Total	236	Total	272

Parasitism based on larvae, pupae and phthisergates (272/508)	53.5%
Parasitism excluding phthisergates (38/274)	13.8%
Parasitism based on ant larvae and immatures of *Orasema* (38/70)	54.2%
Parasitism based on mature larvae and pupae (22/258)	8.5%

meters from parasitized colonies, and yet colonies of *S. invicta* within the same distance from the plant host were unparasitized. In the laboratory, planidia of this *Orasema* sp. were induced to burrow into larvae of *Solenopsis* and survived on the host for over a month, but pupation of the host did not occur, and development could not proceed. Therefore, an unknown factor, possibly host foraging strategy, may result in differential parasitism of hosts in the field.

Pest or parasite?

Species of *Orasema* are interesting in that they may be considered as both a beneficial parasitoid of economically important ants, or as an economic pest that causes injury to plant tissue. The pest status of eucharitids has been documented for three species, *O. aenea* Gahan, *O. assectator* Kerrich and *O. costaricensis*. In India, *O. assectator* deposits its eggs into incisions in the leaf tissue of tea bushes. The incisions are formed by the ovipositor and secondary fungal infections of the incisions cause "sewing leaf blight" (Das 1963, Kerrich 1963). The extent of damage to tea has not been documented since the initial report, and it could have been an isolated phenomenon. In South America, *Orasema* have earned the common name "*bicho costureiro*" or seamstress insect for their characteristic oviposition marks on leaves of various plants (Tocchetto 1942). In Argentina, in a case similar to the Indian species, *O. aenea* was identified as a transmitter of the bacterium, *Pseudomonas savastanol* (E. Smith), that caused tuburcular infections on leaves of olive (Nicolini 1950). *O. costaricensis* was also reported as causing "brown spot" on young banana fingers in Costa Rica as a result of their oviposition punctures (Tocchetto 1942, Roberts 1958, Evans 1966). Wojcik (1990) mistakenly reported that insecticidal baits were used to control *Orasema* in banana plantations, when in fact populations of *Orasema* and its host, *Pheidole*, increased as a result of baits applied against *Solenopsis* (Wojcik, pers. comm.). Because of their intermediate association with thrips, it is possible that *Orasema* was a secondary pest of bananas. With the presence of thrips acting as an oviposition attractant for *Orasema* (P. Hansen, University of Costa Rica, personal communication). There have been no further reports on the pest status of any of these species and they are probably very minor pests.

Is there a potential for an introduced species of *Orasema* becoming a pest in the Galapagos Islands or in the United States? *Orasema costaricensis* is already present on the island of Santa Cruz in the Galapagos and is relatively common in Ecuador, but it has not been reported as a pest in either location. *Orasema* sp. [C1] may be the most closely related species to *O. costaricensis* and, along with another undescribed species, is common throughout the gulf coast region of the United States. Again, these two species have not been reported as pests. Thus, species which are most likely to be damaging have never been observed causing economic problems. Even when oviposition punctures are extremely dense on leaves or stems and the plant tissue becomes heavily scarified, recovery is apparently rapid and older leaf tissue shows almost no signs of damage (Heraty 1990).

Orasema minutissima has been collected from *Gynerium sagittatum* (Aubl.) Beauv. *(Poaceae)* and *Chamissoa altissima* (Jacq.) H.B.K. *(Amaranthaceae)* in Jamaica, but the oviposition habits of *O. minutissima* and *O. xanthopus* are unknown. Among species that are considered closely related, oviposition has been observed for *O. aenea* into leaves of *Iliaceae, Ranunculaceae, Polygonaceae, Passifloraceae* and *Bignoniaceae* (Parker 1942, Tocchetto 1942, Heraty 1990), for *O. costaricensis* into leaves of *Anacardiaceae* (mango) and fruit of *Musaceae* (banana) (Heraty and Darling 1984, Heraty 1990), and for *O.* sp. [C1] into leaves of *Fagaceae* and *Myricaceae* (Heraty 1990). Oviposition into leaves versus involucral bracts of undeveloped flower buds is ancestral and could be the plant structure of choice for all members of this group.

Host mortality

Under some conditions, some species of *Orasema* can be common and obviously have an impact on the host species but, as pointed out by Johnson (1988), there are few studies on the impact of Eucharitidae on the host. One study on *Pseudometagea schwarzii* (Ashmead) *(Eucharitinae)* in Ontario suggested they parasitize only the relatively unimportant overwintering brood (Ayre 1962). However, species of *Orasema* can undergo several generations per year, and comparisons cannot be made with a univoltine, temperate species. There are certain difficulties that need to be overcome in assessing mortality caused by eucharitids. In particular, the continual turnover of immature stages in an ant colony make it difficult to assess the rate of parasitism. A single colony of *P. dentata* Mayr had a total of 508 immature stages, of which 272 showed signs of parasitism by *O.* sp. [C1] (Table 9.2). The rate of parasitism can exceed 50%, if based on all larvae, pupae and phthisergates (53.8%) or only on larval host and immature *Orasema* (54.2%). However, if ant pupae are included, or only phthisergates excluded, the rate of parasitism drops dramatically (8.5 and 13.8%, respectively). The range of larval stages from planidia to pupae, and an additional 27 adults also found in the nest, indicate continual immigration and emigration of *Orasema* into or from the nest. Attempts to "freeze" a moment in time to assess the true rate of parasitism will continue to be difficult.

Orasema xanthopus was found in 38.7% of 1502 colonies of *S. invicta* that were collected in the Mato Grosso and Mato Grosso do Sul provinces of Brazil (Wojcik 1988). Numbers of mature larvae, pupae and adults averaged 17.5 per nest, but one nest contained 598 *O. xanthopus* (Wojcik et al. 1987). In total, 7225 specimens of *Orasema* were collected, representing 80.5% of all insect parasites or nest inquilines (Wojcik et al. 1987). There is no information on parasitism rates of species of *Wasmannia* by *O. minutissima*.

The number of host immatures consumed by an *Orasema* larva is an important consideration for parasite effectiveness. Because of the relatively constant size of the parasite larva and the variable size of the host larva, Wojcik (1989, 1990) hypothesized that *O. xanthopus* needs more than one host pupa to complete its development. In my study of *O.* sp. [C1] from Huntsville, Texas, mature ant larvae with internal planidia were relatively constant in size (1.82 ± 0.09 mm long, 0.84 ± 0.04 mm broad, n=10). Mature larvae of *O.* sp. [C1] were almost equal in size to parasitized host larvae (1.80 ± 0.10 mm long, 0.88 ± 0.06 mm broad, n=10) and were not significantly different in length (Student's t, $P<0.05$). Parasitized larvae were always much larger in size than pupae of minor workers of *P. dentata,* suggesting that parasitoid larvae are either selecting for larvae of major workers or somehow affecting the development of host larvae. Growth of the host may be affected by feeding of the internal first-instar of species of *Orasema*. Additionally, deformation of ant pupae is dramatic and is apparent as soon as the first instar becomes external. If mature larvae fed on more than one host, parasitoid larvae would be associated with undeformed ant pupae; this has not been observed. Thus, it seems unlikely that a single *Orasema* parasitoid feeds on more than one host pupa.

Distribution

The present distribution of *O. minutissima* and *O. xanthopus* can offer insights into their ability for dispersal to or establishment in new areas, as well as possible host associations. *Orasema minutissima* is widespread in the Greater and Lesser Antilles and is known from sporadic collections in Costa Rica and northern South America (Figure 9.6).

Notably, a series of specimens was collected in Ecuador at about the same latitude as the Galapagos Islands. However, the dispersal abilities of this species may be limited. Although the host ant is known from south Florida and the Bahamas, *O. minutissima* is unknown from either location. It is not known whether this absence is due to poorer dispersal capabilites of this species (less than its host), or if these areas are beyond the northern limits of the parasitoid. Adults of *O. minutissima* have been

Figure 9.6. Distribution of *Orasema minutissima.*

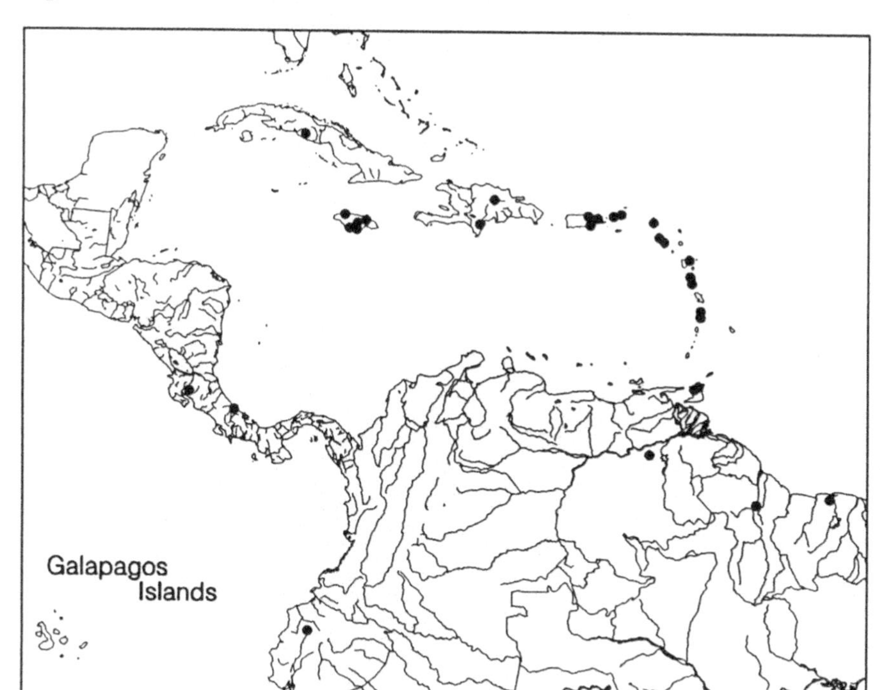

collected from elevations ranging from sea level to 1200 meters, and habitat types recorded from label data are rainforests, coffee plantations, and citrus groves. *W. auropunctata* occupies a wide variety of habitats from dry open areas to heavily shaded areas (Creighton 1950). Further collections of *O. minutissima* are necessary to observe how well it can adapt to the same ecological conditions.

Orasema xanthopus is widespread in South America (Figure 9.7). If the host is indeed restricted to species in the *Solenopsis saevissima*-subcomplex, then collections are indicative of a widespread distribution that is directly comparable to its hosts (Wilson 1952, cf. fig. 1; Buren 1972; Buren et al. 1974). The *S. saevissima*-subcomplex has been divided into nine species that are predominant in certain regions: *S. interrupta* Santschi in western Argentina, *S. richteri* in southeastern Brazil to east-central Argentina, *S. invicta* in southern Brazil to north-central Argentina, and *S. saevissima* F. Smith in Amazonia and southeastern Brazil (Trager 1991). Although the distribution of these ants may overlap, the occurence of *O. xanthopus* over the entire range (Figure 9.7) suggests that it may not be specific to any one species within this sub-complex of *Solenopsis*. However, collections of *S. richteri* in Argentina were parasitized by two other species of *Orasema* and not by *O. xanthopus* (Wojcik and Heraty, unpublished).

Figure 9.7. Distribution of *Orasema xanthopus* (solid square). Line delimits distribution of the *Solenopsis saevissima*-complex according to Wilson (1952).

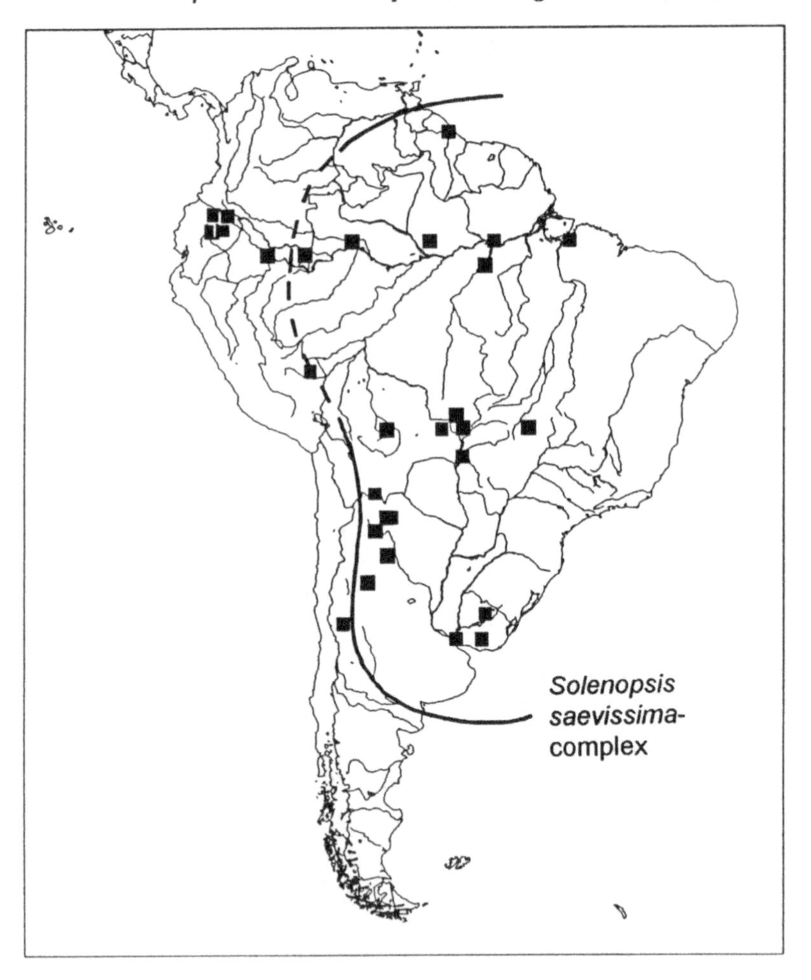

The distribution of *S. invicta* and *S. richteri* in the United States occurs between 26° and 36° north of the equator. The same latitudinal range in the southern hemisphere includes populations of *O. xanthopus* in Argentina and Uruguay (Figure 9.7). So far, the biology of *O. xanthopus* has been studied only in the Mato Grosso region of Brazil, which is the type locality for *S. invicta*; this location occurs much closer to the equator than any of the North American populations. Mato Grosso populations were sampled from disturbed cerrado, pantanal and Campo Limpo vegetation zones (Wojcik, pers. comm.). Habitat types for other South American populations are unknown.

Conclusions

Can, or should, *Orasema* be considered for the biological control of species of *Wasmannia* or *Solenopsis*? Leaving aside the philosophical problems associated with introducing any organism for biological control, I believe that species of *Orasema* do have potential as biological control agents and deserve more study. I have tried to address some of the major issues associated with host-ant and host-plant specificity, effect on host populations, distribution, habitat and their potential for being pests. There are obvious gaps in our knowledge of these groups that need to be addressed, but these could be overcome through study of *O. minutissima* and *O. xanthopus* in their native habitat and areas of relevant distribution.

The Galapagos Islands are unique not only for their flora and fauna, but also for their historical importance in the development of current evolutionary theory. Organisms that were introduced recently threaten the existence of endemic species and considerable effort has been expended to control plant and vertebrate invaders. However, introduced insects, such as as *W. auropunctata* and *Polistes versicolor versicolor* (Oliver), present new problems that need to be addressed in order to protect this fragile ecosystem. The establishment of these insects in the Galapagos Islands, without their complement of natural enemies, has allowed their populations to explode and will result in a continued decline in the diversity among all groups of organisms. Biological control, through the introduction of carefully evaluated insect parasitoids, can be used against introduced species without having adverse effects on native organisms. Similar arguments for the biological control of *S. invicta* and *S. richteri* in North America can be made, but there is not the same sense of urgency as is needed to protect the fauna of the Galapagos Islands.

The introduction of *O. minutissma* to the Galapagos Islands for the control of *W. auropunctata* has the greatest potential for success based on the following qualities:

1. The range of habitat types occupied by *O. minutissima* is limited. Present information suggests its distribution is restricted to moist, shaded habitats (citrus, coffee, rain forest), but this information is based on few collections. *Wasmannia* occurs in high densities and in a variety of habitats in the Galapagos Islands. If *O. minutissima* occurs in the Caribbean in habitats equivalent to the "transition forest" of the Galapagos Islands, then it has the highest potential to be effective.
2. *Orasema minutissima* parasitizes at least two species of *Wasmannia* but is not known to parasitize any other ant genus. The host of its most closely related species, *O. smithi*, is unknown, but is likely to be *Pheidole*. It is possible that *O. minutissima* may parasitize the endemic Galapagos species, *P. williamsi* Wheeler, but this may be a host of *O. costaricensis* which already occurs in the Galapagos Islands. Other *Myrmicinae*, the most likely alternate hosts, are not endemic to the Galapagos Islands. Therefore, there are no other endemic ant species known from the Galapagos Islands that would be parasitized by species of *Orasema*. *W. auropunctata* is already having devastating effects on the native ant fauna (Clark et al. 1982) and other insects (personal observations), and *O. minutissima*, which is only known to attack *Wasmannia*, is not likely to be damaging to other ant species.
3. The two plant genera, *Gynerium* and *Chammisoa*, that are possible oviposition hosts for *O. minutissima*, are not found on the Galapagos Islands. However, their respective plant families, *Poaceae* and *Amaranthaceae*, are represented by several other genera. It is important to know the particular plant structure that is preferred for oviposition by this species in order to evaluate the presence of similar plant forms on the Galapagos Islands. Because *O. costaricensis* has been implicated as a pest of bananas and *O. minutissima* belong to the same species group, *O. minutissima* needs to be evaluated in its native habitat to determine any potential problems caused by its oviposition. However, *O. costaricensis* is already present in the Galapagos Islands and is not a reported pest.
4. *Orasema minutissima* is common on islands such as in the Lesser Antilles (Figure 9.6) which are comparable in size to the Galapagos Islands. In addition, its presence on the western slopes of the Andes in Ecuador indicates that there is no problem of intro-

ducing a new parasite back to the mainland after establishment on the islands.

If *O. minutissima* is imported, the Galapagos Islands offer a unique opportunity for observing the establishment of a biological control agent. *W. auropunctata* is widely distributed and often extremely abundant; if ecological conditions for the host and parasite can be matched, establishment should be relatively certain. Perhaps more importantly, the islands of the Galapagos Archipelago, by the nature of their size, distribution and variable habitats, could be treated as independent experimental units for use in evaluating the spread and effectiveness of an introduced parasitoid. If successful and effective, *O. minutissima* could be used to assess the feasibility of future studies on *O. xanthopus* for control of *Solenopsis* in North America.

Admittedly, we still do not have a tremendous amount of information on species of *Orasema*. There are only 38 names available for the 80-100 species that I have recognized, and at the moment very few of the named species can be identified without comparison to type material. Of these same 100 species, we have sporadic biological information for only 18. This does not mean that we cannot make valid inferences on hosts, behavior and patterns of distribution. If taxa can be placed into hierarchical, monophyletic groups that reflect their phylogeny, then it is easier to make predictions regarding host specificity, habitat preference, or other biological attributes. Museum collections, and revisionary studies based upon them, also provide information on host, habitiat and distribution that are not otherwise made available through the literature. Here, information from both aspects is combined to summarize biological information for the genus *Orasema* and to focus attention on two species that may be useful for biological control of pest ants.

Acknowledgments

I would like to thank J. M. Cumming and J. T. Huber (Biological Resources Division, Agriculture Canada, Ottawa), D. P. Wojcik (Medical and Veterinary Entomology Research Laboratory, Gainesville, FL), and Laura Heraty for their comments and criticisms. Immature stages of Orasema *associated with* Solenopsis *and* Wasmannia *were made available by D. Wojcik and D. P. Jouvenaz (MAVERL) and Roy Snelling (Los Angeles County Museum, Los Angeles, CA), respectively. Voucher specimens of* O. xanthopus *and undescribed species were deposited in the Canadian National Collection of Insects, Ottawa. This work was supported by a National Sciences and Engineering Research Council of Canada postdoctoral fellowship to the author.*

Resumen

El *Eucharitidae* es uno de los grupos mas exitosos en la parasitización de los estados inmaduros de hormigas. Los adultos depositan sus huevos en el tejido de la planta, lejos de su hospero, y el primer instar larvario penetra la colonia dado su comportamiento forético. La larva del eucharitido parasita la larva del hospedero, pero completa su desarrollo en la pupa. El empupamiento y emergencia de los parasitoides ocurre dentro de la colonia. La relación con los hospederos esta altamente correlacionada: especies de *Oraseminae* son casi todas parasitos de *Myrmicinae,* y las especies de *Eucharitinae* son casi todas parasitoides de *Formicinae* y *Ponerinae.* Las especies de *Orasema* son parasitoides de *Pheidole, Solenopsis, Wasmannia* y *Tetramorium. Orasema minutissima* es un parasitoide de *Wasmannia auropunctacta,* y *O. xanthopus* es un parasitoide de *Solenopsis invicta.* Se discute en este capítulo los problemas asociados con estos parasitoides como agentes de control biologico en la islas Galápagos y en Norte America.

La información biologica está limitada solamente a unas pocas especies de *Orasema,* pero las relaciones filogeneticas pueden ser utilizadas para hacer predicciones concernientes a su comportamiento y su especificidad a ciertas plantas y hormigas. Las especies de *Orasema* pueden ser plagas debido a problemas asociados con la perforación de la planta durante la oviposición, pero las especies que causan estos problemas estan presentes en las areas de interés y no han sido reportadas como plaga. *Orasema minutissima* y *O. xanthopus* pertenecen a especies grupo dentro *Orasema* los cuales son solamente parasitoides de especies relacionadas de *Wasmannia* o *Solenopsis.* Se presenta evidencia que únicamente un solo individuo de *Orasema* se desarrolla por pupa del huesped. Se han efectuado unos estudios limitados en la tasa de parasitismo, pero bajo ciertas condiciones, las especies de *Orasema* pueden ser comunes y efectivas. Tanto *O. minutissima* y *O. xanthopus* pueden tener potencial como agentes de control biologico. Se sugiere que *O. minutissima* debe ser usado para control biologico de *W. auropunctata* en las islas Galápagos. Resultados de una liberación en una o varias de las islas pueden ser usados como un enfoque experimental para determinar el trabajo futuro en *O. xanthopus.*

References

Ayre, G. L. 1962. *Pseudometagea schwarzii* (Ashm.) (Eucharitidae: Hymenoptera), a parasite of *Lasius neoniger* Emery (Formicidae: Hymenoptera). *Canadian J. Zool.* 40: 157-164.

Beshear, R. J. 1974. A chalcidoid planidium on thrips larvae in Georgia. *J. Georgia Entomol. Soc.* 9: 265-266.

Boucek, Z. 1988. Australasian Chalcidoidea (Hymenoptera). C.A.B. International, Wallingford.

Buren, W. F. 1972. Revisionary studies on the taxonomy of the imported fire ants. *J. Georgia Entomol. Soc.* 7: 1-26.

Buren, W. F., G. E. Allen, W. H. Whitcomb, F. E. Lennartz and R. N. Williams. Zoogeography of the imported fire ants. *J. New York Entomol.* Soc. 82: 113-124.

Clark, D. B., C. Guayasamín, O. Pazmiño, C. Donoso, and Y. Páez de Villacís. 1982. The tramp ant, *Wasmannia auropunctata*: autecology and effects on ant diversity and distribution on Santa Cruz Island, Galapagos. *Biotropica* 14: 196-207.

Clausen, C. P. 1940a. The immature stages of the Eucharidae (Hymenoptera). *Proc. Entomol. Soc. Washington.* 42: 161-170.

——— . 1940b. The oviposition habits of the Eucharidae (Hymenoptera). *J. Wahington Acad. Sci.* 30: 504-516.

_____ . 1941. The habits of the Eucharidae. *Psyche* 48: 57-69.

Creighton, W. S. 1950. The ants of North America. *Bull. Mus. Comp. Zool.* 104: 1-585, plates 1-57 .

Das, G. M. 1963. Preliminary studies on the biology of *Orasema assectator* Kerrich (Hymenoptera: Eucharitidae) parasitic on *Pheidole* and causing damage to leaves of tea in Assam. *Bull. Entomol. Res.* 54: 393-398.

Evans, H. E. 1966. *Life on a Little Known Planet.* Dell, New York.

Gahan, A. B. 1940. A contribution to the knowledge of Eucharidae (Hymenoptera: Chalcidoidea). *Proc. U.S. Nat'l Mus.* 88 : 425-458.

Gemignani, E. V. 1933. La familia Eucharidae (Hym. Chalcidoidea) en la republica Argentina. *Ann. Mus. Nac. Hist. Nat. Buenos Aires* 37: 277-294.

Girault, A. A. 1913. New genera and species of chalcidoid Hymenoptera in the South Australian Museum, Adelaide. *Trans. Roy. Soc. S. Australia* 37: 67-115.

Heraty, J. M. 1990. Classification and evolution of the Oraseminae (Hymenoptera: Eucharitidae). Unpub. Ph.D. thesis, Texas A&M Univ. [Dissert. Abstr. Int. B 52: 632].

Heraty, J. M. and D. C. Darling. 1984. Comparative morphology of the planidial larvae of Eucharitidae and Perilampidae (Hymenoptera: Chalcidoidea). *Syst. Entomol.* 9: 309-328.

Johnson, D. W. 1988. Eucharitidae (Hymenoptera: Chalcidoidea): biology and potential for biological control. *Florida Entomol.* 71: 528-537.

Johnson, J. B., T. D. Miller, J. M. Heraty and F. W. Merickel. 1986. Observations on the biology of two species of *Orasema* (Hymenoptera: Eucharitidae). *Proc. Entomol. Soc. Wahington* 88: 542-549.

Kempf, W. W. 1972. Catálogo Abreviado das Formigas da Regiao Neotropical (Hymenotera: Formicidae). *Studia Entomol.* 15: 3-344.

Kerrich, G. J. 1963. Descriptions of two species of Eucharitidae damaging tea, with comparative notes on other species (Hymenoptera: Chalcidoidea). *Bull Entomol. Res.* 54: 365-371.

Kistner, D. H. 1982. The social insect's bestiary. Pp. 1-244. In: H.R. Hermann [ed.]. *Social Insects, Vol 3.* Academic Press, New York.

Mann, W. M. 1914. Some myrmecophilous insects from Mexico. *Psyche* 21: 171-184.

Nicolini, J. C. 1950. La avispa costurera y la tuberculosis del olivo. *Rev. Agron., Porto Alegro* 14: 20.

Parker, H. L. 1942. Oviposition habits and early stages of *Orasema* sp. *Proc. Entomol. Soc. Wahington* 44: 142-145.

Reichensperger, A. 1913. Zur Kenntnis von Myrmecophilen aus Abessinen. *I. Zool. Jahrb. Syst.* 35: 185-218.

Roberts, F. S. 1958. Insects affecting banana production in Central America. *Proc. X Intern. Congr. Entomol.* 3: 411-415.

Smith, D. R. 1979. Family Formicidae. Pp. 1323-1468. In: Krombein, K.V., P.D. Hurd, D.R. Smith, and B.D. Burks [eds.]. *Catalog of Hymenoptera in America North of Mexico, Vol. 2.* Smithsonian Inst. Press, Washington.

Tocchetto, A. 1942. Bicho costuriero. *Rev. Agron., Porto Alegre* 6: 587-588.

Trager, J. C. 1991. A revision of the fire ants, *Solenopsis geminata* group (Hymenoptera: Formicidae: Myrmicinae). *J. New York Entomol. Soc.* 99: 141-198.

Van Pelt, A. F. 1950. *Orasema* in nest of *Pheidole dentata* Mayr (Hymenoptera: Formicidae). *Entomol. News* 41: 161-163.

Vander Meer, R. K., D. P. Jouvenaz and D. P. Wojcik. 1989. Chemical mimicry in a parasitoid (Hymenoptera: Eucharitidae) of fire ants (Hymenoptera: Formicidae). *J. Chem. Ecol.* 15: 2247-2261.

Wheeler, G. C. and E. W. Wheeler. 1937. New hymenopterous parasites of ants (Chalcidoidea: Eucharidae). *Ann. Entomol. Soc. Amer.* 30: 163-173, plates 1, 2.

Wheeler, W. M. 1907. The polymorphism of ants with an account of some singular abnormalities due to parasitism. *Bull. Amer. Mus. Nat. Hist.* 23: 1-100.

Williams, R. N. 1980. Insect natural enemies of fire ants in South America with several new records. *Proc. Tall Timbers Conf. Ecol. Anim. Cont. Habitat Manage.* 7: 123-134.

Williams, R. N. and W. H. Whitcomb. 1973. Parasites of fire ants in South America. *Proc. Tall Timbers Conf. Ecol. Anim. Cont. Habitat Manage.* 5: 49-59.

Wilson, E. O. 1952. O complexo *Solenopsis saevissima* na America do Sul (Hymenoptera: Formicidae). *Mem. Inst. Oswaldo Cruz* 50: 49-59. (Eng. ver., Pp. 60-68).

Wilson, T. H. and T. A. Cooley. 1972. A chalcidoid planidium and an entomophilic nematode associated with the western flower thrips. *Ann. Entomol. Soc. Amer.* 65: 414-418.

Wojcik, D. P. 1988. Survey for biocontrol agents in Brazil–a final report, with comments on preliminary research in Argentina. *Proc. Imported Fire Ant Conf.* 50-62.

_____ . 1989. Behavioral interactions between ants and their parasites. *Florida Entomol.* 72: 43-51.

_____ . 1990. Behavioral interactions of fire ants and their parasites, predators and inquilines. Pp. 329-344. In: R. K. Vander Meer, K. Jaffe and A. Cedeno [eds.]. *Applied Myrmecology, A World Perspective.* Westview Press, Boulder, CO.

Wojcik, D. P., D. P. Jouvenaz, W. A. Banks and A. C. Pereira. 1987. Biological control agents of fire ants in Brasil. Pp. 627-628. In: J. Eder and H. Rembold [eds.]. *Chemistry and Biology of Social Insects.* Verlag J. Peperny, München.

10

Impact of *Paratrechina Fulva* on Other Ant Species

Ingeborg Zenner-Polania

Introduction

The ant species, *Paratrechina fulva* (Mayr) ("hormiga loca") was introduced into Colombia in 1985 to control poisonous snakes at sawmills in the Carare-Opon (Santander) natural forests. Three years later, several farms of the nearby municipality of Cimitarra were found to have been accidentally colonized. By 1990, the infested area had increased from 5 ha to 5000 ha.

The economic importance of *P. fulva* has been well documented (Zenner-Polania 1990a, 1990b; Zenner-Polania and Ruiz Bolanos 1982, 1983); also, certain aspects of its relationship with other arthropods are known (Zenner-Polania and Ruiz Bolanos 1985). But the ecological impact of the invasion on the native ant fauna has not been evaluated.

The great majority of introduced animal and plant species do not establish or have only minor effects on the native communities. However, some exotic species considerably alter the structure of the community (Elton 1958). Ants are considered to be one of the most ubiquitous and destructive invaders. Zimmerman (1970), cited by Lubin (1984), documents the changes suffered by the invertebrate fauna of the Hawaiian low forests due to the introduction of the African ant, *Pheidole megacephala*. The Argentine ant, *Linepithema humile,* which was accidentally introduced into California over 80 years ago, is not only found in disturbed areas, but also in natural forests where it has reduced species richness (Ward 1987). The fire ant, *Solenopsis invicta,* has caused inestimable ecological problems by changing the abundance and diversity of native arthropods (Porter and Savignano 1990).

The accidental introduction of plant and animal pests to the Galapagos Islands has caused dramatic changes in the composition and diversity of species. In this case, the little fire ant, *Wasmannia auropunctata* and the tropical fire ant, *Solenopsis geminata,* have caused some of the greatest problems. Lubin (1984) reported that *W. auropunctata* affected 17 out of the 28 genera of native ants in one way or another.

The present research analyzed the impact of the introduction and invasion of *P. fulva* on native ants in the vicinity of Cimitarra (Sant.).

Materials and Methods

The field work was conducted over a period of one year at Cimitarra (Sant.) which is located in the median part of the Magdalena Valley. It is about 200m above sea level. The mean temperature is 30°C and the mean rainfall is 3.833mm/year. For sampling purposes the area was divided into two ecological strata. Stratum I consisted of hilly landscape with primary forest and perennial crops (cocoa, plantain) and stratum II consisted of flat cattle ranches, surrounded by primary and secondary forest. Within each stratum four localities were chosen and within each locality three sampling sites with three levels of *P. fulva* infestation were deliniated: (1) border infestation, (2) heavily infested and (3) free of *P. fulva.*

Ant specimens were obtained by sampling for 3 hours at each site on soil surface, low vegetation, fallen trees and tree trunks up to a height of 1-½m. The ants were stored in vials containing 70% alcohol. Later, species were separated, counted and adequate numbers mounted. After species determinations, they were deposited at the Coleccion Entomologica "Luis Ma. Murillo", ICA-Tibaitata. Identifications were made by using the key developed by Baroni-Urbani (1983) and by comparison with identified species in the museum collection.

The data obtained were submitted to a two-way ANOVA (stratum, infestation level) using a transformation of $\sqrt{X}$. Duncan's multiple range test was performed as well as Shannon's index of species diversity and richness (Odum 1972).

Results and Discussion

The native ant fauna of the study area is very rich. A total of 21 genera were collected out of the 98 registered by Brown (1973) for the northern neotropical region. The total number of species found was 38. These data exclude the majority of arboreal ants and those with nocturnal habits. Only *Monomorium floricola* resisted displacement by *P. fulva* and shared the habitat with the invader. Its survival might be due to its very small size. *Dolichoderus diversus* was found once within one of the completely infested areas. This ant colony constructed its nest within a rolled and sealed cocoa leaf that protected the brood. None of the other native ant species were found within the completely infested territory of *P. fulva.*

The difference in species and populations of ants collected in areas free and on the border of invasion are shown in Figures 10.1, 10.2 and 10.3. These figures do not include the third level of infestation, since the 2 species mentioned in the prior paragraph were the only ones found. Also in the graphs, the data of the two strata were combined, since both ANOVA and Duncan's multiple range test showed no significant difference between the two strata.

Figure 10.1 shows that on the border of the invaded area only five of the 14 native myrmicine ants shared the territory with *P. fulva*. The last one to disappear will probably be *S. geminata,* which has habits similar to *P. fulva*. It is also the predominant ant in the free area and is considered the most aggressive and abundant species, mainly in disturbed ecosystems.

The effects on the three Attini, *Cyphomyrmex, Acromyrmex* and *Atta* were variable. In the free areas, *Atta columbica* prevailed. It was occasionally found foraging, but not nesting, in the border of infestation. *Acromyrmex octospinosus* was found rarely and disappeared as soon as *P. fulva* started invading. *C. rimosus* was collected only in the localities free of "hormiga loca". All three species nest in the soil and *P. fulva* was observed using their brood as food.

Crematogaster spp. compete with *P. fulva* for honeydew secreted by homopterans; their best defense appears to be their arboreal nesting habit. Another abundant species in the free area was *Pheidole biconstricta* which nests in the soil. It also competes with *P. fulva* for honeydew.

From the agricultural point of view, the displacement of the myrmicine species could be considered beneficial since within this subfamily there are many ant species that tend and protect homopteran pests. Furthermore, the myrmicine subfamily includes the leaf-cutting ants and fire ants, which are pests of economic importance. However the displacement of *S. geminata* may not be beneficial since long-term studies in agroecosystems in Mexico showed that in spite of its potential negative impact, its total impact is beneficial (Risch and Carroll 1982, Carroll and Risch 1983). This is based on the finding that besides being a predator, it harvests seeds, especially those of many gramineous weeds.

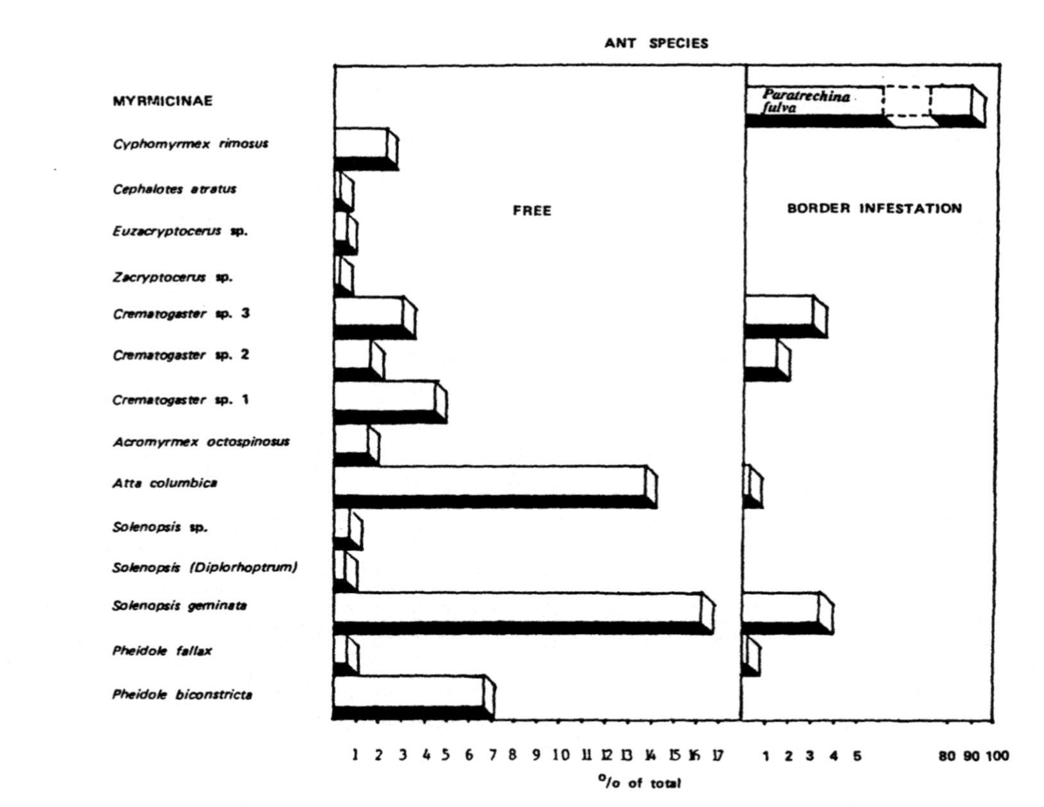

FIGURE 10.1. Composition of native species of *Myrmicinae* in areas free of and on the border of the *P. fulva* infestation.

The general scarceness of homopteran insects in the whole study area stands out. Apparently the majority of the ants, which normally require carbohydrates secreted by sapfeeding homopterans, are capable of exploiting substances produced by different plant and tree organs, such as Beltian bodies, extrafloral nectaries and Müllerian bodies. The flora of the study area is particularly rich in *Leguminosae* and *Euphorbiae*, which possess one type or another of these organs or structures that are rich in oils and protein and are used as food by the ants, including *P. fulva*.

The genera included in the subfamilies *Ponerinae*, *Pseudomyrmecinae* and *Dorylinae* are almost all considered beneficial, since a high percentage of their diet is comprised of arthropods. As can be seen in Figure 10.2, only 3 of the ten species found in the area free of *P. fulva* survived in the border of the infestation. All of them are absent in the area heavily infested with *P. fulva*.

Within perennial ecosystems, including cultivated land, *Pachycondyla*, *Ectatomma* and *Odontomachus* play an important role in the regulation of pest populations. Often their beneficial action is not directly observed, since some of them forage at night (Brown 1976). Furthermore observers may only take into account those hunting at a given moment and forget the existence of the nest and the needs of its potential population. In Colombia their efficacy has not been determined, but all species are known as predators of lepidopteran pests of cotton, cocoa, coffee and oil-palm. Lachaud et al. (1984) studied the foraging strategies of three Ponerinae species in mixed coffee-cocoa plantations and found that 65.03% of the solid food collected by *Ectatoma ruidum* are arthropods. Perfecto (1990) indicates that, in the traditional maize fields of Nicaragua, *E. ruidum* is an excellent predator of *Spodoptera frugiperda* pupae.

Pseudomyrmex spp. disappeared completely as soon as the invasion of the "hormiga loca" started. This is an interesting genus since most of its members have a constant and mutually beneficial relationship with their host plant. The ants protect their host against herbivores and are capable of physically destroying plants which compete for space and light with the tree they inhabit. In turn, the host tree provides food and shelter (Brown 1960, Janzen 1966, 1967; cited by Wilson 1971). The most abundant species found was *P. gebelli* on a *Citrus* sp. which grows wild in pasture land and adjacent roads. *P. elongatus* lives on *Inga* sp., while *P. termitarius* is only found on the ground. According to Dr. Ph. S. Ward (pers. comm.) the three species collect their food on leaves and have general feeding habits such as predation on small insects, collection of dead or dying insects, and visitation to extrafloral nectaries. Therefore, they can be considered beneficial.

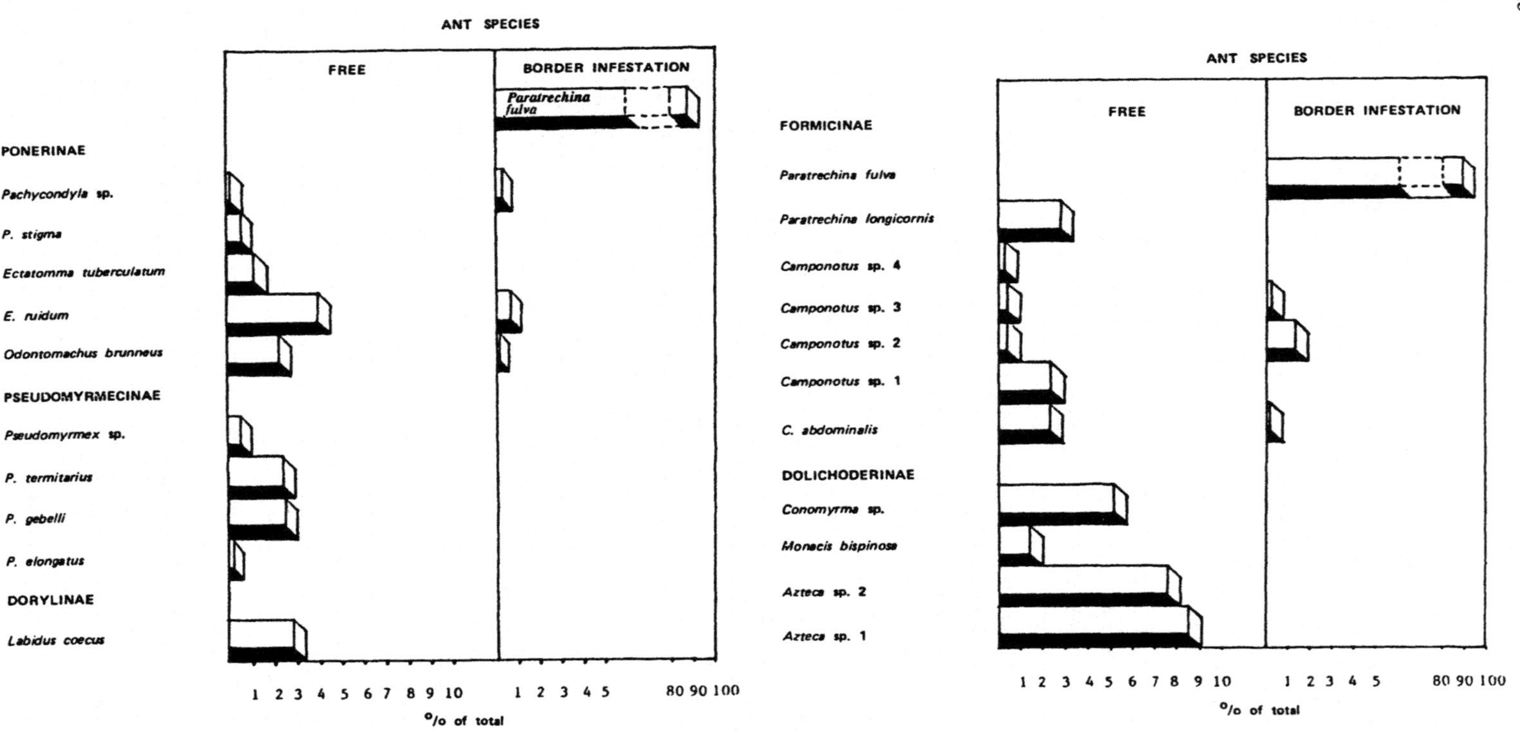

FIGURE 10.2. Composition of native species of *Ponerinae, Pseudomyrmicinae,* and *Dorylinae* in areas free of and on the border of the *P. fulva* infestation.

FIGURE 10.3. Composition of native species of *Formicinae* and *Dolichoderinae* in areas free of and on the border of the *P. fulva* infestation.

Data in Figure 10.3 also show that three of the 10 species of ants belonging to the subfamilies *Formicinae* and *Dolichoderinae* were displaced. The three that persisted, only in low numbers at the border of the infestation, were all *Camponotus* spp.

Some very interesting observations were associated with the disappearance of *Azteca* spp. Both species constructed elaborate carton nests on branches of *Citrus* sp. and *Psidium guavae*. In territory completely colonized by *P. fulva*, these nests were occupied by the invader. This means that even these aggressive and specialized ants, which form huge colonies, are not able to resist the invasions of *P. fulva*.

The Duncan's multiple range tests showed a highly significant difference in population levels of the differerent ant species and the three infestation levels of *P. fulva*.

In all cases in which *P. fulva* displaced other ants, competition seems to be the primary cause. The excessively high populations of the "hormiga loca" virtually makes it impossible for other ants to compete for food and nesting sites.

The results of the calculation of Shannon's index of species diversity are represented graphically in Figure 10.4 for all variables (stratum and level within each locality). At all localities, the abundance of native ants in territory free of *P. fulva* (level 1), as compared to territory completely colonized by *P. fulva* (level 3), is obvious. The same results were obtained for the species richness parameter.

A comparison of the ant species diversity and total number of ants within the three levels for the whole study area (Figure 10.5) shows the negative influence of the invasion of *P. fulva*. The species diversity index of 0.763318 in the *P. fulva* free area drops to 0.091931 at the border of the infestation and reaches 0.008795 for the completely infested area. This means a decrease in the number of ant species of 98.85%. Not even the behavior of the imported fire ant, *S. invicta*, which caused a drop of 70% in the species richness of ants in infested areas (Porter and Savignano 1990) is comparable to this result. A similar influence of an introduced species was, however, observed by Lubin (1984) in the Galapagos Islands, where the little fire ant, *W. auropunctata*, displaced almost all other ant species in the colonized area.

The total number of ants almost doubled in the area highly infested with *P. fulva* when compared with the area free of these ants. However, more than 95% of the individual ants were *P. fulva*. Porter and Savignano (1990) reported a 10- to 30-fold increase in the total number of ants at sites infested by *S. invicta*, with over 99% of the individual ants being imported fire ants.

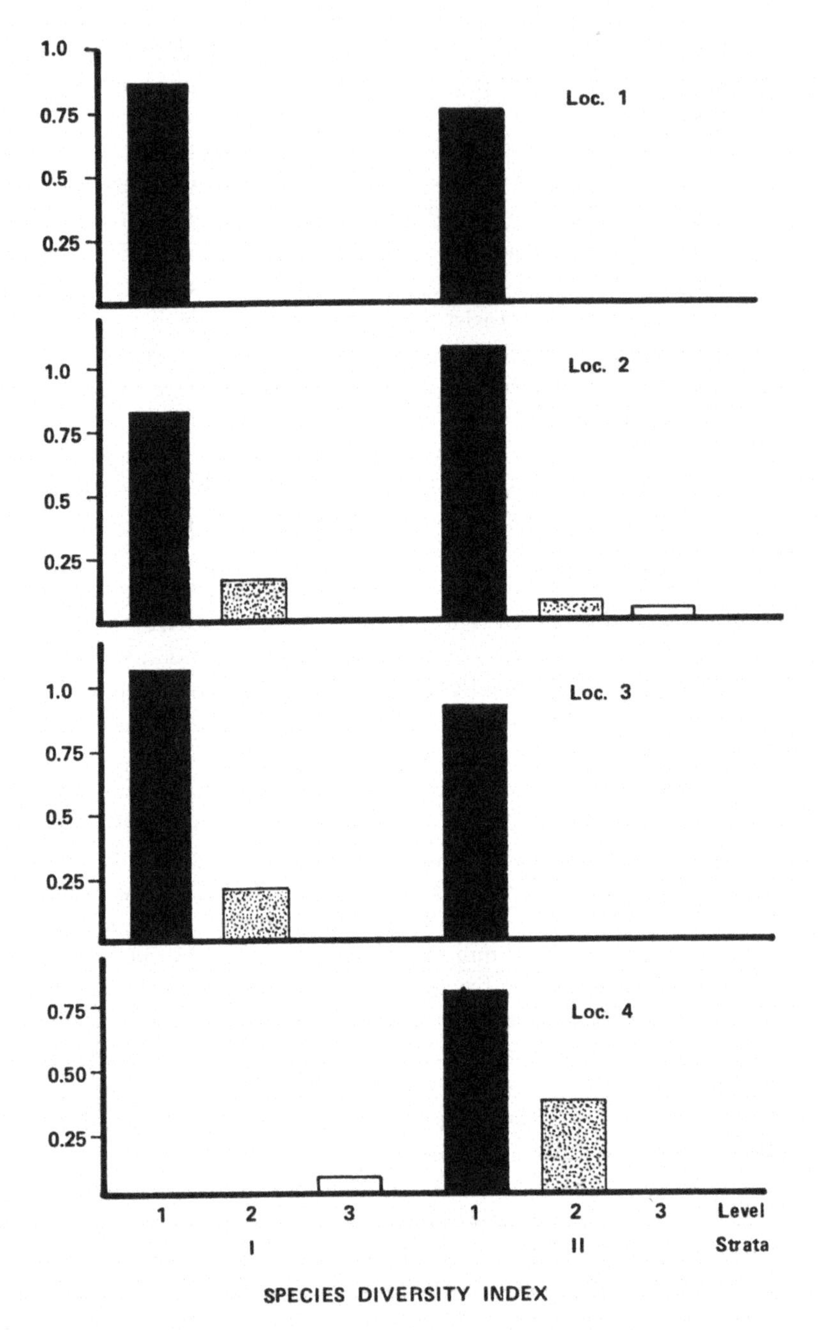

FIGURE 10.4 Impact of *P. fulva* on native ant species richness in 4 localities at Cimitarra (Santander, Colombia).

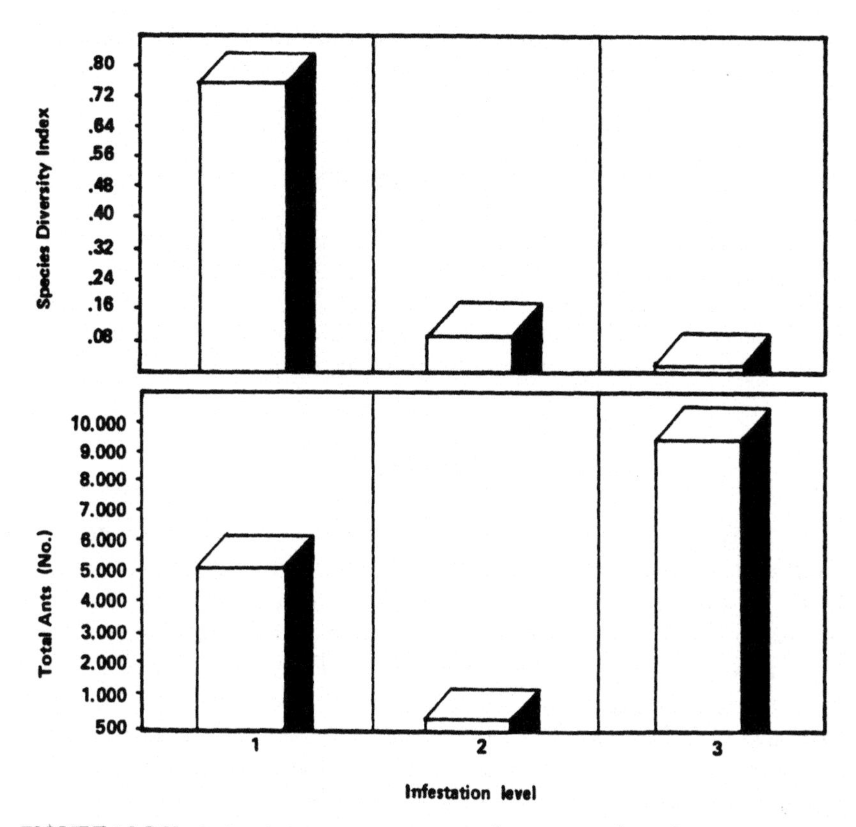

FIGURE 10.5 Variation in native ant species diversity and total ant populations at 3 levels of *P. fulva* infestation.

It can be assumed that *P. fulva* reached this population density because of the favorable ecological conditions: abundance of food, high rainfall which favors its reproduction rate, unlimited nesting sites and absence of natural enemies. Furthermore, the smaller size of the workers, compared with most of the native ants, the aggressiveness and agility they show in detecting food sources and recruiting fellow workers, gives them incomparable advantage over the other ant species.

Conclusions

Colonization by *P. fulva* produced major changes in the abundance and diversity of ant species. The absence of beneficial ant fauna in the invaded areas could induce a change in the native flora. The ability of *P. fulva* to displace native ants is based on its high fecundity, large number of workers, and its efficiency in procuring food. Finally, the results of this and other related studies should be taken into account before ants are introduced as beneficial control agents.

Acknowledgments

The following taxonomists collaborated in the identifications: Dr. D. R. Smith, SEL, PSI, USDA (spp. of Cyphomyrmex, Solenopsis (Diplorhoptrum), Dolichoderus, Monomorium, Pachycondyla *and* Conomyrma*); Dr. Ph. S. Ward, U. of California, Davis (*Pseudomyrmex *spp.) and Dr. E. O. Wilson, Harvard University (*Pheidole *spp.). Their collaboration is highly appreciated.*

Resumen

La importancia económica en Colombia de la hormiga loca, *Paratrechina fulva,* es bien conocida, sin embargo, su impacto sobre la fauna de hormigas en áreas invadidas por ella se desconoce. Con el propósito de proveer información básica en este aspecto se realizaron colecciones directas y observaciones de *Formicidae* durante cuatro períodos en el lapso de un año, en el municipio de Cimitarra (Santander, Colombia). Se encontró que la invasión de la hormiga loca afecta en forma notoria la fauna local de hormigas, disminuyendo la riqueza original de especies en un 98,85%. Del total de 38 especies de hormigas encontradas, solamente *Monomorium floricola* y *Dolichoderus diversus,* sobrevivieron en zonas colonizadas por *P. fulva.* El índice de diversidad de especies de hormigas, calculado en 0,763318 para áreas libres, disminuyó a 0,008795 para áreas invadidas.

En general, los resultados señalan la amenaza seria que constituye la hormiga loca para la biodiversidad de hormigas que habitan dentro o sobre el suelo en un área dada, e inclusive permiten deducir que la invasión ocasiona cambios ecológicos, sólo reversibles a muy largo plazo.

References

Baroni-Urbani, C. 1983. Clave para la determinacion de los generos de hormigas neotropicales. *Graellsia* 39:73-82.

Brown, W. L., Jr. 1960. Ants, acacias and browsing mammals. *Ecology* 41: 587-592.

_____ . 1973. A comparison of the Hylean and Congo-West African rain forest ant faunas. Pp.161-185. In: E. J. Meggers, A. S. Ayenus, and W. D. Duckworth [eds.]. *Tropical Forest Ecosystems in Africa and South America. A comparative Review.* Smithsonian Institution Press, Washington, D.C.

_____ . 1976. Contribution toward a reclassification of the Formicidae. Part VI. Ponerinae, Tribe Ponerini, Subtribe Odontomachiti. Section A. Introduction, subtribal characters; Genus Odontomachus. *Studia Entomol.* 19: 67-171.

Carroll, C. R. and S. J. Risch. 1983. Tropical Annual Cropping Systems: Ant Ecology. Environmental Management. (USA). v.7, no.1. pp.51-57.

Elton, C. S. 1958. The ecology of invasions by animals and plants. Methuen, London, England.

Janzen, D. H. 1966. Coevolution of mutualism between ants and acacias in Central America. *Evolution* 20: 249-275.

_____ . 1967. Interaction of the bull's-horn acacia (*Acacia cornigera* L.) with an ant inhabitant (*Pseudomyrmex ferruginea* F. Smith) in eastern Mexcio. *Kansas University Science Bull.* 47: 315-558.

Lachaud, J. P., D. Fresnau, and J. Garcia-Perez. 1984. Etude des strategies d'approvisiomement chez 3 especies de fourmis ponerines (Hymenoptera: Formicidae). *Folia Entomol. Mexicana* 61: 159-177.

Lubin Y. D. 1984. Changes in the native fauna of the Galapagos Islands following invasion by the little red fire ant, *Wasmannia auropunctata. Biol. J. Linn. Soc.* 21: 229-242.

Odum, E. 1972. Ecologia. Nueva Editorial Interamericana, Mexico.

Perfecto, I. 1990. Indirect and direct effects in a tropical agroecosystem: The maize-pest-ant system in Nicaragua. *Ecology 71: 2125-2134.*

Porter, S. D. and D. A. Savignano. 1990. Invasion of polygyne fire ants decimates native ants and disrupts arthropod community. *Ecology 71: 2095-2106.*

Risch, S. J. and C. R. Carroll. 1982. The ecological role of ants in two Mexican Agroecosystems. *Oecologia* 55: 114-119.

Ward Ph. S. 1987. Distribution of the introduced Argentine ant (*Iridomyrmex humilis*) in natural habitats of the lower Sacramento valley and its effects on the indigenous ant fauna. *Hilgardia* 55(2): 1-16.

Wilson E. O. 1971. *The Insect Societies.* Belknap Press of Harvard University Press, Cambridge, Mass.

Zenner de Polania, I. and N. Ruíz Bolaños. 1982. Uso de cebos contra la hormiga loca, *Nylanderia fulva* (Mayr) (Hymenoptera-Formicidae). *Revista Colombiana Entomol.* 8: 24-31.

_____ . 1983. Control quimico de la hormiga loca, *Nylanderia fulva* (Mayr). Revista ICA (Colombia) 18: 241-250.

_____ . 1985. Habitos alimenticios y relaciones simbioticas de la "hormiga loca" *Nylanderia fulva* con otros artropodos. *Revista Colombiana Entomol.* 11: 3-10.

Zenner de Polania, I. 1990a. Biological aspects of the "hormiga loca", *Paratrechina (Nylanderia) fulva* (Mayr), in Colombia. Pp. 290-297. In: R.K. Vander Meer, K. Jaffe and A. Cedeno [eds.]. *Applied Myrmecology: a World Perspective.* Westview Press, Boulder, CO.

_____ . 1990b. Management of the "hormiga loca", *Paratrechina (Nylanderia) fulva* (Mayr) in Colombia. Pp. 701-707. In: R.K. Vander Meer, K. Jaffe and A.

Cedeno [eds.]. *Applied Myrmecology: a World Perspective*. Westview Press, Boulder, CO.
Zimmerman, E. C. 1970. Adaptive radiation in Hawaii with special reference to insects. *Biotropica* 2: 32-38.

11

The Ecology and Distribution of *Myrmicaria Opaciventris*

J. P. Suzzoni, M. Kenne and A. Dejean

Introduction

The ant genus *Myrmicaria,* characterized by an important sexual dimorphism, is widely distributed throughout Africa (except in Madagascar) and Indo-Malaysia south of the 15th degree of latitude North. The greatest diversity of species is found in East Africa, so this region is considered as the center of the area of distribution of the genus. Generally, they are ground-dwelling except for the arboreal-nesting species *Myrmicaria brunnea* (Arnold 1926, Lévieux 1983). The best known species, *Myrmicaria eumenoides* Gerstaeker [=*Myrmicaria natalensis* Smith], has been studied in the Ivory Coast, Kenya and South Africa (Samways 1982, 1983a and b; Lévieux 1983; Dittebrand and Kaib 1987; Kaib and Dittebrand 1990). This species is found infrequently in the Guinean savannah and it is never found in the rain forest in the Ivory Coast (Lévieux 1983).

Although *M. eumenoides* is of economic importance on fruit crops because of its intensive predation on pests (Lévieux 1982, 1983; Samways 1983a and b), its biology has been poorly studied. Of interest are the studies of Dittebrand and Kaib (1987) and Kaib and Dittebrand (1990) which show the relationship between the size of the poison gland reservoir (very large for a myrmecine ant) and age polyethism.

Data on the ecology and distribution of *Myrmicaria opaciventris* (Emery) are scarce. Its area of distribution is probably bounded in the North, East and South by that of *M. eumenoides* which corresponds to Central Africa (Stitz 1910, Wheeler 1922). From this information, we can hypothesize that *M. opaciventris* is adapted to forest zones; however, our observations suggest some qualification of this statement. In equatorial areas of Cameroon, the species seems to occupy areas of human habitation where it spreads out very quickly into newly cleared areas.

Can the mode of colony foundation explain the repartition of an ant species associated with newly cleared areas in an equatorial country?

We propose here to verify this hypothesis by studying (1) the distribution of nests in different biotopes, (2) swarming behavior, (3) nest selection sites, (4) the manner in which new colonies are founded, and (5) the evolution of incipient colonies in the laboratory.

Myrmicaria opaciventris is a ground-nesting species. Its large colonies are polydomous with large nests (with a mound surface from 0.25 to more than one square meters) (Wheeler 1922). The mounds are connected together by trunk routes that become trenches which later form galleries. This behavior suggested that it played an important ecological role. Thus, a long-term study was organized by one of the co-authors to determine the species impact on agrosystems. According to unpublished data (Mavita 1984, Kilaba 1985, Nimi 1985, Kenne 1990, Agendia 1991), it appears that this species is a very efficacious predator, able to capture large prey items because of its short and long range ability to recruit nestmates.

On the other hand, the workers tend many species of Homopterans. Thus, we need to consider the possibility of disease transmission since several species of these sap-feeding insects are suspected of being vectors of plant diseases.

Material and Methods

Ecology. Ecological data were recorded from November 1987 to June 1991. Observations in forests (forest reserves of Ebodjié, Campo, Korup and old secondary forest) and cocoa plantations were made by two to four people along roads situated at a distance of 30m from the tree line (forest edge). In the dirt road, we chose a 5-meter-wide zone including the road itself; along roadsides, we examined a 2-meter-wide band. The ant mounds were easy to locate visually because of their size and because they consist of soil particles dumped by the workers, even when the colonies are young.

Swarming. In1988, we observed by chance two mating swarms on March 13th and March 22nd, respectively. They occurred in the morning following a rain at the beginning of the rainy season. Consequently, in 1990 and 1991, we made a series of observations of known colony nests at the University of Yaoundé campus and in the nearby neighborhood (49 nests in 1990 and 30 in 1991). A daily circuit was made from one nest to another between 10:00 and 11:00 a.m., thus, it was soon possible to correlate swarming with rainfall. This allowed us to predict the occurrences of the mating swarms and describe the behavior of the ants.

Colony-founding habitat. Numerous empirical observations led us to believe that founding queens landed and shed their wings in areas where there was no groundcover, or on land recently cleared by farmers. To test the validity of this deduction, we observed queens in the afternoon following a swarm and compared four areas each along a dirt road 500m long and 2.5m wide (Figure 11.1):

(A) a dirt road where the ground was well trampled (test area 1),
(B) a freshly-plowed field where farmers were planning to plant cassava (test area 2),
(C) an area situated behind the dirt road occupied by large Graminaceae (control area 1), and
(D) a stretch of ground situated along the other side of the road, which was bordered by bushes separating it from the forest edge (control area 2).

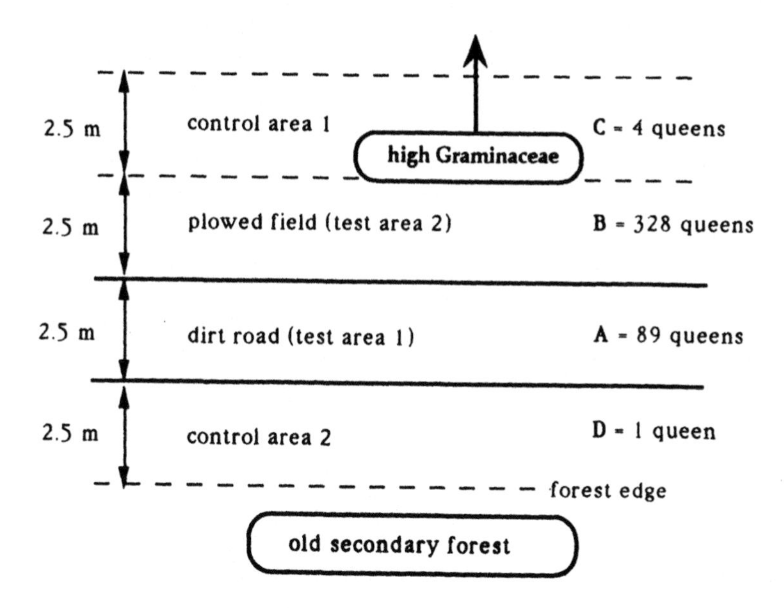

Figure 11.1. Preferences of *M. opaciventris* between different areas for digging their first shelter (Ottotomo, 13th March 1988). Bold letters indicate vegetation type. Statistical comparisons:

A and B: ε=11.7>6.1; P<10^{-9}
A and C: ε=17.7>6.1; P<10^{-9}
A and D: ε=17.9>6.1; P<10^{-9}
B and C: ε=8.7>6.1;P<10^{-9}
B and D: ε=9.1>6.1; P<10^{-9}
C and D: X^2C=0.8>3.8; NS

Observations were made in each test and control area of the number of holes excavated by the queens and the number of queens employed to excavate.

Tests for polygyny. Adult colonies consist of several nests connected by trunk routes, trenches or galleries. In the latter case, we marked workers by spraying them with different colors of paint. This allowed us to determine which nests belonged to the same colony.

Excavations were made between May and June in 1988 and in 1989 in two such polydomous colonies to test whether the different nests of the same colony had none, one or several queens (the observations were easier to make during this period because alate females are rarely found).

Laboratory queen rearing. In 1988 and 1989 we captured queens engaged in excavating their founding-nest chambers. In 1990, we captured alate females at the very moment they left their nest for the nuptial flight. All of the queens were divided into four groups whose composition was as follows: one alate female (31 cases), one queen (72 cases), two queens (40 cases) and three queens (24 cases).

The ants were reared in glass test tubes (1.0cm x 7.5cm) supplied with water and kept in a laboratory where the temperature varied from 23.5° C to 25° C. No food was provided. We followed the development of the incipient colonies by periodic readings on the number of eggs, larvae, nymphs and workers.

Statistical comparisons. For comparison of results, we used 2x2 contingency tables and the Chi-square (X^2) test. When necessary (a value <5), we used the Yates correction. The contingency tables corresponded to one degree of freedom, with a particular case where the square root of Chi-square was equal to ε, allowing the use of ε tables for low values of probability. For comparisons of brood or worker production in these incipient colonies, the t test was used. The mean values are given with the standard error.

Results

Type of habitat

Among the four types of habitat in the forest zones, *M. opaciventris* nests were most frequently found on roadsides. In this case (and for dirt roads), workers foraged into the neighboring areas. Dirt roads in the cultivated regions were in second place, followed by cocoa plantations and, in last place, the forests themselves.

The importance of the effect of man's activity on the distribution of *M. opaciventris* nests is also illustrated when we compare zones of forest reserves (Ebodjié, Campo, Korup) to the second series consisting of old secondary forests situated in cultivated regions (Table 11.1). *M. opaciventris* nests were significantly higher in the last region ($X^2C = 5.21 > 5.02$; $P < 0.025$ for forests, including dirt roads; $X^2C = 15.8 > 15.1$; $P < 10^{-4}$ for the roadsides). Comparison with cocoa plantations is irrelevant because we know of only one (small size) in the forest reserve territory and there is no *M. opaciventris* present even in the periphery.

The results obtained from the fallow land located in the University's valley probably correspond to the optimum density of nests (1.85/ha). The difference is significant with the forest zones and with the cocoa plantations, but not with the plots at the University of Yaoundé.

TABLE 11.1. Density of *M. opaciventris* nests in 4 kinds of habitats in the forest zones of 8 geographic regions in southern Cameroon compared with the density of fallow land (University Valley) and 3 plotted zones on campus. The climate of these regions is equatorial. *S=surface in ha; N=number of nests; D=density of nests/ha*

		A forest			B forest-track			C cocoa-plantat.			D road side		
		S	N	D	S	N	D	S	N	D	S	N	D
forest reserves	Ebodjie	6	0	0	0.25	0	0	-	-	-	0.4	0	0
	Campo	0.5	0	0	0.1	0	0	1	0	0	0.2	1	5
	Korup	2	0	0	0.25	0	0	-	-	-	-	-	-
secondary forests in cultivated regions	Pan Pan	8	0	0	0.25	4	16	3	1	0.3	0.2	6	30
	Kala	2	0	0	0.25	0	0	1	0	0	0.8	28	35
	Ndupe	3	1	0.3	0.25	3	12	-	-	-	0.4	16	40
	Bertoua	1	0	0	0.25	1	4	1	1	1	0.2	5	25
	Ottotomo	2	0	0	1	5	5	1	0	0	0.4	12	30
TOTAL		24	1	0.04	2.6	13	5	7	2	0.28	2.6	68	26.1

	E Fallow land		
	S	N	D
Valley of the University	18.4	34	1.85
	F Plots of the University		
	S	N	D
Plot A	3.7	4	1.08
Plot B	7.1	6	0.84
Plot C	6	5	0.83
TOTAL	16.8	15	0.89

For the results obtained above we compared, by twos, a 50/50 distribution using Chi-square (Yates correction):.

A and B: $X^2C=10.8>6.6$; $P<0.01$
A and C: $X^2C=0.1<3.8$; NS
A and D: $\varepsilon=8.1>6.1$; $P<10^{-9}$
A and E: $\varepsilon=5.4>5.3$; $P<10^{-7}$
A and F: $X^2C=11.2>10.8$; $P<0.001$
B and C: $X^2C=10.3>6.6$;$P<0.01$
B and D: $\varepsilon=6.0>5.7$; $P<10^{-8}$
B and E: $X^2C=2.9<3.8$; NS
B and F: $X^2C=6.1>5.4$; $P<0.02$
C and D: $\varepsilon=7.9>6.1$; $P<10^{-9}$
C and E: $X^2C=6.6>5.4$; $P<0.02$
C and F: $X^2C=1.27<3.8$; NS
D and E: $\varepsilon= 7.2>6.1$; $P<10^{-9}$
D and F: $X^2C=7.7>6.1$; $P<10^{-9}$
E and F: $X^2C=2.3<3.8$; NS

Description of the swarms

The results presented in Table 11.2 show that each time swarming took place, it followed a rainy day and happened during the morning between 8:00 and 11:30 a.m. The swarming behavior was composed of 4 steps:

1. excited workers exited the nest (from 8:20 to 10:00 a.m.);
2. the first males exited and took flight between 9:00 and 10:40 a.m.;
3. male swarming stopped and was followed 10 to 30 minutes later by the beginning of the female swarm (at this time, a few copulations were observed around the nest);
4. alate females exited the nest and took flight. This last step began about 9:15 to 10:30 a.m. and the swarming ended between 11:00 a.m. and 1:00 p.m.

Of 138 swarms observed, the males and the females exited the nests in 125 cases, while for the 13 other cases, only the males exited. An important difference was observed between 1990 and 1991. In 1990, except in 2 cases, each nest swarmed only once; while in 1991, several nests swarmed 2, 3 or even 4 times. In both cases, the number of males was obviously higher than the number of females.

All males and females that participated in a swarm and then landed close to their nests were usually attacked and killed by hunting workers. In contrast, those individuals that exited and then returned to their nest without a nuptial flight, were always accepted. Of 64 males landing less than 5m from the nest, 58 were captured; all of the females (37) were captured.

Colony-founding queens. The experimental design (Figure 11.1) permitted us to test the hypothesis that the queens exhibited a choice in the type of habitat where they dug their first shelter. The data show that founding queens preferentially excavated their first shelter in recently cleared and plowed fields or directly on the dirt road. However, there was a highly significant difference between these 2 habitats. Very few queens excavated their founding-nest chambers on the control plots.

Following a swarming flight on March 22, 1989 in the valley of the University of Yaoundé, we captured 527 queens on a recently plowed plot of 2,500m^2. A systematic search by 4 people on two adjacent plots of the same area resulted in capture of only 13 queens in a lawn area and 72 in a zone covered by high Graminaceae. The differences between the test plots and the 2 controls were very highly significant (ε=18.5> 6.1; $P<10^{-9}$; ε=22.0>6.1; $P<10^{-9}$). The 2 control plots also showed a highly significant difference (ε=4.44>4.41; $P<10^{-5}$), which indicated the queens were attracted to areas where soil appeared between the tufts of high Graminaceae.

TABLE 11.2. Swarming activity of nests of *M. opaciventris* at the University of Yaoundé and vicinity. Between January 25, 1990 (the first rain) and March 31, 1990 (the last swarm) and between February 8, 1991 and March 22, 1991, we recorded 25 pluvious days, 12 of them followed by swarming and 83 days without rain, never followed by swarming.
(N)= night; (D)= day; (M)= morning; (E)= evening. Statistical comparison: $X^2C=44.2$; $\mathcal{E}=6.6>6.1$; $P<10^{-9}$

1990 : 49 nests			1991 : 30 nests		
date of rains	date of swarms	% nests in swarm	date of rains	date of swarms	% nests in swarm
25 Jan. (N)	26 Jan. (M) *	0	08 Feb. (N)	09 Feb. (M) *	0
01 Mar. (D)	02 Mar. (M)	06.1	15 Feb. (E)	-	0
04 Mar. (N)	-	0	21 Feb. (N)	22 Feb. (M)	30
05 Mar. (N)	06 Mar. (M)	20.4	24 Feb. (E)	25 Feb. (M)	60
07 Mar. (D)	08 Mar. (M)	57.1	26 Feb. (E)	-	0
16 Mar.(M)	-	0	01 Mar. (N)	02 Mar. (M)	3.3
17 Mar. (D)	18 Mar. (M)	4.1	02 Mar. (M)	-	0
18 Mar. (E)	-	0	05 Mar. (E)	-	0
26 Mar. (N)	-	0	10 Mar. (D)	11 Mar. (M)	80
29 Mar. (D)	-	0	14 Mar. (E)	15 Mar. (M)	66.7 **
30 Mar.(M)	31 Mar. (M)	16.3 *	15 Mar. (N)	16 Mar. (M)	36.7 **
			17 Mar. (N)	18 Mar. (M)	10
			18 Mar. (N)	-	0
			22 Mar. (E)	-	0

*Alate queens originating from other nests were observed digging their founding-nest chamber.
**Certain nests emit only males: March 31, 1990: 2/8 nests; March 15, 1991: 3/20 nests; March 16, 1991: 8/11 nests.

Behavior of the queens. As soon as the alate females had landed after the nuptial flight, they shed their wings by raking their middle and posterior legs forward in repetitive movements until the wings broke off. Of 121 colony-founding queens observed during swarming on March 13, 1988, 66 excavated their nest chamber alone, while in 55 cases, 2 to 6 queens joined in the excavation as follows: 2 queens, 32 cases; 3 queens, 20 cases; 4 queens, 2 cases and 6 queens, 1 case. Similarly, on March 22, 1989, 127 nest chambers, were excavated with the following results: 1 queen, 85 cases; 2 queens, 24 cases; 3 queens, 13 cases and 4 queens, 5 cases. Subsequently, we marked a series of new nest chambers with a match and examined them 7 days later. We found 88 queens alone, 25 chambers with 2 queens, 10 with 3 queens and 4 with 4 queens. In all cases, numerous eggs and first instars larvae were observed.

When two or more queens establish a nest, they synchronize their excavating activity. Thus, the duration of the excavation activity is less than that of single queens. We compared the digging activity of 55 groups of queens with that of 66 single queens. At the moment when the 55 groups of queens had finished their digging activity, only 12 out of 66 single queens had finished ($\varepsilon=8.8>6.1$; $P<10^{-9}$).

Behavior of the first workers in nature. We observed by chance in two different areas that nanitic workers of incipient colonies attacked conspecific young nests and raided their brood. These observations could not be generalized because of a lack of sufficient replications; however, scientific reports with other ant species support the idea of small colonies being eliminated by ones which start with a larger initial worker force (see Bartz and Hölldobler 1982, Rissing and Pollock 1986).

Ecological control of polygyny in adult colonies. Three polydomous colonies observed in 1988 and 1989 were found to contain 4 queens in a group of 6 nests, 5 queens in a group of 4 nests and 7 queens in a group of 7 nests. In the latter case, 2 queens were present in one nest and one nest had no queen.

Other excavations made in 1990 and 1991 without a systematic search of the number of queens by nest and by colony confirm the above results. In this case, 3 queens were found in a 3-nest colony, 2 queens in a 2-nest colony and 3 queens in a 4-nest colony. Excavations of 15 other nests (without testing whether they were members of the same colony) resulted in 12 queens.

Laboratory rearing

Survival of the queens before the emergence of the first workers. In monogynous foundings, 18.1% of the queens died before the emergence of the first workers (13/72 queens). In contrast, 8.8% of the digynous queens (7/80) and 8.3% of the trigynous queens (6/72) died prior to the first worker production. A comparison between haplometrosis and pleometrosis survival showed significant differences (13/72 queens compared to 13/152 queens; $X^2 = 4.3>3.8$; $P<0.05$). The difference between the two groups of pleometrotic queens was very low ($X^2=0.007<0.015$; $P<0.9$).

Survival of new colonies. After 100 days of rearing, 22.2% of the haplometrotic colonies had died (16/72) compared to 12.5% of the digynous colonies (5/40) and 20.8% of the trigynous colonies (5/24). The comparison between haplometrotic and pleometrotic colonies (16/72 and 10/64 respectively) did not show a significant difference ($X^2=0.9<3.8$; NS).

After 100 days of rearing, we also recorded that multiple queen colonies were reduced to a single egg-laying queen in 47.7% (16/35) of the digynous colonies and in 27.3% (5/19) of trigynous ones. In 8 out of the 19 digynous colonies (42.1%), 2 queens remained. The supernumerary queens were killed fighting among themselves in one of the digynous colonies; while in the other digynous colonies and in all of trigynous colonies, the queens were decapitated and dismembered by the first workers.

Brood and worker production. Only one of 31 isolated alate females produced eggs (a maximum of 165 eggs 45 days later); however, no eggs hatched and they were, ultimately, eaten by the female.

Laboratory rearing of inseminated queens permitted us to verify that egg production in pleometrotic foundings is higher than that of haplometrotic ones ($P<0.001$). The maximum number of eggs laid per colony was recorded between the 30th and the 40th day after the nuptial flight (Figure 11.2). The data reveal that compared to haplometrotic colonies, the production of eggs in pleometrotic colonies was not proportional to the number of queens. We compared the number of eggs laid by digynic and trigynic queens to the theoretical number obtained by multiplying the egg production of the single queens by 2 or 3. We obtained significant differences for 2 queens (fd=59; $t=2.1>1.9$; $P< 0.05$) and for 3 queens (fd=57; $t=2.4>2.3$; $P<0.02$).

Duration between the nuptial flight and the emergence of the first brood. The first larvae hatched about the 30th day after the nuptial flight (Figure 11.3). The larvae received regurgitated food from the queen and also consumed other eggs. They also attacked other larvae. The maximum number of larvae was recorded between the 50th and the 60th day after the nuptial flight, pleometrotic colonies produced more larvae then haplometrotic ones ($P<0.01$).

Figure 11.2. Egg production ± SE in haplo- and pleometrotic foundings of *M. opaciventris*.

Statistical comparisons (30 days after nuptial flight):
1 queen (44.9±22.8) and 2 queens (68.4±19.9): fd=59; t=4.1; P<0.001
1 queen and 3 queens (97.0±36.0): fd=57; t=6.7; P<0.001
2 queens and 3 queens fd=46; t=3.37; P<0.001

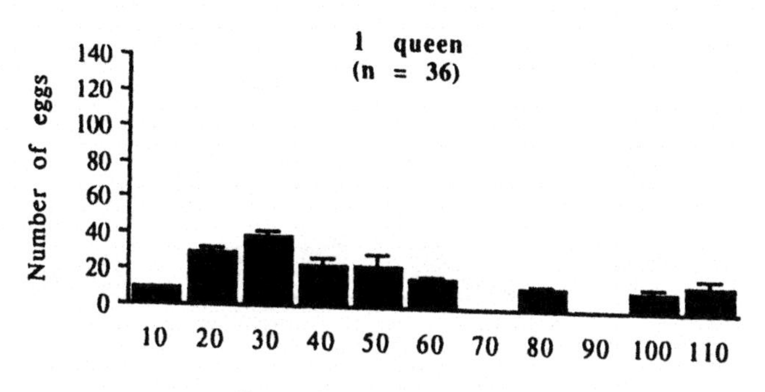

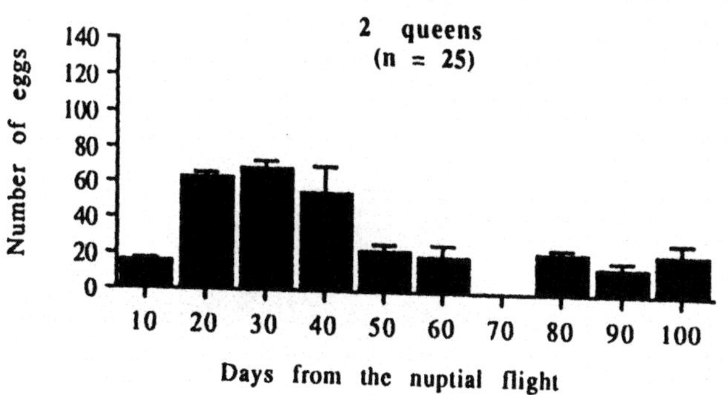

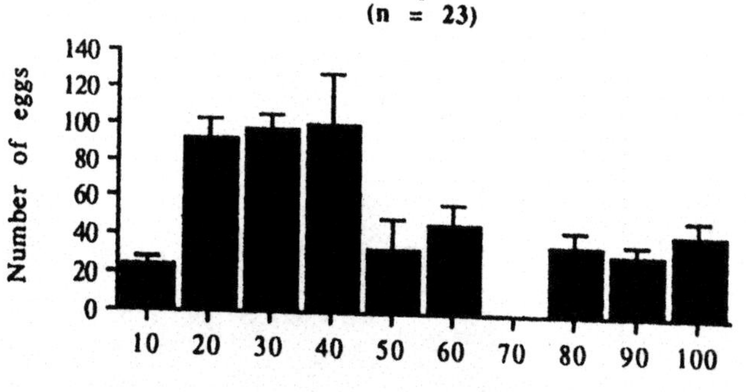

Figure 11.3. Larvae production $\pm$ SE haplo- and pleometrotic foundings of *M. opaciventris*.

Statistical comparisons (60 days after the nuptial flight):

1 queen (12.0 ± 0.8) and 2 queens (24.7 ± 2.6): fd=94; t=5.6; $P<0.001$

1 queen and 3 queens (29.0 ± 3.0): fd=80; t=7.7; $P<0.001$

2 queens and 3 queens fd=58; t=1.07; NS

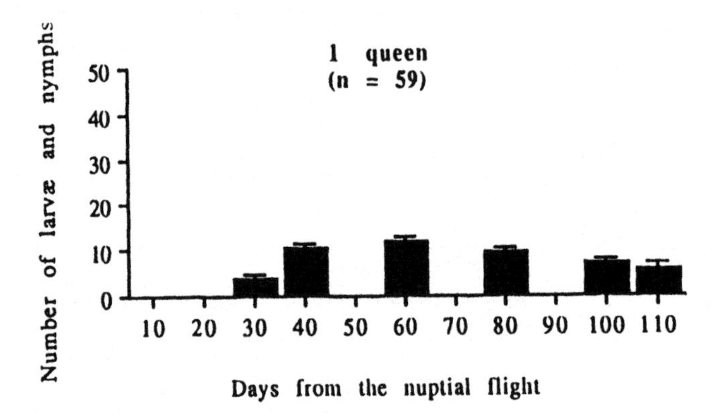

2 queens
(n = 37)
Number of larvæ and nymphs
0
10
20
30
40
50
10 20 30 40 50 60 70 80 90 100 110
Days from the nuptial flight

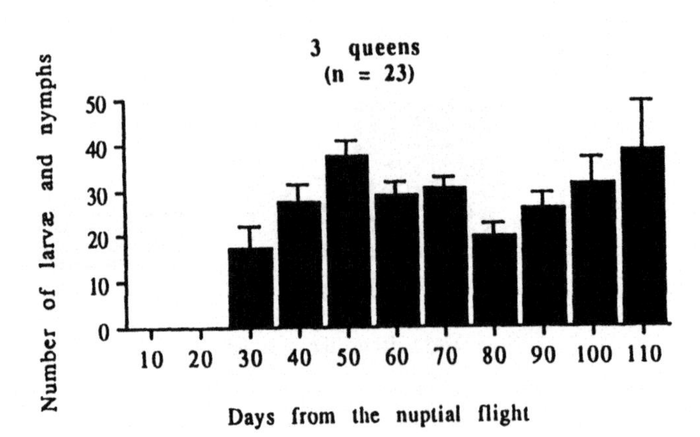

Figure 11.4. Worker production $\pm$ SE in haplo- and pleometrotic foundings of *M. opaciventris*.

Statistical comparisons (70 days after the nuptial flight):

1 queen (4.3 ± 3.7) and 2 queens (7.8 ± 4.6): fd=90; t=3.8; $P<0.001$

1 queen and 3 queens (14.3 ± 16.8): fd=76; t=4.2; $P<0.001$

2 queens and 3 queens fd=46; t=1.99; $P<0.05$

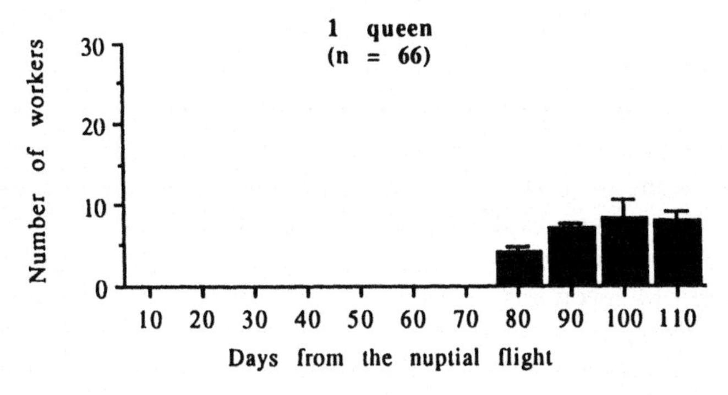

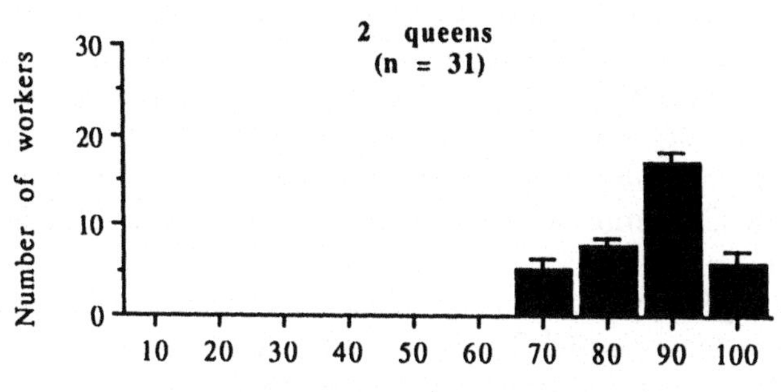

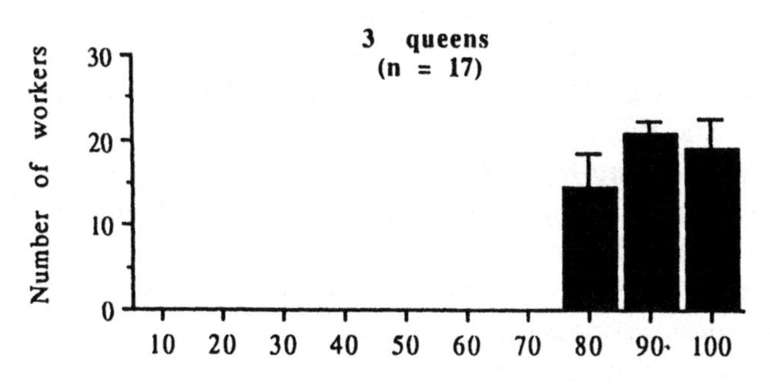

The first workers emerged about the 75th day after the nuptial flight (Figure 11.4). The pleometrotic colonies had an immediate advantage in larval numbers over the haplometrotic ones ($P<0.001$). The difference between di- and trigynic colonies was significant ($P<0.05$).

Discussion

Type of habitat. The natural vegetation of southern Cameroon is equatorial rain forest, an unlikely habitat for *M. opaciventris,* which occupies preferentially dirt roads or roadsides, two habitats resulting from man's activity. However, the density of nests is significantly lower in the forest reserve zones (primary forests, very low density of inhabitants) than in the old secondary forests (situated in regions with numerous cultivated areas).

One could argue in favor of the colonies extending their range into zones cleared by man. One could also deduce a decline in the populations of the species when secondary forest expands or in cocoa tree plantations. It seems that the roadsides permit the species to disperse rapidly because of a very good adaptation to this kind of habitat.

Little is known about the origin of *M. opaciventris* but it is obvious that southern Cameroon with its equatorial climate is not the homeland of the species.

Nuptial flights. Numerous species of *Camponotus, Crematogaster* and *Pheidole* have their nuptial flights just after the first rains of the rainy season. *Tetramorium aculeatum* and *Oecophylla longinoda* have a greater number of nuptial flights than *M. opaciventris,* but *M. opaciventris* differs from these species by having diurnal swarms. Synchronized exiting from the nest by excited workers followed by males and then females during the nuptial flight seems to be general characteristics of ants (Hölldobler and Wilson 1990).

Activity of colony-founding queens. It is interesting to note that the colony-founding queens preferentially excavate their founding-nest chambers on recently plowed areas or less frequently on areas where the soil is not covered by vegetation. In southern Cameroon the farmers clear the ground during the end of the dry season in view of sowing their fields just before the first rains.

The attraction of numerous queens to recently cleared areas permits the species to rapidly occupy these areas, especially as the period of swarming corresponds to the period of clearing and plowing of the soil by the farmers. The occupation of cleared ground is therefore very rapid so that *M. opaciventris* has the status of the pioneer species.

Excavation of founding nest-chambers by queens occurs in numerous species (Hölldobler and Wilson 1990). The synchronization of the activity by several queens in *M. opaciventris* permits a more rapid construction of the founding-nest chamber. This offers greater protection of the queens since they are out of view of potential predators faster.

The success of this species in occupying newly cleared areas is also reinforced by the proportion of pleometrosis (97/248 cases, 39.1%). This is a moderate value compared to that of other species: 80% for *Lasius flavus*, 82% for *Acromyrmex versicolor*, 89% for *Veromessor pergandei* and 97% for *Myrmecocystus versicolor* (Hölldobler and Wilson 1990). The importance of this phenomenon is discussed below.

Ecological control of polygyny in adult colonies. It seems that each nest has one queen (rarely two), so these large colonies establish a system where each nest is relatively autonomous with its own brood production. There is therefore the possibility of comparison with the idea of a supercolony, for example, *Formica lugubris* (Chérix 1980). However, there are two major differences. The *F. lugubris* super-colony occupies a very large area and each nest is heavily polygynic.

The super-colony allows a species to occupy large areas where it can exploit many food sources (extrafloral nectaries, fruits, Homoptera, dead animals and invertebrate prey).

We have not determined if the different *M. opaciventris* nests result from the same nuptial flight, or, if young queens are accepted by their nest after copulation. This last hypothesis merits testing because we have observed hunting workers attack males and females of their own colony as intruders (and prey). The same observation was made for *Iridomyrmex purpureus*, but some queens succeeded in digging their nest chambers in the vicinity of the mature colony and most of them were attended and protected and even helped by the workers in the digging (Hölldobler and Carlin 1985, Hölldobler and Wilson 1990). We never observed this behavior in the case of *M. opaciventris*, but the hypothesis remains to be tested. Although few females are inseminated by their nestmate males, some of these could return to their nest without a nuptial flight.

Laboratory rearing. The advantages of pleometrosis are discussed by Hölldobler and Wilson (1990) and Passera et al. (1991), and they consist of:

- better survival of the queens,
- larger and faster development of the initial brood,
- larger initial worker force, produced in less time,
- less weight loss per individual queen,
- a head start in the case of intercolonial competition between young colonies (often situated closely to one another).

Bartz and Hölldobler (1982) observed *Myrmecocystus mimicus* and verified that the optimum group size in nature (between 3 and 4 queens) corresponded closely to theory and that for this value mortality among founding queens was the lowest. Nonacs (1990) showed that queens of *Lasius pallitarsis* prefer to join lighter ones and that queens with a large size advantage avoided others. The association of queens during founding is not correlated with kin selection. For instance, foundress queens of *M. mimicus* and *Messor pergandei* associate without displaying aggressive behavior even when originating from different localities.

The hypothesis of a better survival rate of colonies started by 2 or 3 queens before emergence of the first workers was confirmed in *M. opaciventris* but, on the other hand, we did not find any consequential effect at the founding-colony level after 100 or 110 days. At this time, only one queen remained alive in most of the polygynous colonies. Haplometrotic foundings seem advantageous because of investment of only one queen versus two or three in the pleometrotic foundings. So, the true impact must be controlled at the brood and worker productivity level.

Production of more brood (eggs and larvae) in pleometrotic foundations is also confirmed in *M. opaciventris*. However, compared to haplometrosis, the production of eggs by pleometrotic queens is not proportional, each pleometrotic queen producing less eggs than haplometrotic ones. This last result, by deduction, is linked to the observation of smaller weight loss per individual queen in pleometrosis.

The first workers for single and multiple queen colonies appear at about the same time but the number of workers is immediately greater in the pleometrotic foundings resulting in a larger initial worker force. This can be advantageous in case of intraspecific competition with other incipient colonies situated in the vicinity.

Resumen

Investigaciones de campo permiten de poner en evidencia que las colonias de *M. opaciventris* se establecen en zonas de claros hechas por el hombre, en el sur de Camerun. Estos resultados son confirmados por observaciones directas del comportamiento de las hembras fundadoras, las cuales cavan de preferencia, su camara de fundacion en zonas donde la tierra ha sido trabajada recientemente por los campecinos para sembrar, antes de las primeras lluvias. La pleometrosis fue observada en 39% de las fundaciones.

Las sociedades adultas estan constituidas de varios nidos en interconexion por pistas, trincheras y galerias. Cada nido contiene una reina (rara vez dos), de manera que las sociedades adultas presentan un tipo particular de poliginia.

Cultivos en laboratorio permiten de evaluar las ventajas de la pleometrosis y de poner en evidencia una mejor suprevivencia de las reinas fundadoras antes de la emergencia de las primeras obreras y una produccion inicial de larvas y de obreras mas importante.

References

Agendia, V. 1991. Colony founding in *Myrmicaria opaciventris* (Formicidae, Myrmicinae). Expression of the genotype during the alimentary behavior of workers of incipient societies. Dissertation E.N.S. Yaoundë, Cameroon.

Arnold, G. 1926. A monograph of the Formicidae of South Africa. *Ann. South Africa Mus.* 23: 191-295.

Bartz, S.H. and B. Hölldobler. 1982. Colony founding in *Myrmecocystus mimicus* Wheeler (Hymenoptera:Formicidae) and the evolution of the foundress associations. *Behav. Ecol. Sociobiol.* 10:137-147.

Chérix, D. 1980. Note préliminaire sur la structure, la phénologie et le régime alimentaire d'une super-colonie de *Formica lugubris* Zett. *Insect Soc.* 27: 226-236.

Dittebrand, H. and M. Kaib. 1987. Food recruitment in the Ant *Myrmicaria eumenoides*, influenced by the defense of *Schedorhinotermes lamanianus*. P. 414. In: J. Eder and H. Rembold [eds.]. *Chemistry and Biology of Social Insects.* Verlag J. Peperny, München.

Hölldobler, B. and N. F. Carlin. 1985. Colony founding, queen dominance and oligogyny in the Australian meat ant, *Iridomyrmex purpureus. Behav. Ecol. Sociobiol.* 18: 45-58.

Hölldobler, B. and E. O. Wilson. 1990. *The Ants.* Springer-Verlag, Berlin.

Kaib, M. and H. Dittebrand. 1990. The poison gland of the ant *Myrmicaria eumenoides* and its role in recruitment communication. *Chemoecology* 1: 3-11.

Kenne, M. 1990. Etude éco-éthologique de la prédation chez *Myrmicaria opaciventris* (Emery) (Formicidae Myrmicinae). Mémoire de maitrise, Yaoundé, Cameroon.

Kilaba, C. 1985. Contribution a l'étude du comportement alimentaire chez *Myrmicaria opaciventris* (Formicidae Myrmicinae). Memoire de graduat. Kikwit, Zaïre.

Lévieux, J. 1982. Quelques observations sur l'activité de nutrition en saison séche de la Fourmi *Myrmicaria striata* Stitz (Hymenoptera, Formicidae, Myrmicinae) dans une savane préforestière de Côte d'Ivoire. *Rev. Ecol. Biol. Sol.* 19: 439-444.

_____ . 1983. Mode d'exploitation des ressources alimentaires épigées de savanes africaines par la Fourmi *Myrmicaria eumenoides* Gerstaecker. *Insect Soc.* 30: 165-176.

Mavita, T. 1984. Evaluation de la surface du champ trophoporique chez trois espéces de Fourmis (*Myrmicaria opaciventris, Odontomachus troglodytes* et *Brachyponera senaarensis*). Mémoire de graduat. Kikwit, Zaïre.

Nimi, N. 1985. Contribution a l'étude du comportement alimentaire chez les Fourmis: la recherche, la capture des proies et le retour au nid chez *Myrmicaria opaciventris* (Formicidae, Myrmicinae). Mémoire de graduat. Kikwit, Zaïre.

Nonacs, P. 1990. Size preferences in the choice of pleometrotic partners: competition in the peaceable kingdom of foundress ant queens? Pp. 240-241. In: G. K. Veeresch, B. Malick, and C. A. Viraktamath [eds.]. *Social Insects and the Environment.* Oxford and I.B.H. Publishing Co.

Passera, L., E. L. Vargo and L. Keller. 1991. Le nombre de reines chez les Fourmis et sa conséquence sur l'organisation sociale. *Ann. Biol.* 30: 137-173.

Rissing, S. W. and G. B. Pollock. 1986. Social interaction among the pleometrotic queens of *Veromessor pergandei* (Hymenoptera: Formicidae) during colony foundations. *Anim. Behav.* 34: 226-233.

Samways, M. J. 1982. Soil dumping by *Myrmicaria natalensis* (Smith) (Hymenoptera: Formicidae) as a competitive advantage over other ant species. *Phytophylactica.* 14: 3-5.

_____ . 1983a. Asymmetrical competition and amensalism through soil dumping by the ant, *Myrmicaria natalensis. Ecol. Entomol.* 8: 191-194.

_____ . 1983b. Community structure of ants (Hymenoptera: Formicidae) in a series of habitats associated with Citrus. *J. Appl. Ecol.* 20: 833-847.

Stitz, H. 1910. West Afrikanische Ameisen. *I. Mitt. Zool. Mus. Berlin.* 5: 1-133.

Wheeler, W. M. 1922. Ants of the American Museum Congo expedition. A contribution to the myrmecology of Africa. *Bull. Am. Mus. Nat. Hist.* 45: 1-1139.

12

Exotic Ants and Community Simplification in Brazil: A Review of the Impact of Exotic Ants on Native Ant Assemblages

Harold G. Fowler, Marcelo N. Schlindwein and Maria Alice de Medeiros

Introduction

Elton (1958) called attention to the strategies and possible consequences of the invasion of exotic organisms into foreign ecosystems. However, it is only recently that the true impact of these exotic organisms on the functioning of natural systems gained concentrated scientific attention (Mooney and Drake 1986, Vitousek et al. 1987). The introduction of an exotic species does not mean that they will always establish viable populations and if they do become established, it does not necessarily mean they will increase their population densities to the point of becoming pests. However, a few species will attain pest status resulting in the displacement of native floral and faunal elements (Elton 1958, Bond and Slingsby 1984, Mooney and Drake 1986).

Although the number of ant species world-wide probably exceeds 8,000, relatively few species are consistently registered as pests following their introduction (Table 12.1). Most of these species are characterized by small body size, mass-recruitment, and polygynous social organization. Polygyny permits the formation of large unicolonial populations, as well as population growth by budding, either directly or indirectly as a result of man's activities. Small worker size permits these ants to

TABLE 12.1. Principal exotic ant pest species and their colonial reproductive structure and assumed geographic origins.

Species	*Polygyne (P)/Monogyne (M)*	*Origin*
L. humile (Mayr)	P	Brazil, Argentina
S. geminata (F.)	P/M	Tropical Americas
S. invicta Buren	P/M	Brazil
Solenopsis richteri	M	Argentina
M. pharaonis (L.)	P	Tropical Africa
T. melanocephalum (L.)	P	Tropical Africa
P. megacephala (L.)	P	Tropical Africa
W. auropunctata	P	Tropical America(?)
M. destructor	P	India
M. floricola	P	Subtropical N. America
T. simillumium	P	Tropical Africa
T. caespitum	P	Europe
T. bicarinatum	P	S.E. Asia
P. longicornis	P	Tropical Africa
P. fulva	P	Brazil
A. longipes	P	Tropical Asia (?)

?= Origin questionable

explore small spaces, and as most are not limited by nest sites, their small body size allows them to use a large number of nesting sites. The larger species are highly aggressive, but are much less apt to be consistently troublesome pests. Mass recruitment allows these ants to rapidly mobilize workers to retrieve food resources quickly, as well as to exclude other potential competitors from the same resources.

Wilson (1959) first called attention to the introduction of a number of tramp species in the distant Melanesian islands, and to the fact that most were restricted to marginal habitats. Although little is known of island fauna, Wilson and Taylor (1967a, 1967b) have documented that Hawaii has no endemic ants. The same situation occurs in the Galapagos. Why ants, in spite of their great taxonomic diversity, have not been able to colonize distant islands without man's help is still not clear. However, island systems, due to their poorly formed trophic levels, are susceptible to the introduction and establishment of small predators, such as ants, which are able to exploit poorly protected habitats. For these systems, the impact of exotic ants on native fauna is better seen in groups other than ants. These organisms have evolved in space free of ant predators and may be rapidly decimated (e.g., in the Galapagos).

What happens to native ant fauna in the face of exotic ant migrants? Among the ants, only a few exotic species have gained notoriety because of significant population increases, and only a few studies have examined their impact on the native ant fauna. There are, however, numerous examples of exotic ants having significant impact upon island and continental biotas. We call attention to two cases documented in Brazil, and then review what is known of the effects of exotic ants on native ant assemblages.

Pheidole megacephala *in Northeastern Brazil*

In the São Francisco River valley, between the states of Bahia and Pernambuco, near the state border of Alagoas, the construction of a hydroelectric dam resulted in a large influx of material and man-power over a period of 10 years. One uninvited immigrant, *Pheidole megacephala,* probably arrived at the hydroelectric dam site through shipments of heavy equipment. Collections reveal that, at present, *P. megacephala* is restricted to the city of Itapirica and has not spread to neighboring cities or to the dry caatinga (Figure 12.1). The distruction of the riverine vegetation following dam closure, makes it unlikely that *P. megacephala* could survive the harsh conditions of the caatinga nor disperse naturally. It is, however, highly likely that this ant will be quickly dispersed through human commerce, as Itapirica is likely to become a major center of economic activity due to the presence of the hydroelectric dam. It is also extremely likely that this ant will colonize irrigated portions of the dry caatinga associated with the dam and then become an agricultural pest, much as they have in Hawaii (Fluker and Beardsley 1970). The fact they will be linked to Itapirica through high-intensity human transport increases the possibilities of this scenario.

Our studies in this area, which were related to a wildlife rescue project following dam closure, revealed foraging trails of *P. megacephala* in virtually all of Itapirica, whether on the ground or upon and within human structures and vegetation. Many vertebrates which were rescued, especially reptiles, were placed under quarantine and then confined to cages, where they were lost to the ants because of their weakened condition. During the two weeks in which these rescue attempts were made two boa constrictors were killed, as were many smaller snakes and lizards. Within the city, we detected only a few geckos and no lizards were seen during the day in areas which would typically harbor populations.

Because of their abundance, *P. megacephala* was also a nuisance pest in human habitation. No kitchens that we examined were free of sizeable foraging trails. In some regions of the city, the legs of cribs, beds and storage cabinets had been placed in water tins to retard ant access.

FIGURE 12.1. Ant species of the São Francisco River Valley and surrounding areas. Itapirica was the only locality with *P. megacephala*.

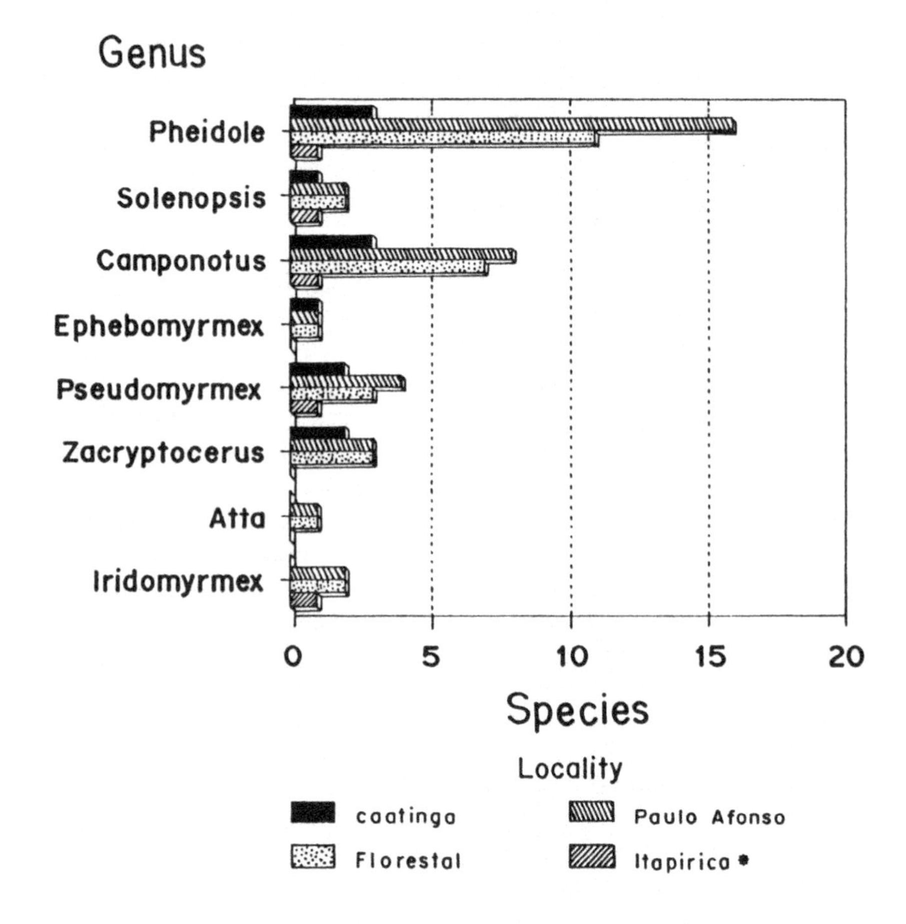

Baiting studies (Figure 12.1) revealed that there was a drastic reduction of ant species in Itapirica when compared with the species richness that was expected based upon sampling in comparable nearby areas. This reduction was even more pronounced in the the numerically dominant genus *Pheidole*. Where *P. megacephala* was found, we were unable to attract any native *Pheidole* to the baits. Only one arboreal species each of *Pseudomyrmex* and *Camponotus* were collected, although more were expected (Figure 12.1). These results are even more convincing because our collecting intensity was at least three times as great in Itapirica due to the fact that our base of operations was housed there.

Tapinoma melanocephalum *and* **Paratrechina longicornis** *in Southeastern Brazil*

In the southern portion of the State of São Paulo near the border with the State of Parana, there is still an extensive native Atlantic coastal forest. However, portions of it have been deforested and it is now the largest banana-producing area of Brazil. Banana plantations are established in areas of cleared forest and are thus much like islands within a matrix of native tropical forest.

Of the 80 banana plantations surveyed, 31 had *Paratrechina longicornis*, 18 had *Tapinoma melanocephalum*, but none had both species. In plantations with *P. longicornis*, an average of 7±1.8 species were present, while in plantations with *T. melanocephalum*, only 2±0.5 other ant species were found. In the remaining banana plantations with neither *T. melanocephalum* nor *P. longicornis* present, there was an average of 13±4.3 ant species registered.

Because of varying cultural practices in the banana plantations, it is difficult to ascertain if ant species reduction occurred due to one of the exotic species colonizing a species-reduced environment, or whether the exotic species colonized and caused the reduction. However, in fields with other crops (tea and cocoa), collections failed to detect either of these exotic species. Nevertheless, both *T. melanocephalum* and *P. longicornis* have been found at low levels in native vegetation of nearby islands (Fowler and Pesquero 1992). This attests to the fact that they are capable of establishing beach-heads, at least on islands with few species. We have not, however, found either of these species during our intensive collections of surrounding mainland native forest.

The significant reduction of ant species numbers found associated with these ants, as well as their differential and mutually exclusive distribution, suggests that perhaps competitive interactions are important. Dispersal of both species is probably assisted by man which may well explain their checkerboard distribution.

If populations of these ants explode in these areas, they also could pose additional risks to the overall ecology and species-rich fauna of the Brazilian Atlantic coastal forest (Fowler et al. 1992a), since we have shown that they are capable of invading marginal habitats of this endangered region (Fowler and Pesquero 1992). Our observations in the state of São Paulo are less surprising than those of *P. megacephala* from the caatinga of the arid northeast of Brazil, due to the fact that the majority of plant and animal species now found within the state of São Paulo are exotic introductions (Fowler et al. 1992b).

Exotic Ants and Their Impact upon Native Ant Fauna

Are the prior cases rare or common among other known examples of exotic ant introductions? Do we have enough information to infer possible outcomes? A review of the literature (Table 12.2) suggests that some patterns are apparent. Although this listing is not complete, it represents most of what is known of exotic ants and their impact on local ant fauna.

TABLE 12.2. Examples of species replacement following establishment of exotic ants.

Invader	*Location*	*Ant Species Replacement*	*Source*
L. humile			
	Madeira	+#	Stoll 1898
	Bermuda	+#	Haskins and Haskins 1965
			Crowel 1968
			Lieberburg et al. 1975
	Hawaii	+#	Fluker and Beardsley 1970
	France	+	Bernard 1983
	USA	+	Erickson 1972
	Alabama, USA	?	Barber 1916
			Newell and Barber 1913
	Mississippi, USA	+	Haugh 1934
	California, USA	+	Ward 1987
	Western Australia	+	Majer 1992
	South Africa	+	Bond and Slingsby 1984
W. auropunctata			
	Congo	?	Meer Mohr 1927
	Galapagos	+#	Clark et al. 1982
			Lubin 1984
P. fulva			
	Colombia	+	Zenner-Polania 1990
A. longipes			
	India	+	Soans and Soans 1971
	Seychelles	?	Haines et al. (Chap.18)
P. longicornis			
	Para, Brazil	?	Brown 1954
	São Paulo, Brazil	+	This study

Invader	*Location*	*Ant Species Replacement*	*Source*
M. pharaonis			
	USA	?	Smith 1965
	Germany	?	
	Pará, Brazil	?	Brown 1954
	São Paulo, Brazil	+#	Fowler et al. 1992c
T. melanocephalum			
	Pará, Brazil	?	Brown 1954
	São Paulo, Brazil	+	This study
P. megacephala			
	Pernambuco, Brazil	+	This study
S. invicta			
	USA	+	Porter and Savignano 1990 Camilo and Phillips 1990 Wojcik (Chap.23)

+=Species displaced #=Non-native species displaced ?=Information not available

First, it is apparent from documented cases in island systems that exotic ants generally displace other endemic ants. Moreover, the best documented cases involve only a small subset of known tramp species (Table 12.1). In particular, *P. megacephala,* and *Linepithema humile [=Iridomyrmex humilis]* have a disproportionate representation in this sample, both for island and continental studies. This is probably due primarily to the association of these ants with agricultural habitats in which their presence draws more attention. Because of the limited numbers of documented examples and due to the ecological void of most distant islands, it seems likely that exotic ants, once introduced, will have a great impact upon other faunal elements that have not experienced the predatory pressure of ants during their evolutionary history.

Some of the potential exotic invaders listed in Table 12.1 can colonize human structures, in particular *Monomorium pharaonis, T. melanocephalum, P. longicornis, L. humile* and *P. megacephala.* The native ant fauna associated with human structures are depauperate, and interactions with other exotic species are generally stronger (see Bueno and Fowler, Chap. 16).

Continental ant fauna have revealed few significant population reductions (Table 12.2), but once again, this is not an unbiased sample. The impact of the fire ant, *Solenopsis invicta,* has been best documented, although its expansion and domination is not characteristic of other documented cases, with the exception of *L. humile* in Africa and California (Woodworth 1908, De Kock and Giliomee 1989), and *Wasmannia auropunctata* in Africa and the Galapagos. We performed rarefaction analysis of documented cases of faunal changes following introduction of *S. invicta* in Texas and *L. humile* in California and Mississippi (Table 12.2). As an example, the data from Mississippi (Table 12.2), demonstrates a significant reduction of species richness. We found this in all of the other cases we evaluated by rarefaction techniques. *S. invicta* in Texas (Table 12.2), produced the most drastic species reductions. Our data on *L. humile* in Figure 12.2 represents the least drastic of all of our rarefaction estimates. Thus, even in these subtle cases, the clear impact of exotic species on native ant fauna is significant.

FIGURE 12.2. The impact of the Argentine ant, *L. humile*, on the diversity of the native ant fauna of Mississippi as ascertained through rarefaction. Data are reinterpreted from Haugh 1934.

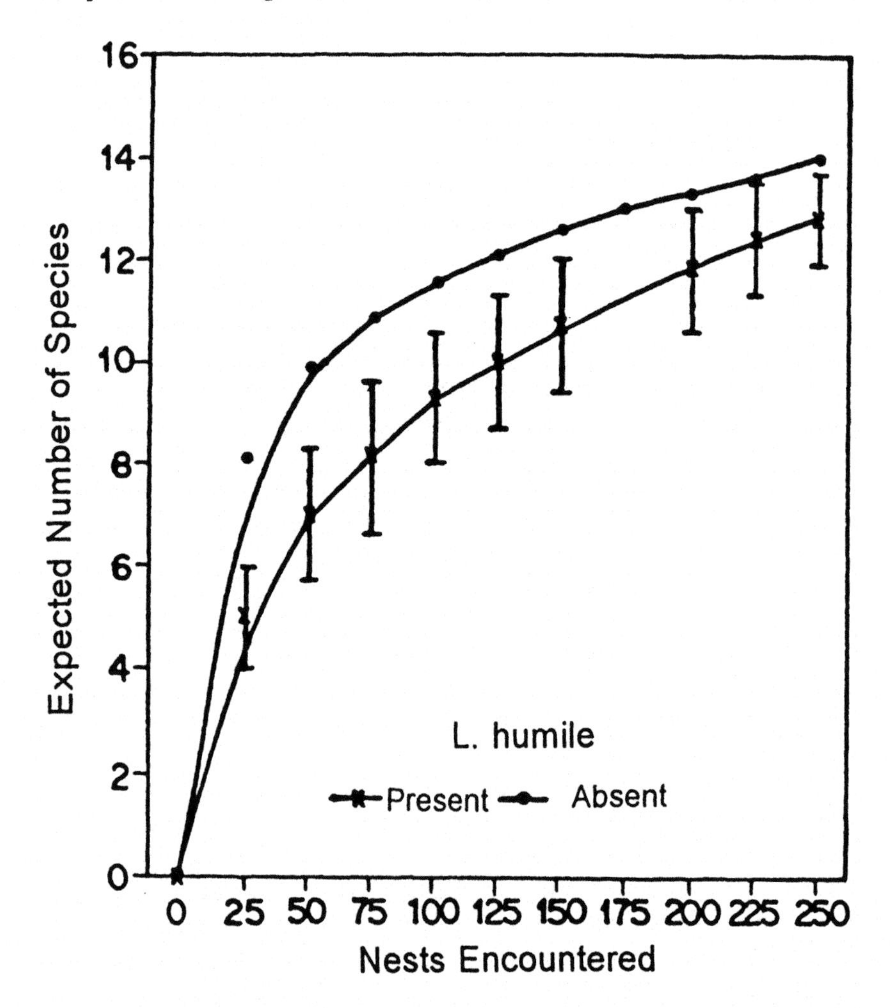

Ants are one of the principal groups of terrestrial predators. Their communities are diverse and compete with other predators, often through complex and indirect pathways. Unfortunately, ecological theory and data bases do not allow a proper assessment of the impact of these introduced species. This is even more obvious when the relationships between ants and plants is included (Meer Mohr 1927, Bond and Slingsby 1984). Exotic ants, once established, can become potential agricultural, urban, and ecological pests. Methods of classical biological control have proved unfruitful in the control of these introduced social insects (Fowler and Romagnano 1992), principally due to their social behavior. New insights and theories are needed, as well as increased quarantine efficiency.

Conclusion

Based upon our knowledge of ecology, studies are needed to devise strategies that increase the competitiveness of native ants over exotic ants. This may involve agents of classical biological control (Fowler and Romagnano 1992). However, the severe ecological or economic pressure produced by the rapid expansion of invading exotic ants (such as found in the Galapagos) makes it necessary to reduce further damage through chemical control. This last-ditch effort will always have to be invoked until the organization of communities is better understood and placed in a practical management scheme. This necessarily involves the ablity to predict competitive interactions, and development of strategies that can weaken the competitive advantages of the invading species. Even for the best-studied invader, *S. invicta*, we are still far from this point, and further studies on risk-reducing means of chemical control must take precedence if fragile island fauna are to be conserved, and further economic and agricultural damages are to be minimized.

Acknowledgments

Portions of this work was supported by the wildlife rescue and translocation program of the Itapirica Hydroelectric Dam by the Companía Hidroelectria do São Francisco (CHESF), and by the Conselho Nacional de Desenvolvimento Científico e Tecnólogico (CNPq) through grant 300171/88-9. D. F. Williams was instrumental through his constant stimulus in completing this work, and his assistance in attending the Galapagos Ant Conference is graciously acknowledged. Luis C. Forti, Ana Maria Dias de Aguiar, Suzana Keutelheute, Yasmine Neptune, Rosemary Vieira, Arioldo Cruz Neto and José Eduardo Zaia assisted in field collections through support from the

Projeto Rondon, and their participation is graciously recognized. Nozor Pinto kindly inked the figures.

Resumen

En este capítulo, informamos acerca del impacto de las hormigas exoticas en la fauna del Brazil. El primer caso se refiere a *Pheidole megacephala* en el noreste del Brazil y *Tapinoma malanocephalum* y *Paratrechina longicornis* en el sureste del Brazil. Tambien revisamos lo que se conoce hasta ahora de los efectos de las hormigas exóticas en la fauna de hormigas nativas.

Los estudios de muestreo con cebos demostraron que en la ciudad de Itapirica, hubo una reducción de las especies de hormigas comparado con la gran variedad de las especies de hormigas en otras areas vecinas. Esta reducción fue mayor en *Pheidole* el cual es el genero dominante del area. Cuando se encontró *Pheidole megacephala,* no se pudo encontrar especies nativas del genero *Pheidole* en los cebos. En realidad, los rastros de esta hormiga fueron encontrados en toda la ciudad. Muchos vertebrados, especialmente reptiles mantenidos en cuarentena, fueron consumidos por hormigas localizadas dentro de su jaula. Dentro de la ciudad, solo pudimos encontrar pocas salamanquesas, pero no se pudieron encontrar lagartijas. Dada su abundancia, *P. megacephala* fueron un problema en habitaciones, y en todas las cocinas inspeccionadas.

En el otro extremo del Brazil en la parte sureña del estado de Sao Paulo, se efectuaron muestreos en el area con gran producción de bananos. De 80 plantaciones de banana muestreadas, 31 tuvieron *P. longicornis,* 18 tuvieron *T. melanocephalum,* pero en ninguna se encontraron las dos especies al mismo tiempo. En las plantaciones con *P. longicornis,* se encontró un promedio de 7 especies, mientras que en otras plantaciones con *T. melanocephalum,* se encontraron unicamente 2 especies. En aquellas plantaciones de banano sin *T. melanocephalum* o *P. longicornis,* hubo un promedio de 13 especies de hormigas.

Cuando las hormigas exoticas se han establecido, pueden convertirse en plagas ecologicas, plagas de zonas agricolas y urbanas. Los metodos de control biologico clasico no han dado muchos resultados para el control de insectos sociales. Basados en nuestros conocimientos de ecología, el control biologico clasico necesita basarse en estrategias que favorezcan las hormigas nativas. Presentemente, dada las presiones ecologicas y economicas producidas por la rapida expansión de las hormigas exoticas invasoras, tales como las encontradas en las islas Galápagos, el unico control que pudiera proveer mas daño es el quimico.

References

Barber, E. R. 1916. The Argentine ant: distribution and control in the United States. *U.S. Dept. Agric. Bull.* 377: 1-23.

Bernard, F. 1983. *Les Fourmis et leur Milieu en France Mediterranéene.* Editions Lechavalier, Paris.

Bond, W. J., and P. Slingsby. 1984. Collapse of an ant-plant mutualism: the Argentine ant (*Iridomyrmex humilis*) and myrmecochorous Proteaceae. *Ecology* 65: 1031-1037.

Brown, W. L., Jr. 1954. Some tramp ants of Old World origin collected in tropical Brazil. *Entomol. News* 75: 14-15.

Camilo, G. R., and S. A. Phillips, Jr. 1990. Evolution of ant communities in response to invasion by the fire ant *Solenopsis invicta.* In: R. K. Vander Meer, K. Jaffe and A. Cedeño-Leon [eds.]. *Applied Myrmecology: a World Perspective.* Westview Press, Boulder, CO.

Clark, D. B., C. Guayasamín, O. Pazmiño, C. Donoso, and Y. Páez de Villacís. 1982. The tramp ant *Wasmannia auropunctata*: Autecology and effects on ant diversity and distribution on Santa Cruz Island, Galapagos. *Biotropica* 14: 196-207.

Crowell, K. L. 1968. Rates of competitive exclusion by the Argentine ant in Bermuda. *Ecology* 49: 551-555.

De Kock, A. E., and J. H. Giliomee. 1989. A survey of the Argentine ant, *Iridomyrmex humilis* (Mayr) (Hymenoptera: Formicidae), in South African fynbos. *J. Entomol. Soc. S. Africa* 52: 157-164.

Elton, C. S. 1958. *The ecology of invasions by animals and plants.* Methuen, London.

Erickson, J. 1972. The displacement of native ant species by the introduced Argentine ant *Iridomyrmex humilis* Mayr. *Psyche* 78: 257-268.

Fluker, S. S., and J. W. Beardsley. 1970. Sympatric associations of three ants: *Iridomyrmex humilis, Pheidole megacephala,* and *Anoplolepis longipes* in Hawaii. *Ann. Entomol. Soc. Am.* 63: 1290-1296.

Fowler, H. G., A. M. D. de Aguiar, M. A. Pesquero, and J. H. C. Delabie. 1992a. A preliminary assessment of ant assemblage (Hymenoptera: Formicidae) from the Brazilian Atlantic Coastal Forest. *Stud. Neotrop. Fauna Environ.* (in press).

Fowler, H. G., S. Campiolo, and M. A. Pesquero. 1992b. O exôtico estado de São Paulo. *Ciencia Hoje* 15: 18-23.

Fowler, H. G., F. Anaruma, and O. C. Bueno. 1992c. Seasonal space usage by the exotic Pharaoh's ant, *Monomorium pharaonis,* in an institutional setting in Brazil. *J. Appl. Entomol.* (in press).

Fowler, H. G., and M. A. Pesquero. 1992d. Ant communities of the Ilha do Cardosa State Park and their relation with vegetation. *Rev. Brasil Biol.* (in press).

Fowler, H. G., L. F. T. de Romagnano 1992e. Ecological bases for biological control. *Rev. Brasil Agropec.* (in press).

Haskins, C. P., and E. F. Haskins. 1965. *Pheidole megacephala* and *Iridomyrmex humilis* in Bermuda--equilibrium or slow replacement. *Ecology* 46: 736-740.

Haugh, G. W. 1934. Effect of Argentine ant poison on the ant fauna of Mississippi. *Ann. Entomol. Soc. Am.* 27: 621-632.

Lieberburg, I., P. M. Kranz, and A. Seip. 1975. Bermudan ants revisited: the status and interaction of *Pheidole megacephala* and *Iridomyrmex humilis*. *Ecology* 56: 473-478.

Lubin, Y. D. 1984. Changes in the native fauna of the Galapagos Islands following invasion by the little red fire ant, *Wasmannia auropunctata*. *Biol. J. Linn. Soc.* 21: 229-242.

Majer, J. D. 1992. Dispersão da formiga argentina, *Iridomyrmex humilis*, com referença especial a Western Australia. In: H. G. Fowler, and O. C. Bueno. [eds.]. *Importancia Economico e Manejo de Formigas Praga*. Editora da UNESP, São Paulo. (in press).

Meer Mohr, J. C. 1927. Au sujet du role de certaines fourmis dans les plantations coloniales. *Bull. Agric. Congo Belge*. 31: 97-106.

Mooney, H. A., and J. A. Drake [eds.]. 1986. *The ecology of biological invasions of North America and Hawaii*. Springer-Verlag, New York.

Newell, W., and T. C Barber. 1913. The Argentine ant. *U.S. Dept. Agric. Bull.* 122: 1-98.

Porter, S. D., and D. A. Savignano. 1990. Invasion of polygyne fire ants decimates native ants and disrupts arthropod community. *Ecology* 71: 2095-2106.

Smith, M. R. 1965. House infesting ants of the eastern United States, their recognition, biology and economic importance. *U.S. Dept. Agric. Tech. Bull.* 1326: 1-105.

Soans, A. B. and J. S. Soans. 1971. A case of intergeneric competition and replacement in the ants *Oecophylla smaragdina* (Fab.) and *Anoplolepis longipes* (Jerdon) (Hymenoptera: Formicidae). *J. Bombay Nat. Hist. Soc.* 68: 289-290.

Stoll, I. 1898. Zur Kenntnis der geographischen Verbreitung der Ameisen. *Mitteil. Schweiz. Entomol. Gesellsch.* 10: 120-126.

Vitousek, P. M., L. R. Walker, L. D. Whiteaker, D. Mueller-Dounbois, and P.A. Matson. 1987. Biological invasion by *Myrica faya* alters ecosystem development in Hawaii. *Science* 238: 802-804.

Ward, P. S. 1987. Distribution of the introduced Argentine ant (*Iridomyrmex humilis*) in natural habitats of the lower Sacramiento Valley and its effect on the indigenous ant fauna. *Hilgardia* 55(2): 1-16.

Wilson, E. O. 1959. Adaptive shift and dispersal in a tropical ant fauna. *Evolution* 13: 122-144.

Wilson, E. O., and R. W. Taylor. 1967a. The ants of Polynesia (Hymenoptera: Formicidae). *Pacific Insects Mon.* 14: 1-109.

_____ . 1967b. An estimate of the potential evolutionary increase in species density in the Polynesian ant fauna. *Evolution* 21: 1-10.

Woodworth, C. W. 1908. The Argentine ant in California. *Univ. California Agric. Exp. Sta. Circ.* 38: 1-11.

Zenner-Polania, I. 1990. Biological aspects of the "hormiga loca", *Paratrechina (Nylanderia) fulva* (Mayr), in Colombia. Pp. 290-297. In: R. K. Vander Meer, K. Jaffe and A. Cedeño-Leon [eds.]. *Applied Myrmecology: a World Perspective*. Westview Press, Boulder, CO.

13

Spread of Argentine Ants (*Linepithema Humile*), with Special Reference to Western Australia

J. D. Majer

Introduction

The Argentine ant (*Linepithema humile*) (genus revised by Shattuck 1992), which is believed to be native to Brazil and Argentina, has spread throughout the World as a result of trade movements. The regions where the ant has been most successful in establishing itself are in the 30°-36° latitude belts of both the Northern and Southern hemispheres (Fluker and Beardsley 1970, Lieberburg et al. 1975).

It generally replaces much of the resident ant fauna in areas where it becomes established. The documented instances where *L. humile* has replaced some, or all, of the resident ant fauna in areas where it has been introduced are recorded in Table 13.1. A number of conclusions are evident from this Table and from reading the papers from which it was constructed.

First, *L. humile* has a great capacity to replace other ants which formerly occupied the area. This appears to be related to the fact that it is omnivorous with generalist feeding and nesting requirements and, as such, it overlaps the niches of many other ant species.

Second, the ant has a large capacity for colony growth as a result of the large number of queens in each colony and colony fission. This allows the ant to replace the existing resident ants, be they native species or, like itself, a previously introduced ant. With the exception of *Tapinoma sessile* and *Liometopum occidentale* in California (Ward 1987), the main ant species which *L. humile* has replaced are taxonomically unrelated.

TABLE 13.1. Locations where the Argentine ant *(Linepithema humile)* has replaced other species as a result of its introduction.

Country	*Outcome of Introduction*	*Main Species Replaced*	*Reference*
France (Cote d'Azur)	Colonized large areas and eliminated most native ants	*Pheidole* spp. *Plagiolepis* spp.	Bernard (1983)
Madeira	Colonized most of island, eliminating main resident ant	*Pheidole megacephala**	Stoll (1898)
Bermuda	Rapid initial spread, which later slowed as a result of equilibrium in preferred range	*Pheidole megacephala**	Haskins and Haskins (1965) Crowell (1968) Lieberburg et al. (1975)
Hawaii	Tended to colonize areas above 100m but not hotter, lower areas where *P. megacephala* was in its preferred range	*Pheidole megacephala** *Anoplolepis longipes**	Fluker and Beardsley (1970)
U.S.A. (Alabama)	Colonized large areas	*Solenopsis saevissima richteri***	Wilson (1951)
(California)	Colonized large areas of field	*Pogonomyrmex californicus* *Pheidole capensis* *Veromessor pergandii*	Erickson (1972)
(California)	Only colonized riparian woodland sites with permanent water sources and that tend to be environmentally degraded	*Liometopum occidentale* *Tapinoma sessile* *Formica occidua*	Ward (1987)
South Africa (SW Cape)	Colonized disturbed areas of fynbos (heathland)	*Anoplolepis custodiens* *Pheidole capensis*	Donnelly and Giliomee (1985) De Kock and Giliomee (1989)
North Africa	Rare or absent, apparently being outcompeted by the native *Tapinoma simrothi*	None	Bernard (1983)

*Denotes non-native ant. ** Separated into two species: *Solenopsis invicta* and *Solenopsis richteri.*

Third, many authors have observed that *L. humile* has a tendency to thrive in areas which have been disturbed to some degree. Possibly this is associated with a prior simplification of the local ant fauna, rendering it less able to resist invasion by the exotic species.

A number of authors have pointed out that *L. humile is* generally found in large numbers only in moist areas. This was noted in California by Erickson (1972). Ward (1987) found the ant only in riparian woodland, failing to find it in the drier woodlands and chaparral. In the Côte d'Azur in France, it is generally found only within the 25 km wide littoral zone (Bernard 1983).

Finally, even when introduced, *L. humile* does not always replace the resident ant. In Hawaii it totally replaced *Pheidole megacephala* only at the cooler, higher elevations; at lower elevations *P. megacephala* persisted, apparently being better adapted to the more tropical climate. *L. humile* does not appear to have become abundant in North Africa and, although I think that low moisture levels could be a factor here, competition with the related native dolichoderine, *Tapinoma simrothi is* advanced by Bernard (1983) as the likely reason for its inability to spread.

Abundance of Argentine ants in Western Australia vs. Brazil

Pitfall trap data of ant communities in disturbed ecosystems in Brazil, where *L. humile is* thought to be a native ant, indicate that it makes up only a minor component of the community (Table 13.2, columns a and b); the majority of ants in the two examples given being members of the subfamilies *Myrmicinae* and *Formicinae*. My own observations in secondary forest at Viçosa, Minas Gerais, Brazil, indicate that the ant is not common there either. Similarly, in areas of undisturbed woodland or forest around Perth, Western Australia, the ant is absent (Table 13.2, column c), even though the ant community shown in the example contained eight other species of the related genus, *Iridomyrmex*. However, in disturbed areas near water, outbreaks of *L. humile* occur (Table 13.2, column d).

In disturbed areas of Western Australia, the diversity of other ants is greatly reduced. However, in disturbed areas where *L. humile is* present, diversity is far lower than could be expected from the impact of habitat disturbance alone. In such areas, *L. humile* may make up over 80% of the individuals in the ant community (Table 13.2, column d). In addition, where *L. humile* is present, the number of native *Iridomyrmex* is very low and the introduced ant species is spread across most of the ground surface where the outbreak occurs (Majer and Flugge 1984).

TABLE 13.2. Comparison of the numbers of species and percentage of individuals in samples of ground-dwelling ants taken in Brazil and Western Australia, showing the contribution to numbers made by the Argentine ant (*L. humile*).

	Brazil				*Western Australia*			
	(a) *L. humile* present		*(b)* *L. humile* present		*(c)* *L. humile* absent		*(d)* *L. humile* present	
Subfamily	No. of spp.	% of indiv.	No. of spp.	% of indiv.	No. of spp.	% of indiv.	No. of spp.	% of indiv.
Ponerinae	3	0.15	5	0.16	4	48.60	-	-
Ecitoninae	1	2.20	-	-	-	-	-	-
Pseudomyrmecinae	-	-	5	0.02	-	-	-	-
Myrmicinae	10	57.87	16	93.24	13	25.90	3	15.60
Formicinae	3	39.68	10	6.18	10	16.40	-	-
Dolichoderinae								
L. humile	1	0.07	1	0.23	-	-	1	82.90
Iridomyrmex or other *Linepithema*	-	-	1	0.06	8	8.80	-	-
other genera	-	-	3	0.07	2	0.30	-	-

(a) Agroecosystems in Viçosa, M.G. (From Della Lucia et al. 1982)
(b) Agroecosystem and cerrado in Sete Lagoas, M.G. (From Castro and Queiroz 1987)
(c) Eucalyptus open-woodland in Perth, W.A. (From Rossbach and Majer 1983)
(d) Old refuse dump in Perth, W.A. (From Majer and Flugge 1984)

Why is the Argentine ant not more widespread in Western Australia?

Figure 13.1 shows the locations where Argentine ants have been found since 1941. It is immediately evident that the locations are not confined to the higher rainfall coastal strip. They are, however, more common in the higher rainfall belt and, in all areas, are often close to rivers, lakes or swamps. Moisture is therefore likely to be one factor which restricts where outbreaks may occur.

Second, although this has not been quantified, the areas where infestations occur are generally disturbed areas such as townsites, market gardens, grassed areas around lakes and rubbish dumps. Such areas are known to have a simplified ant fauna, even before *L. humile* colonized the area (Majer 1978). In the 15 years that I have been studying ants in pristine areas of the south-west of Western Australia, I have never encountered *L. humile* in undisturbed vegetation. Furthermore, in an outbreak area near Karridale, in the south of Western Australia, transects of pitfall traps along forest-townsite boundaries have revealed that *L. humile* does not occur in the adjoining undisturbed native *Eucalyptus* forest (J. van Schagen, pers. comm.).

There are a number of possible reasons why *L. humile* thrives in disturbed areas, including increased insolation of the soil, changed moisture regimes or the simplification of the native ant community. It is the latter two reasons, and in particular the reduction in richness of native ants, which I suggest may be the major factor in the spread of *L. humile* within such areas.

Much of the current Australian ant fauna radiated from the original ground-dwelling ants resident in the heaths and shrublands of southern Australia prior to separation of the continent from other land masses. This has accounted for the extremely high richness of ants within the genera *Camponotus, Melophorus* and *Iridomyrmex* within Australia (Greenslade 1979). Indeed, the terrain over much of Australia is covered with a mosaic of different *Iridomyrmex* spp (Greenslade 1987).

In a comparison of study sites throughout the mediterranean regions of the world, Majer (1989) found that Australia is unique in possessing a high richness of *Iridomyrmex* spp (Table 13.3). It is postulated here that the existence of this diverse and ubiquitous fauna of *Iridomyrmex* spp in Western Australia and, indeed, throughout much of the country has contributed to the containment of *L. humile* outbreaks. The mechanism might operate (1) through competition with the resident dominant *Iridomyrmex,* which would probably have similar food needs and, to a lesser extent, nesting requirements or (2) by diffuse competition from the array of *Iridomyrmex* spp and other native genera occuring within the area.

FIGURE 13.1 Locations in Western Australia where Argentine ants have been discovered since 1941. An infestation has also been found in Esperance which is further east along the south coast (adapted from Whitehouse 1988).

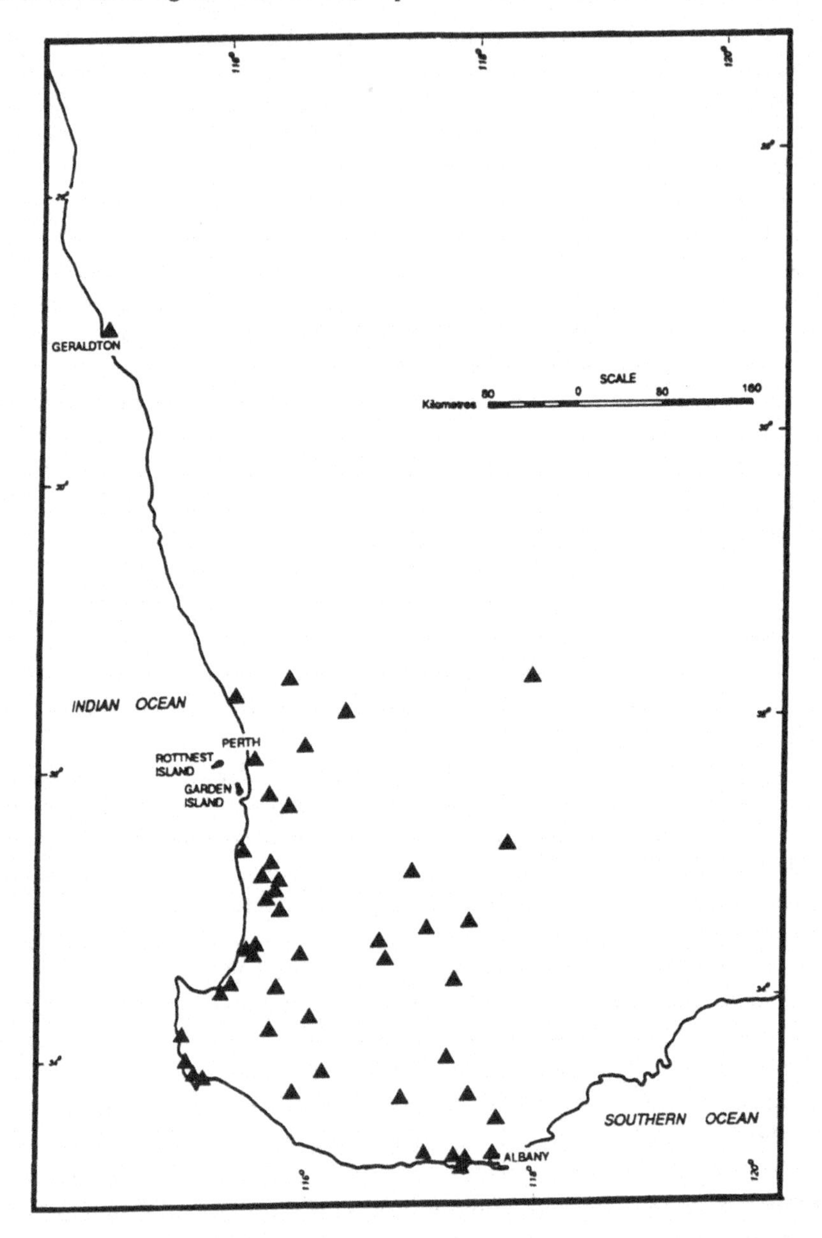

TABLE 13.3. Number of ant species per subfamily collected from study sites in mediterranean regions throughout the world. The number of species of *Iridomyrmex* and *Tapinoma* per study site are also shown (from Majer 1989).

Subfamily	*Western Australian forest*	*Western Australian heath*	*South Australian forest*	*South African fynbos*	*Israeli vegetation transect*	*Lebanon whole country*	*Californian vegetation transect*	*Chilean vegetation transect*
Myrmeciinae	3		4					
Ponerinae	9	5	7	1		1		
Ecitoninae							2	
Dorylinae				1		1		
Pseudomyrmecinae							1	1
Myrmicinae	24	22	15	27	31	58	21	8
Formicinae	22	18	12	13	16	36	21	9
Dolichoderinae								
All	15	12	6	2	2	4	5	5
*Iridomyrmex**	13	10	5	1			1	
Tapinoma	1	1			2	2	1	1

*Note: This analysis was prepared prior to Shattuck's (1992) splitting of *Iridomyrmex* into four genera.

An opportunity for control of Argentine ants

Majer and Flugge (1984) reported on a field trial in which part of an Argentine ant outbreak was eliminated by spraying with heptachlor. They found that this territory was then occupied by a native ant, *Iridomyrmex greensladei* (Figure 13.2). Unfortunately, this effect was short-lived, since *L. humile* regained its original territory within 2 years of the spraying (Figure 13.2). The apparent reason was that the untreated part of the original *L. humile* outbreak area eventually re-spread into the sprayed area.

The result of this trial demonstrates the mutual exclusivity of *L. humile* and native *Iridomyrmex* spp and also points to a possible way in which Argentine ants may be controlled. If the environment could be manipulated to give a selective advantage to native *Iridomyrmex* spp, or the *L. humile* could in some way be weakened, it might be possible to create areas where native ants could maintain territory against invasions by *L. humile*. Some of the ways in which this might be done are discussed in Majer (1990).

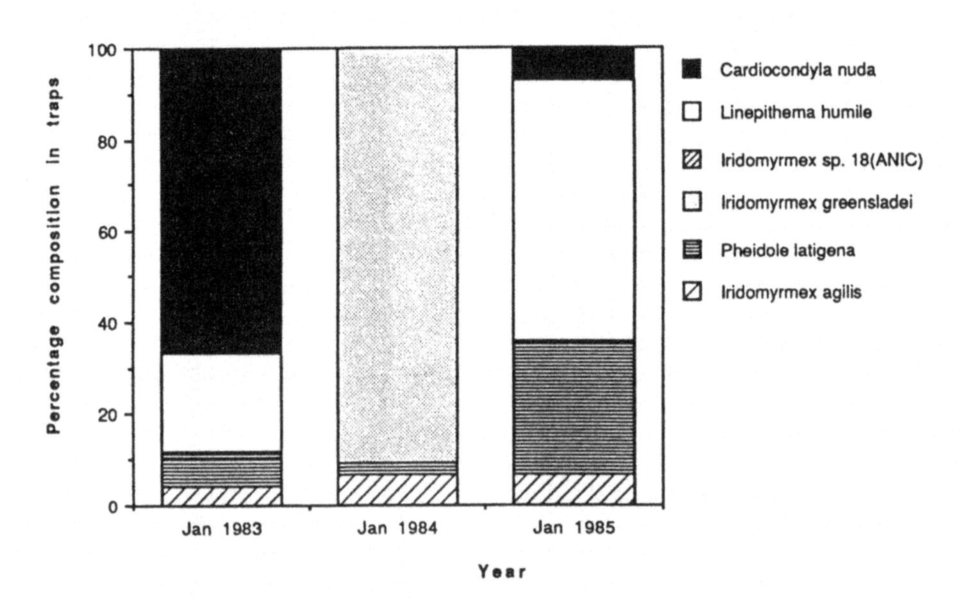

FIGURE 13.2 Percentage composition of pitfall trap catches in an area of Argentine ant infestation prior to February 1983, and also 11 and 23 months after spraying with heptachlor. (Adapted from Majer and Flugge 1984, Flugge 1985).

Acknowledgments

I would like to thank the Centro de Pesquisas do Cacao (CEPLAC) for providing support and facilities while I prepared this paper. I also thank Peter Davis and John Van Schagen for commenting on an earlier draft of this paper. Disclaimer: *The views expressed in this paper are my own and do not necessarily reflect those of the workers who are investigating means of Argentine ant control in Western Australia.*

Resumen

Desde su origen en Sur America, la hormiga Argentina, *Linepithema humile* [=*Iridomyrmex humilis*] se ha distribuído por todo el mundo. Se ha establecido en muchas regiones, pero mas eficazmente en aquellas regiones del hemisferio sur y norte con temperaturas limites entre los 30-36 grados.

En este capítulo se revisan los reemplazos que han ocurrido a las especies de hormigas cuando *L. humile* se ha establecido y preguntamos el porqué esta hormiga se ha expandido poco en el sur de Australia occidental. Hay muchas explicaciones, incluyendo el caso de que un efectivo programa de control se ha establecido desde 1954. Aquí, estudia la hipotesis adicional de que el esparcimiento de *L. humile* ha sido limitado por la competencia con la fauna nativa y diversa de *Iridomyrmex*. Esta evidencia se establece al comparar la composición de las comunidades de hormigas nativas en otras partes mediterraneas del mundo por medio de la observación de la respuesta de las hormigas Australianas occidentales nativas a la eliminación quimica de *L. humile*.

References

Bernard, F. 1983. *Les Fourmis et leur Milieu en France Mediterranéene.* Editions Lechevalier S.A.R.L., Paris.

Castro, A. G. and M. V. B. Queiroz. 1987. Estrutura e organação de uma comunidade de formigas em agro-ecossistema neotropical. *An. da Soc. Entomol. Brasil* 16: 363-375.

Crowell, K. L. 1968. Rates of competitive exclusion by the Argentine ant in Bermuda. *Ecology* 49: 551-555.

De Kock, A. E. and J. H. Giliomee. 1989. A survey of the Argentine ant, *Iridomyrmex humilis* (Mayr), (Hymenoptera: Formicidae) in South African fynbos. *J. Entomol. Soc. South Africa* 52: 157-164.

Della Lucia, T. M. C., M. C. Loureira., L. Chandler, J. A. H. Freire, J. D. Galvão, and B. Fernandes 1982. Ordenação de comunidades de Formicidae em quatro agroecossistemas em Viçosa, Minas Gerais. *Experientiae* 28: 67-94.

Donnelly, D. and J. H. Giliomee. 1985. Community structure of epigaeic ants (Hymenoptera: Formicidae) in fynbos vegetation in the Jonkershoek Valley. *J. Entomol. Soc. South Africa* 48: 247-257.

Erickson, J. 1972. The displacement of native ant species by the introduced Argentine ant *Iridomyrmex humilis* Mayr. *Psyche* 78: 257-266.

Flugge, R. 1985. Post spray succession of ant fauna at Burswood Island. approximately two years after spraying for Argentine ants. Unpublished report. The Western Australian Environmental Protection Authority.

Fluker, S. S. and J. W. Beardsley. 1970. Sympatric associations of three ants: *Iridomyrmex humilis, Pheidole megacephala,* and *Anoplolepis longipes* in Hawaii. *Ann. Entomol. Soc. Am.* 63: 1290-1296.

Greenslade, P. J. M. 1979. *A Guide to Ants of South Australia.* South Australian Museum, Adelaide.

_____. 1987. Environment and competition as determinants of local geographical distribution of five meat ants, *Iridomyrmex purpureus* and allied species (Hymenoptera: Formicidae). *Austral. J. Zool.* 35: 259-273.

Haskins, C. P. and E. F. Haskins. 1965. *Pheidole megacephala* and *Iridomyrmex humilis* in Bermuda--equilibrium or slow replacement. *Ecology* 46: 736-740.

Lieberburg, I., P. M. Kranz, and A. Seip. 1975. Bermudan ants revisited: the status and interaction of *Pheidole megacephala* and *Iridomyrmex humilis. Ecology* 56: 473-478.

Majer, J. D. 1978. Preliminary survey of the epigaeic invertebrate fauna with particular reference to ants, in areas of different land use at Dwellingup, West. *Austral. Forest Ecol. Manage.* 1: 321-334.

_____. 1989. Pp. 219-224.The Formicidae. In: R. L. Specht, [ed.] *Mediterranean--type ecosystems.* Kluwer Academic Publications, Dordrecht.

_____. 1990. Control of Argentine ants *(Iridomyrmex humilis)* in Western Australia: the past, present and future. Paper presented at the International Course Update in Pest Ant Control, Campo Grande, Mato Grosso do Sul 12-14 October 1989 (in press).

Majer, J. D. and R. Flugge. 1984. The influence of spraying on Argentine *(Iridomyrmex humilis)* and native ants (Hymenoptera: Formicidae). *West. Austral. Inst. Tech. School Biol. Bull.* No. 8.

Rossbach, M. H. and J. D. Majer. 1983. A preliminary survey of the ant fauna of the Darling Plateau and Swan Coastal Plain near Perth, Western Australia. *J. Roy. Soc. West. Australia* 66: 85-90.

Shattuck, S. O. 1992. Review of the dolichoderine ant genus *Iridomyrmex* Mayr with description of three new genera (Hymenoptera: Formicidae). *J. Austral. Entomol. Soc.* 31:13-18.

Stoll, I. 1898. Zur Kenntnis de geographischen Verbreitung de Ameisen. *Mitt. Schweiz. Entomol. Gesellsch.* 10: 120-126.

Ward, P. S. 1987. Distribution of the introduced Argentine ant *(Iridomyrmex humilis)* in natural habitats of the Lower Sacramento Valley and its effect on the indigenous ant fauna. *Hilgardia* 55(2): 1-16.

Whitehouse, S. 1988. A review of Argentine ant control in Western Australia. *West. Austral. Environmental Protection Authority Bull.* No. 325.

Wilson, E. O. 1951. Variation and adaptation in the imported fire ant. *Evolution* 5: 68-79.

14

Ant Pests of Western Australia, with Particular Reference to the Argentine Ant (*Linepithema Humile*)

J. J. van Schagen, P. R. Davis and M. A. Widmer

Introduction

The Argentine ant, *Linepithema humile* [*Iridomyrmex humilis* Mayr], originated in South America; however, with the unintentional aid of humans, the ant has now been distributed to many regions of the world including the United States, South Africa, Europe and Australia. Most infestations occur between the 30° and 36° latitudes (north and south).

Domestically, the ants are a great nuisance. They are very invasive, particularly in houses, and build up in large numbers in gardens. Agriculturally, they cause damage by tending homopterans, especially on citrus trees. Populations of these pest insects can subsequently increase if the Argentine ants protect them from their predators, some of which were originally introduced as biological control agents (Woodworth 1908, Woglum and Borden 1921). Environmental damage can result from the Argentine ant's ability to displace the native ant fauna (Erickson 1971, Tremper 1976, Bond and Slingsby 1984, Ward 1987, Davis et al. unpublished data, Van Schagen et al. unpublished data). Even tramp species such as *Pheidole megacephala* (the coastal brown ant), which may have already displaced native ant fauna, have on occasions been displaced by Argentine ants (Haskins and Haskins 1965, Fluker and Beardsley 1970). Conversely, in their native Brazil, Argentine ants were found to account for only a small component of the ant community (Majer 1990), and did not seem to displace other ants.

In South African fynbos, Bond and Slingsby (1984) demonstrated that the displacement of native ants by Argentine ants resulted in a disruption of myrmecochory and seeds became more vulnerable to predation by vertebrates and invertebrates. This could result in the loss of many plant species in South Africa and in Australia where Berg (1975) recorded some 1500 species of myrmecochorous vascular plants.

Argentine ants were first discovered in Albany, Western Australia (W.A.) in 1941, and rapidly spread to southwestern parts of the State, including the Perth metropolitan area. In 1954, when the ant was well established, an eradication campaign was initiated by the W. A. Department of Agriculture involving the spraying of infested areas initially with dieldrin and later with heptachlor (Forte and Greaves 1953, 1956, Jenkins and Forte 1973). In 1988, the campaign was cancelled following a review of the use of heptachlor for Argentine ant control by the Environmental Protection Authority (E.P.A. 1988). At that stage a total of 31,000 hectares of infestation had been treated and Argentine ants were successfully eradicated from most of this area. However, some 1500 hectares of infestation remained untreated due to its environmental or agricultural sensitivity. This was comprised mainly of wetlands in the Perth metropolitan area and some country locations. Control of the ants is now the responsibility of the individual land holders.

The aim of this chapter is to discuss the main ant pests occurring in W.A. and to give an account of the spread of Argentine ants since the cancellation of the eradication campaign in 1988.

Methods

Infestations of Argentine ants in W.A. are being monitored by the Agriculture Protection Board (A.P.B.). New infestations are discovered mainly as a result of public inquiries about ants that are a nuisance to them. Most members of the public are informed about Argentine ants through the media and with the aid of posters and leaflets produced by the A.P.B. Many inquiries are received with regard to ant pests. Most of these are via telephone, but during the period of August 1989 to July 1991, 372 mail inquiries, complete with specimens, were received. These species were positively identified and this allowed us the opportunity to assess the relative pest status of ants occurring in Western Australia. In view of the fact that this survey was performed immediately after the termination of the Argentine ant eradication campaign, it can be used as a baseline study with which future pest ant assessment can be compared.

Results

L. humile was the major ant pest in W.A., being recorded 125 times (89 in the Perth metropolitan area and 36 from the country towns) (Table 14.1). *P. megacephala* accounted for 72 of the metropolitan inquiries but only two from the country areas. Unlike the Argentine ant, *P. megacephala* generally does not enter homes and is mainly a pest in gardens, where it often nests under slabs and brickpaving. Other ant species which appear to be significant pests include native ants such as *Iridomyrmex glaber*, the black house ant, *Iridomyrmex chasei* and the introduced species, *Technomyrmex albipes*, the white-footed house ant. Other introduced ants included *Paratrechina longicornis*, the crazy ant, *Monomorium pharaonis*, the Pharaoh ant and *Monomorium destructor*, the Singapore ant.

TABLE 14.1. Numbers of inquiries received via post from members of the public in the Perth metropolitan area and country towns in southwestern W. A. regarding ant pests on their properties. Where the species is not specified it refers to a native ant of that genus.

Ant species	*Metropolitan area*	*Country towns*
PONERINAE		
Brachyponera lutea	1	
Rhytidoponera violacea		1
MYRMICINAE		
Aphaenogaster sp.	1	2
Crematogaster sp.	4	1
Monomorium destructor	2	5
M. pharaonis	5	
M. spp.	4	4
Pheidole megacephala	72	2
P. sp.	6	1
Solenopsis sp.	1	
Tetramorium spp.	7	1
DOLICHODERINAE		
Iridomyrmex chasei	20	6
I. glaber	24	8
L. humile [= I. humilis]	89	36
I. purpureus	1	1
I. spp.	5	3
Technomyrmex albipes	30	12
FORMICINAE		
Camponotus spp.	10	2
Paratrechina longicornis	5	

The Argentine ant has been spread to most major towns in southwestern Western Australia. By June 30, 1991, the total known area of infestation was 2932 hectares (A.P.B. unpubl. records); however, this is an underestimation because several new infestations remain to be surveyed, and it does not take into account the expansion of known infestations, which have not been re-surveyed since 1988. There are now more than 40 separate infestations of Argentine ants in the Perth metropolitan area, ranging in size from 350 hectares at Herdsman Lake (wetlands) to only 0.5 hectare. Though distributed widely, the ants generally tend to concentrate near permanent water sources (including rivers, lakes and well irrigated gardens).

P. megacephala, on the other hand, was found in a much more defined area, namely on the southern side of the Swan river from Fremantle to Perth. This is where most of the inquiries originated. No known infestations of Argentine ants occur in this area.

Discussion

Results suggest that Argentine ants are the major ant pest of southwestern W.A. This is surprising given that a highly successful eradication campaign was terminated only three years prior to the end of this study. Although this study was biased towards Argentine ants as a result of the pre-existing public awareness of the ant, the expansion of infestations from 1500 hectares into a total area of almost 3000 hectares in less than three years is an indication of its dominance over other ants. Many new infestations have been discovered since 1988, spreading throughout the metropolitan area and many country locations. Apart from the radial expansion of the colonies, Argentine ants are spread mainly through transport by humans via a wide range of commodities, including soil, pot plants, foodstuffs and garbage. Several plant nurseries within the metropolitan area are infested with Argentine ants and present a spread risk. The A.P.B. is currently trying to minimise the spread by informing the public and owners of infested businesses. Apart from this advice, few active operations are being carried out to prevent the spread of Argentine ants.

A rapid increase in the infested area will result in Argentine ants becoming a greater nuisance with a concomitant increase in insecticide usage for its control. Argentine ants threaten both the environment and, from a more economic point of view, agriculture in W.A.

It has been demonstrated that Argentine ants displace native ants in South Africa (Bond and Slingsby 1984, De Kock and Giliomee 1989) where their spread into the 'fynbos' areas was linked mainly with disturbance and accessibility by man. Similar findings were recorded in California by Erickson (1971) and Ward (1987), and in W.A. by Davis et al. (unpublished data) and Van Schagen et al. (unpublished data). A spread of Argentine ants into the forests of southwestern W.A. could occur from the many popular tourist attractions and picnic sites located therein. If our native ant fauna is displaced in those areas by the Argentine ants, a similar situation to the South African 'fynbos' could occur.

Agriculturally, there is the risk of economic loss in the horticultural, viticultural and apicultural industries, both directly and indirectly through an increase in scales and aphids and the raiding of bee hives. Since the ants occur primarily in the higher rainfall coastal areas, export facilities are also put at risk. Contamination of export produce could result in the refusal of goods in overseas ports, or costly decontamination procedures.

W.A.'s main port at Fremantle is still free of Argentine ants, possibly due to the abundance of *P. megacephala* in the area which compete with the Argentine ants for space and therefore exclude them. Now that Argentine ants are not being actively controlled or contained, it is likely that they will eventually displace *P. megacephala* from the Fremantle area.

A major ant pest in northern W.A. which is not reflected in this survey, is the Singapore ant, *M. destructor*, which is mainly a domestic pest. The reason for the lack of public inquiries regarding this ant is that people in the area are well aware of its existence and identity, and individual cases are mainly dealt with by professional pest control operators (P.C.O.s). Similarly, we believe that *M. pharaonis* has established in the Perth Central Business District (pers. obs.), however this is also not reflected in this study because these cases are mainly dealt with by P.C.O.s. In this case, the total extent of the *M. pharaonis* infestation is not known in detail, however, 5 inquiries were received from throughout the Central Business District.

It is desirable that some form of control be exercised to minimize the spread of Argentine ants and to control them in situations were they pose a risk to the environment or agriculture. The A.P.B. is conducting research into alternative methods of controlling the ant, especially the use of toxic baits, but it seems that Argentine ants have the potential of spreading rapidly and becoming a significant pest throughout southwestern Western Australia.

Acknowledgments

The authors like to thank Jim Giovinazzo for his help with the surveying of infestations; Tara Craven for her help with the preparation of this manuscript; and Assoc. Prof. Jonathan Majer for reviewing an early draft of the manuscript.

Resumen

Se da en este capitulo una lista de las mayores hormigas plaga que ocurren en en suroeste de Australia occidental, incluyendo el area metropolitana de Perth. Los resultados se basan en una encuesta efectuada desde Agosto de 1989 a Julio 1991, cuando el publico enviaba los especimenes de hormigas plaga para identificación. Estos especimenes fueron identificados y se registró el nombre de las especies así como tambien el problema que estas causaban. La hormiga argentina *Linepithema humile* [=*Iridomyrmex humilis*], fué la plaga mas importante en esta parte de Australia occidental, y la hormiga café de la costa, *Pheidole megacephala* fué segunda en importancia.

References

Berg, R. Y. 1975. Myrmecochorous plants in Australia and their dispersal by ants. *Austral. J. Bot.* 23: 475-508.

Bond, W., and P. Slingsby. 1984. Collapse of an ant-plant mutualism--The Argentine ant (*Iridomyrmex humilis*) and myrmecochorous Proteaceae. *Ecology* 65: 1031-1037.

De Kock, A. E., and J. H Giliomee. 1989. A survey of the Argentine ant, *Iridomyrmex humilis* (Mayr), (Hymenoptera: Formicidae) in South-African fynbos. *J. Entomol. Soc. South Africa* 52: 157-164.

Environmental Protection Authority. 1988. A review of the use of heptachlor for the control of Argentine ants and termites in Western Australia. *E.P.A. Bull.* 354. Perth, Australia.

Erickson, J. M. 1971. The displacement of native ant species by the introduced Argentine ant, *Iridomyrmex humilis* Mayr. *Psyche* 78: 257-266.

Fluker, S. S., and J. W. Beardsley. 1970. Sympatric associations of three ants: *Iridomyrmex humilis, Pheidole megacephala* and *Anoplolepis longipes* in Hawaii. *Ann. Entomol. Soc. Am.* 63: 1290-1296.

Forte, P. N., and T. Greaves. 1953. New insecticides for the control of the Argentine ant in Western Australia. *J. Dept. Agric. West. Australia* 2: 267-280.

_____. 1956. The effectiveness of dieldrin for the control of the Argentine ant in Western Australia. *J. Dept. Agric. West. Australia* 5: 85.

Haskins, C. P., and E. F. Haskins. 1965. *Pheidole megacephala* and *Iridomyrmex humilis* in Bermuda--Equilibrium or slow replacement? *Ecology* 46: 736-740.

Jenkins, C. F. H., and P. N. Forte. 1973. Chemicals for Argentine ant control. *J. Agric. West. Australia* 14: 95-96.

Majer, J. D. 1990. Spread of Argentine ants (*Iridomyrmex humilis*), with special reference to Western Australia. *Proc. Third Int. Workshop on Pest Ants*. Campo Grande, Mato do Sul, Brazil.

Tremper, B. S. 1976. Distribution of the Argentine ant, *Iridomyrmex humilis* Mayr, in relation to certain native ants in California: Ecological, physiological and behavioral aspects. *PhD Thesis*. Univ. California, Berkeley.

Ward, P. S. 1987. Distribution of the introduced Argentine ant (*Iridomyrmex humilis* in natural habitats of the Lower Sacramento Valley and its effects on the indigenous ant fauna. *Hilgardia* 55(2): 1-16.

Woglum, R. S., and A. D. Borden. 1921. Control of the Argentine ant in Californian citrus groves. *U.S. Dept. Agric. Bull.* 965: 1-43.

Woodworth, C. W. 1908. The Argentine ant in California. *Univ. California Cir.* 38: 1-11.

15

Ant Fauna of the French and Venezuelan Islands in the Caribbean

Klaus Jaffe and John Lattke

Introduction

Surveys of the ants of the islands in the Caribbean have been made by Wilson (1988) and Jaffe et al. (1991). These studies are somewhat contradictory, as Wilson reports that from the 89 genera and 176 species to be found in the Antilles, 46% can be classified as endemic, while Jaffe et al. found no endemic species after 6 months of intensive collecting in the French islands. Kempf (1972) and later Brandão (1991) report endemic subspecies and varieties from the Caribbean islands in their catalogs of neotropical ants, but given the splintering of the past, it is likely that most, if not all, of these epithets will be synonymized. These latter authors used data taken from various independent collectors who collected at different times in the past. Few comparative studies among islands exist. We report here data from recent collections which were identified by the same person and kept in the same collection (Museo de Ciencias Naturales, Universidad Simon Bolivar, Caracas). This allows a better comparison of the ant fauna on different islands as well as an opportunity to reassess the degree of endemism and the distribution patterns of ants in the Caribbean islands.

The accompanying Table presents the data of this recent effort. Ants were collected by hand using the sampling technique described by Romero and Jaffe (1989). Fifty ant species were found on only one island out of 88 species reported (57%). None of these 50 species can be classified unequivocally as endemic to a single island nor to the Antilles, since 18 species that could be identified to species level are known to

TABLE 15.1. Ant species found on the French islands (St. Martin: SM; Desirade: DE; Guadeloupe: GU; Marie Galante: MG; Les Saintes: LS; Martinique: MR); Venezuelan islands (La Blanquilla: BL; Margarita: MA; La Orchila: OR; Las Aves: AV) and Aruba (AR), in the Caribbean.

Species	SM	DE	GU	MG	LS	MR	BL	MA	OR	AV	AR
MYRMICINAE											
Acromyrmex octospinosus			X					X			
Cardiocondyla wroughtoni	X	X	X	X		X		X			X
Cyphomyrmex rimosus			X	X	X	X					
Crematogaster sp1			X								
Crematogaster sp2			X								
Crematogaster sp3						X					
Crematogaster sp4							X			X	
Crematogaster sp5											X
Crematogaster sp6								X			
Monomorium minimum		X	X	X		X					
Monomorium destructor		X	X		X	X					
Monomorium carbonarium		X	X								
Monomorium floricola		X	X		X	X	X	X			X
Monomorium salomonis			X		X						
Mycetophylax conformis							X	X			X
Mycocepurus smithi			X	X		X					
Pheidole fallax		X	X	X	X	X					
Pheidole megacephala	X	X	X		X			X			
Pheidole sp1							X				X

Species	SM	DE	GU	MG	LS	MR	BL	MA	OR	AV	AR
MYRMICINAE (continued)											
Pheidole sp2			X		X						
Pheidole sp3			X								
Pheidole sp4			X		X						
Pheidole sp5			X								
Pheidole sp6			X								
Pheidole sp7			X								
Pheidole sp9			X								
Pheidole sp10			X								
Pheidole sp11			X								
Pheidole sp12								X			
Pheidole sp13								X			
Quadristruma emmae			X								
Solenopsis geminata		X	X	X	X	X		X			X
Solenopsis sp1			X								
Solenopsis sp2		X	X								
Solenopsis sp3			X								
Solenopsis sp4							X	X		X	
Solenopsis sp5							X				
Solenopsis sp6								X			
Solenopsis sp7								X			
Strumigenys smithi			X								
Smithistruma alberti			X								

continues

Species	SM	DE	GU	MG	LS	MR	BL	MA	OR	AV	AR
MYRMICINAE (*continued*)											
Tetramorium bicairnatum			X				X				
Tetramorium simillimum			X								
Trachymyrmex jamaicensis			X								
Wasmannia auropunctata			X	X	X	X					
Rogeria foreli			X								
Zacryptocerus clypeatus								X			
DOLICHODERINAE											
Azteca delpi antillana		X	X	X							
Conomyrma sp1							X				
Conomyrma sp2							X		X		
Conomyrma sp3											X
Conomyrma sp4								X			
Tapinoma melanocephalum		X	X	X		X					X
Tapinoma litorale		X									
Tapinoma sp			X								
PONERINAE											
Anochetus emarginatus								X			
Anochetus mayri						X					
Ectatomma ruidum			X			X		X			
Gnamptogenys striatula			X								
Hypoponera gleadowi			X								
Hypoponera punctataissima			X			X					
Hypoponera sp8			X								
Leptogenys arcuata			X								
Odontomachus insularis			X	X		X					
Odontomachus bauri			X			X					

Species	SM	DE	GU	MG	LS	MR	BL	MA	OR	AV	AR
PONERINAE (continued)											
Pachycondyla stigma		X	X								
Platythyrea sinuata			X	X		X					
Prionopelta antillana			X			X					
PSEUDOMYRMECINAE											
Pseudomyrmex curacaensis			X								
Pseudomyrmex flavidulaus			X	X							
Pseudomyrmex gracilis								X			
FORMICINAE											
Brachymyrmex sp1							X				
Brachymyrmex sp2			X	X		X					
Brachymyrmex sp3		X	X	X		X					
Brachymyrmex sp4			X			X					
Camponotus abdominalis							X				
Camponotus crassus									X	X	
Camponotus sexguttatus			X								
Camponotus lindigi											X
Camponotus sp1								X			
Camponotus sp2			X			X					
Camponotus sp3			X								
Camponotus sp4			X								
Camponotus sp8			X			X					
Camponotus sp9											X
Camponotus sp10											X
Myrmelachista sp1			X								
Paratrechina longicornis		X	X	X	X	X	X	X	X	X	X
Paratrechina sp1							X				
Paratrechina sp2								X			

exist elsewhere in the continent. The remaining 32 species are only classified as morpho-species belonging to genera which lack recent taxonomic revisions. Thus, the real number of endemic species in the Lesser Antilles is far lower than 36% and probably close or equal to zero. Wilson (1988) reported endemic species from islands close to South America, such as St. Thomas, St. Vincent, St. Lucia and Trinidad but none from the French islands. We confirm the absence of endemic species from the French islands, but suggest that endemism reported from other islands in the Lesser Antilles are probably due to incomplete taxonomic study of the collections rather than true endemism.

Few ant species have been found only on the Caribbean islands (Kempf 1972). The species which are clearly endemic are found only on the large islands of the Greater Antilles (Cuba, Jamaica, Puerto Rico and Hispaniola). Species collected only from the Lesser Antilles are generally distributed throughout the neotropics or have been collected from most parts of the Caribbean basin, including the mainland of the continent. We have found that ant species in our collections can be placed into 3 categories: those which are found (1) on most islands, (2) only on the islands of the southern Caribbean, close to the South American continent, and (3) only on the Lesser Antilles, north of St. Lucia. Among the two latter groups, species in the genus *Conomyrma* appear to be common on islands near the Venezuelan coast, but are absent in the French Antilles. On the other hand, species in the genus *Tapinoma* are common in the French Antilles but absent on Aruba and the Venezuelan Islands. Similar patterns can be found among species in the genera *Crematogaster, Pheidole* and *Solenopsis*. These results would suggest that the Lesser Antilles could be grouped into two distinct biogeographical zones coinciding with the southern and the northern arc of the Lesser Antilles. The limits between them would lay somewhere near St. Lucia. The fauna in the southern arc, including the islands close to Venezuela, would be partially or totally related to the South American fauna, whereas the fauna of the northern arc would be more closely related to that in the Greater Antilles.

As a whole, the Lesser Antilles seem to be biogeographically different from the Greater Antilles, if only because few of the known endemics from the Greater Antilles are found on the Lesser Antilles. Two notable exceptions which exemplify dispersal from the Greater Antilles to the Lesser Antilles are (1) *Trachymyrmex jamaicensis* which is found on Guadeloupe and is also common on islands of the Greater Antilles and (2) *Azteca delpini antillana* which occurs on Guadeloupe, Hispaniola and St. Lucia, but is absent from Venezuela which suggests it dispersed from Hispaniola to the Lesser Antilles.

When data from this study is added to that summarized by Wilson

(1988) and then considered in the context of the biogeographic models discussed by Liebherr (1988), the biogeographic conclusion is that the southern arc of the Lesser Antilles seems to have been colonized by ants dispersing from Venezuela. This coincides with one of the four tracks of colonization proposed by Rosen (1975), the South American Caribbean track. The northern arc seems to have been colonized by ants that dispersed from South America and from the Greater Antilles. No indication as to vicariance (Croizat 1976) in the Lesser Antilles is evident from our data. This confirms suggestions by Liebherr, and others cited in Liebherr (1988), that the insect population of the Lesser Antilles is rather recent and has colonized the islands by dispersing from the mainland or from the Greater Antilles.

One interesting feature is that widely distributed ant species or tramp species; i.e. *Tapinoma melanocephalum, Paratrechina longicornis, Cardiocondyla wroughtoni, Monomorium floricola, Pheidole megacephala, Solenopsis geminata,* and *Wasmannia auropunctata* are never simultaneously found on all the islands sampled. Their absence from some islands is probably not due to sampling errors, as these species, when present, become the dominant ants in their habitat. However, the alternative hypothesis needs to be investigated since tramp species could become sub-dominant, i.e. cease to dominate other species. The lack of tramp species on some islands suggests that in some situations, they can be kept out of islands even in the presence of intense human commerce and activities.

The only ant species present which has been reported as a local pest is *Acromyrmex octospinosus* in Guadeloupe. Other species, although regarded as pests elsewhere do not seem to reach this status in all situations. For example, *W. auropunctata* is known as a stinging ant in Martinique and Guadeloupe but does not seem to be much of a problem to farmers. Old reports of ants becoming pests in the Caribbean do exist. For example, in 1500, an ant species, probably *S. geminata,* was reported as a pest on Hispaniola and Jamaica. Also, in 1760 reports of similar pest problems occurred on Barbados, Grenada and Martinique (Reaumurs report in an unpublished work of Wheeler cited by Holldobler and Wilson 1990). Today, no pest problems caused by *S. geminata* in the Lesser Antilles are known to us. Thus, interestingly, the only known pest ant today is *A. octospinosus,* which was introduced to Guadeloupe around 1954 (Blanche 1961).

The loss of an ant's pest status at some period of time after its initial introduction seems to be a general feature of pests. Two possible explanations are available. Either the ants become less ecologically aggressive, i.e. they disperse less, individual colonies may require larger foraging territories, or the species may specialize in certain ecosystems and thus compete with fewer ant species. Also, humans may habituate to the ants and thus not consider them to be pests any more. For the first hypothesis to be confirmed, comparative behavioral studies of recently introduced ants *vs.* well established populations are needed. Evidence for the second hypothesis exists because many of the pest species in the Caribbean can be managed and can thus be used as biological control agents (Pollard and Persad 1991, Jaffe et al. 1991). The truth is probably a mixture of the two proposed hypothesis which could explain the "natural taming" of pest ants.

As previously mentioned, tramp species, although widely distributed among the islands have not colonized all of them. For example, *W. auropunctata*, a known tramp and pest species, was not found on the southern islands studied and *P. longicornis* was not found on Saint Martin, although it was present on all the other islands studied. *P. megacephala* seems to be absent or is very infrequent on Margarita and *S. geminata* could not be collected on Saint Martin. This pattern is congruent with a colonization model described by island biogeographic theories (MacArthur and Wilson 1967), where impediments to colonization and occasional extinction of some species determine the species distribution on islands.

Some of the islands, such as St. Martin, are poor in ant fauna, whereas the two islands with the largest diversity of ecosystems, Guadeloupe (which was well sampled) and Margarita (which was poorly sampled compared to its size and ecosystem diversity), have islands with the largest number of species. On the other hand, xeric islands such as Blanquilla, Las Aves, and La Orchila, have few species, confirming that the diversity of ecosystems is an important feature in determining the number of species on an island.

Our ant collection completes and corrects the data on the biogeography of the ant fauna in the Antilles (Wilson 1988), including species from Margarita, La Blanquilla, Las Aves and Aruba, and shows an increase in the number of species reported for Guadeloupe and Martinique. Our data also suggests that Wilson underestimated the ant population in this area, possibly due to the fact that intensive collecting had not been done on these islands.

Conclusion

In conclusion, the ant fauna in the Caribbean islands, at least in the Lesser Antilles, appears to be of recent origin, as no clear endemism is apparent. Potential pest species seem to be very good colonizers, causing trouble to humans shortly after introduction to an island.

However, their importance as pests appears to decrease with time, suggesting that aggressive colonizing ant species develop a more passive ecological behavior as they adapt to their new environment.

Resumen

Se colenctaron hormigas por muestreo directo en la islas Francesas y Venezolanas de las Antillas y en Aruba. Los resultados revelan que la fauna de hormigas en el Caribe, al menos en las Antillas Menores, es de origen reciente, ya que no presentan endemismo aparente. Biogeograficamente, las Antillas Menores muestran dos zonas, una al sur de Saint Lucia, con especies provenientes de Venezuela, y otra al norte con especies tanto del continente Suramericano como de las Antillas Mayores. Especies plaga potenciales aparecen como buenas colonizadoras, creando problemas solo durante el periodos inmediatamente despues de su introducción, disminuyendo su importancia como plagas lo que sugiere que especies de hormigas agresivamente colonizadoras desarrollan comportamientos ecologicamente mas pasivos al adaptarse a su nuevo ambiente.

References

Blanche, D. 1961. La fourmi-manioc. *Ministere de l'Agriculture. Service de la Protection des Vegetaux SEP ed.*

Brandão, C. R. 1991. Adendos ao catalogo abreviado das formigas da regiao neotropical. Rev. *Brasil. Entomol.* 35: 319-412.

Croizat, L. 1976. Biogeografia analitica y sintetica (panbiogeografia) de las Americas. *Boletin de la Academia de Ciencias Fisicas, Matematicas y Naturales (Venezuela)* XV and XVI.

Hölldobler, B., and E. O. Wilson. 1990. *The Ants.* Belknap Press, Harvard Univ. Cambridge, Mass.

Jaffe, K., H. Mauleon, and A. Kermarrec. 1991. Qualitative evaluation of ants as biological control agents with special reference to predators of *Diaprepes spp* on citrus groves in Martinique and Guadeloupe. In: C. Pavis and A. Kermarrec [eds.]. *Rencontres Caribes en Lutte biologique.* INRA, Les Colloques Nr. 58, Paris.

Kempf, W. W. 1972. Catálogo abreviado das formigas (Hymenoptera: Formicidae).da região neotropical. *Studia Entomol.* 15: 3-334.

Liebherr, J. K. 1988. Biogeographic patterns of West Indian *Platynus* carabid beetles (Coleoptera). Pp. 121-152. In: J. K. Liebherr [ed.]. *Zoogeography of Caribbean Insects.* Cornell Univ. Press. Ithaca, NY.

MacArthur, R. H., and E. O. Wilson. 1967. *The theory of island biogeography.* Princeton Univ. Press, NJ.

Pollard, G.V., and A. B. Persad . 1991. Some ant predators of insect pests of tree crops in the Caribbean with particular reference to the interaction of *Wasmannia auropunctata* and the leucaena psyllid *Heteropsylla cubana*. In: C. Pavis and A. Kermarrec [eds.]. *Rencontres Caribes en Lutte biologique*. INRA, Les Colloques Nr. 58, Paris.

Romero, H., and K. Jaffe. 1989. A comparison of methods for sampling ants (Hymenoptera: Formicidae) in savannas. *Biotropica* 21: 348-352.

Rosen, D. E. 1975. A vicariance model of Caribbean biogeography. *Syst. Zool.* 24: 431-664.

Wilson, E.O. 1988. The Biogeography of the west Indian Ants (Hymenoptera: Formicidae). Pp 214-230. In: J.K. Liebherr [ed.]. *Zoogeography of Caribbean Insects.* Cornell Univ. Press, Ithaca, NY.

16

Exotic Ants and Native Ant Fauna of Brazilian Hospitals

Odair C. Bueno and Harold G. Fowler

Introduction

Since the early reports by Beatson (1972), the study of ants within hospitals has gained impetus because of their potential for transmitting intra-hospital infections (Green et al. 1953; Beatson 1972; Aleksev et al. 1972; Eichler 1962, 1978, 1990; Edwards and Baker 1981; Sy 1987; Fowler et al. 1992b). The majority of studies have been conducted in Europe and North America, and have documented serious risks associated with the Pharaoh ant, *Monomorium pharaonis* (L.), a tramp ant species which has spread world-wide from the tropical forests of the Old World (Rupes et al. 1983). Through this accidental transport, it has become a major pest, principally urban, in many areas in which it has been introduced (Edwards 1986). Because *M. pharaonis* is a tropical ant that uses any available cavity for nesting, it is commonly found associated with human structures in the cooler temperate regions (Czejkowska 1979; Krzeminska et al. 1979; Eichler 1962, 1978; Berndt and Eichler 1987). Its occurence in human structures in the tropics (Sudd 1962, Fowler et al. 1992a) is probably due to intense competition with the native ant fauna outside of human structures (Edwards 1986).

In the south temperate region of Chile, Ipinza-Regla et al. (1981) documented that the Argentine ant, *Linepithema humile* [=*Iridomyrmex humilis* (Mayr)], was also a potential vector of intra-hospital infections. This study, and that of Fowler et al. (1992b), are the only studies conducted outside of the Northern hemisphere, and only the latter has examined the situation under tropical situations. Fowler et al. (1992b) also showed that a number of ant species were associated with hospitals, and that *Tapinoma melanocephalum, Paratrechina longicornis,* and other ant species could serve as mechanical vectors for bacterial infections in hospitals. These data are extremely important in that they focus attention upon ants as possible disease vectors, especially in light of the fact that Brazil has one of the highest rates of intra-hospital infections in the world, (7 to 20% depending upon the region and the hospital). In this paper, we investigated twenty hospital "islands" in Brazil, and demonstrated that in spite of this nation's having one of the richest ant faunas of the world, hospitals are colonized predominantly by exotic ant species in an apparent random fashion.

Methods

We sampled five hospitals of each of the following types in various locations in Brazil: (1) university teaching hospitals, generally the largest hospitals found in any region (>200 beds), which treat many types of medical problems; (2) municipal hospitals which typically have up to 200 available beds and which treat routine medical cases and conduct minor surgeries; (3) large private hospitals, having up to 100 beds, which generally specialize in certain medical fields such as cardiology, obstetrics, etc.; and (4) small private hospitals with about 20 beds and which treat routine medical problems, particularly those of the more affluent sector of society. One large university hospital, one municipal hospital, and one small private hospital were intensively sampled with bait stations over a one year period in the state of São Paulo. All other hospitals, which varied in geographic extension from the Amazon to São Paulo, were sampled with bait stations only once. We also performed microbiological studies (Fowler et al. 1992b) at the university teaching hospital in São Paulo. These studies are reviewed in this chapter. Further details can be found in Fowler et al. 1992b.

Results

The results of our sampling (Figure 16.1) demonstrated that ant faunas of Brazilian hospitals are extremely variable in size, but all have at least 10 species present and can attain up to 23 different species. Generally, the larger the hospital, the more ant species present.

FIGURE 16.1. Mean, minimum and maximum ant species richness of various types of sampled hospitals in Brazil.

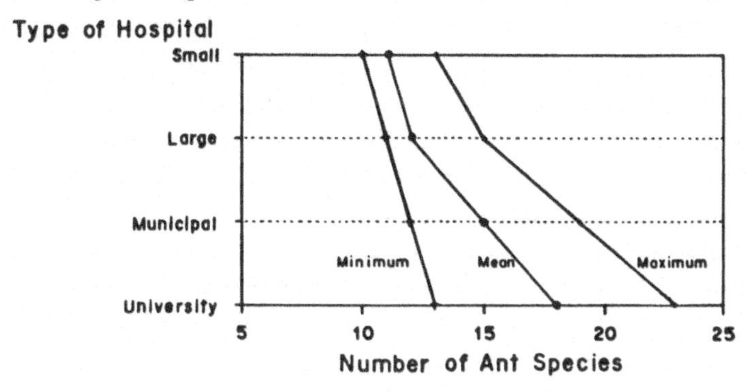

FIGURE 16.2. Proportional representation of the ant fauna of three hospitals sampled in Brazil.

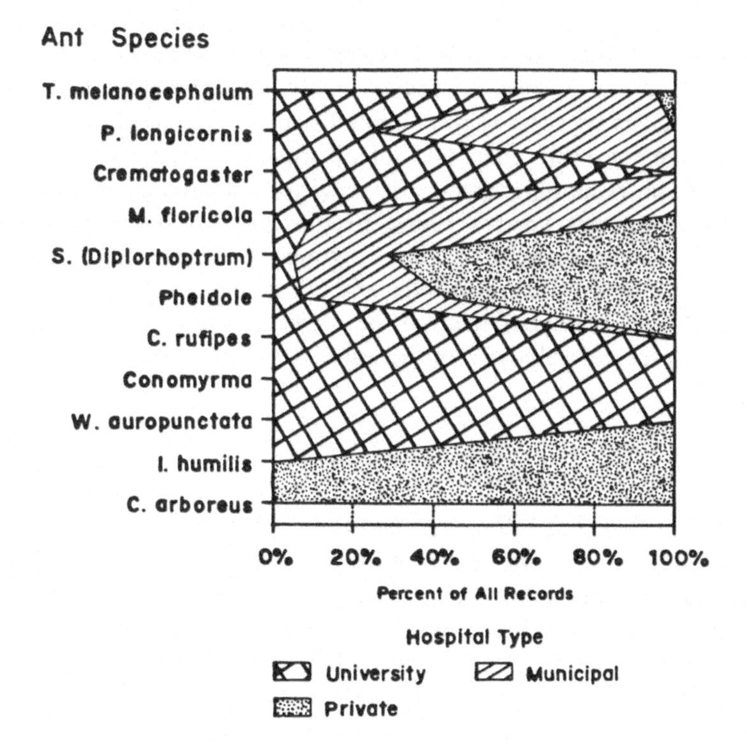

The ant species composition of the 3 hospitals that were intensively surveyed (Figure 16.2) reveals strong differences with no apparent patterns. Of the ant species found in the hospitals, *T. melanocephalum* was found in 12, *P. longicornis* in 9, *M. floricola* in 12, *M. pharaonis* in 5, *Pheidole megacephala* in 3, (all exotic). *Wasmannia auropunctata*, the most consistently found native species, was discovered in 6 hospitals. All hospitals had additional species of *Pheidole*, most had native species of *Iridomyrmex* and *Solenopsis* (*Diplohroptrum*), and many had species of *Camponotus*, especially associated with outer walls. In some of the hospitals, one species dominated a wing or a floor, with another species dominating other portions of the hospital. Native species of *Pheidole* and *Camponotus* were generally restricted to extremely small areas. Most of the 16 species of *Pheidole* we identified were very small, and nested generally in the floor or walls. None of these foraged for more than 10 m. In spite of its potential as a major problem, *L. humile* (Mayr) was found in only 5 hospitals, and in none of these was it the most abundant ant species.

Bacteria associated with ants in intra-hospital infections in the intensively studied university teaching hospital were most commonly encountered on *T. melanocephalum*. Although the percentage of *T. melanocephalum* workers with platable bacteria was not large, they were responsible for more than 60% of all ant-associated bacteria (Figure 16.3). Of the remaining species, *P. longicornis* was next in importance. All the exotic ant species found in the hospitals were polygynous, unicolonial, reproduced by colony fragmentation, and very small. With the exception of *W. auropunctata* and *L. humile*, all of the native ant species found were either large, or if small, such as in some of the *Pheidole*, were not polygynic or unicolonial.

Discussion

These studies have demonstrated that a large number of ant species are found in Brazilian hospitals and all have the potential to mechanically vector bacteria which can produce intra-hospital infections. Exotic species are generally dominant. Contrasted with ant problems in hospitals from the temperate region (Eichler 1990, Ipinza-Regla 1981), management of ant populations is much more difficult, because as one dominant species is eliminated, another will soon occupy its niche.

Surprisingly, the presence and dominance of many exotic ant species in Brazilian hospitals, such as *T. melanocephalum, P. longicornis, M. floricola, P. megacephala* and *M. pharaonis* was indicative of a stronger emphasis on hygiene. As these species nest in any space in the hospital,

FIGURE 16.3. The association of ant species with bacteria related with intra-hospital infections in a university hospital in Brazil.

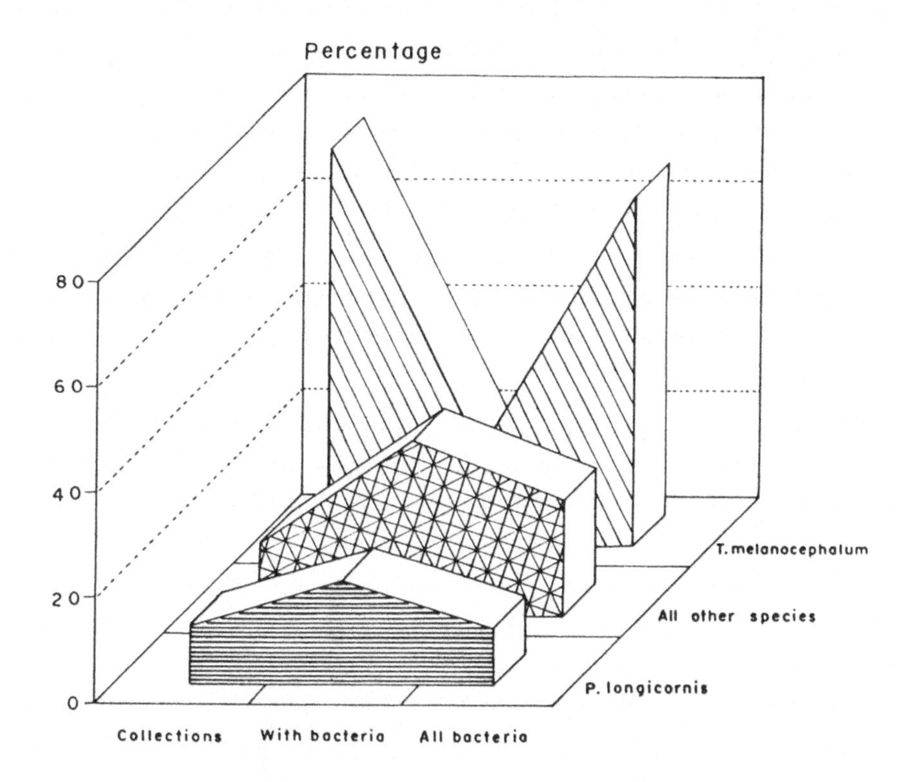

spread through colony fragmentation, and are very small, they escape the attention of maintenance personnel. These same characteristics in biology and social organization make them difficult to control using classical pesticides. In fact, these control measures may cause proliferation because groups of workers and larvae without queens are isolated and then they produce new queens creating new foci of infestations.

With the exception of *W. auropunctata,* which exhibits the same strategy as the exotic species, the native ant fauna are more easily controlled with conventional insecticides. Some species, such as those of the genus *Camponotus,* are indicative of infrastructure deficiencies, especially when found in the interior portions of the hospitals. Most native ants, even the small species of *Pheidole,* are generally found close to the outer walls of the hospitals, and probably rely more upon outside food sources than interior food sources. In the inner portions of the hospitals, only the exotic species, and *W. auropunctata,* are consistently present. These are generally the areas in which patients are located, making them more prone to contract intra-hospital infections.

Even though we have little information as to the contribution of ants to intra-hospital infections in Brazil, we recommend that a multi-stage control and monitoring program be developed. The bait stations we have used, when properly placed and at regular intervals, serve as an early-warning system to identify new foci of infestations. Large ant species, and species associated with outer walls, indicate that maintenance should be performed, and these can be treated with classical structural insecticides. However, even here we strongly recommend using baits. Pelleted bait formulations can be used for most species; but liquid baits are preferred for species of *Camponotus*.

For small, polygynic, unicolonial species, classical insecticides do not work, and often produce larger problems than before treatment. Although we have not yet evaluated products for the majority of these species, it seems likely that insect growth regulators, whether juvenoids or not, are likely to be the most promising control alternative (Rupes et al. 1983, Edwards 1986).

Acknowledgments

Virgilio Pereira da Silva, Francisco Anaruma Filho, Maria Apareciada Freitas and Lourival Antonio Mighelin assisted in collections, and Terue Sadatsune performed the microbiological identifications. Antonio C. Montelli provided logistical support. Financial support was provided by the Secretary of Science and Technology of the State of São Paulo (Grant No. 122/88-9), the Brazilian Council for Scientific Research, CNPq (Grant No. 300171/88-9), and the Foundation for Scientific Support of the State of São Paulo, FAPESP, and the Foundation for the Development of UNESP, FUNDUNESP. We would especially like to thank Dr. D. F. Williams and R. S. Patterson, and the Charles Darwin Research Station for providing us with this forum. We gratefully acknowledge the support provided by W. Eichler, A. Ferreira, S. B. Vinson and D. F. Williams, and Nozor Pinto for kindly rendering the figures.

Resumen

La mirmecofauna de hospitales brasileños, determinada por colecciones en 15 hospitales, varia desde 10 hasta 23 espécies por hospital y es dominada por espécies exóticas. Las espécies principales son *Tapinoma melanocephalum, Monomorium floricola* y *M. pharaonis, Paratrechina longicornis* y *Pheidole megacephala,* todas exóticas. De las espécies nativas encontradas, destaca *Wasmannia auropunctata* y várias espécies de *Pheidole, Camponotus* y *Iridomyrmex,* especialmente *Linepithema humile*

[=*Iridomyrmex humilis*]. Cada hospital tenia una fauna distincta, y el tamaño de la mirmecofauna implica mayores problemas no manejo de hormigas, que son comprobadas vectores mecánicos potenciales de infecciones intra-hospitalares. Discutimos las implicaciones de nuestros datos para programas de vigilancia y control.

References

Aleksev, A. N., V. A. Bibivoka, T. Brikman, and Z. Kantarbaeva. 1972. The persistence of viable plague microbes on the epidermis and in the alimentary tract of *Monomorium pharaonis* in experimental conditions. *Med. Parazit.* 41: 237-239.

Beatson, S. H. 1972. Pharaoh's ants as pathogen vectors in hospitals. *Lancet* 1: 425-427.

Berndt, K. P., and Wd. Eichler. 1987. Die Pharaoameise, *Monomorium pharaonis* L. (Hym., Myrmicidae). *Mitt. Zool. Mus. Berlin* 63: 3-188.

Czejkowska, M. 1979. Wystepowaniei rozprzenianie sie *Monomorium pharaonis* L. *(Hymenoptera, Formicidae)* na terenie Warszawy. *Frag. Faun.* 23: 343-361.

Edwards, J. P. 1986. The biology, economic importance and control of the Pharaoh's ant, *Monomorium pharaonis* L. Pp. 257-271. In: S. B. Vinson [ed.]. *Economic impact and control of social insects.* Praeger Press, New York.

Edwards, J. P., and L. F. Baker. 1981. Distribution and importance of the Pharaoh's ant, *Monomorium pharaonis* L., in national health service hospitals in England. *J. Hosp. Infec.* 2: 249-254.

Eichler, Wd. 1962. Verbreitung und Ausbreitungsterndenzen der Pharaoameise in Mitteleuropa (Nach eigenen Erfahrungen in Deutschland, Osterreich, Polen und Tschechoslowakei berichtet). *Prakt. Schad.* 14: 1-2.

_____. 1978. Die verbreitung der Pharaoameise in Europa. *Memor. Zool.* 29: 31-40.

_____. 1990. Health aspects and control of *Monomorium pharaonis.* Pp. 671-675. In: R.K. Vander Meer, K. Jaffe and A. Cedeno-Leon [eds.]. *Applied Myrmecology: a World Perspective.* Westview Press, Boulder, CO.

Fowler, H. G., F. Anaruma Filho, and O. C. Bueno. 1992a. Seasonal space usage by the introduced Pharaoh's ant, *Monomorium pharaonis* (Hymenoptera: Formicidae), in institutional settings in Brazil and its relation to other structural ant species. *J. Appl. Entomol.* (in press).

Fowler, H. G., O. C. Bueno, T. Sadatsune, and A. C. Montelli. 1992b. Ants as potential vectors of pathogens in hospitals in the State of São Paulo, Brazil. *Ins. Sci. Appl.* (in press).

Green, A. A., M. J. Kane, P. S. Tyler, and D. G. Halstead. 1954. The control of Pharaoh's ants in hospitals. *Pest. Infest. Res.* 1953: 24.

Ipinza-Regla, J., G. Figueroa, and J. Osorio. 1981. *Iridomyrmex humilis,* "hormiga argentina", como vector de infecciones intrahospitalarias. I. Estudio bacteriologico. *Folia Entomol. Mex.* 50: 81-96.

Krzeminska, A., B. Styczynska, H. Monrawska., and B. Sereda. 1979. Zwalczanie mrówek faraona *Monomorium pharaonis* (L.) w mreszhaniach na terenie Warszawy. *Rocz. Wow. Zak. Hig.* 30: 623-631.

Rupes, V., J. Chmela, I. Hrdy, and K.Krecek. 1983. Effectiveness of methoprene-impregnated baits in the control of *Monomorium pharaonis* and populations infesting health establishments and households. *J. Hig. Epidem. Microbiol. Immunol.* 27: 295-303.

Sudd, J. H. 1962. The natural history of *Monomorium pharaonis* (L.) infesting houses in Nigeria. *Entomol. Month. Mag.* 98: 164-166.

Sy, M. 1987. Über einige Probleme bei der Bekämpfung der Pharaoameise *(Monomorium pharaonis)* (L.) mit dem Rinal Pharaoameisenkoder. *Anz. Schad. Planz. Umwelt.* 60: 51-55.

17

Big-Headed Ants, *Pheidole Megacephala:* Interference with the Biological Control of Gray Pineapple Mealybugs

Gary C. Jahn and John W. Beardsley

Introduction

The gray pineapple mealybug, *Dysmicoccus neobrevipes* (Beardsley), is a major pest on pineapple in Hawaii because its feeding is associated with pineapple wilt disease (Rohrbach et al. 1988). As gray pineapple mealybugs feed they produce a waste material known as honeydew that consists of free amino acids and various types of sugars (Gray 1952). Ants are commonly found feeding on the honeydew excreted by gray pineapple mealybugs, in particular the big-headed ant, *Pheidole megacephala* (Fabricius), which is the dominant ant species in Hawaiian pineapple. Its distribution in Hawaii is limited by high rainfall (Reimer et al. 1990). Big-headed ants are of African origin (Illingworth 1917), while gray pineapple mealybugs are of tropical American origin (Rohrbach et al. 1988). It is not known when either species was introduced to Hawaii; however, according to Illingworth (1926a), the earliest report of big-headed ants is that of Blackburn and Kirby (1880), who called big-headed ants *Pheidole pusilla* (Heer).

When big-headed ants are eradicated from a pineapple field the gray pineapple mealybug populations decline and wilt disease is brought under control (Beardsley et al. 1982). One explanation for this phenomenon is that big-headed ants protect gray pineapple mealybugs from natural enemies. Among the natural enemies of gray pineapple mealybugs are the coccinellids *Cryptolaemus montrouzieri* (Mulsant) (Illingworth 1926b) and *Curinus coeruleus* Mulsant (Jahn 1992).

Another explanation of the importance of big-headed ants to gray pineapple mealybug survival is that big-headed ant's consumption of honeydew prevents the gray pineapple mealybug from drowning in its own waste material (Rohrbach et al. 1988).

The evidence for protection

Ant protection of Homopterans is known to occur. For example, Bartlett (1961) has demonstrated that in the absence of Argentine ants, *Linepithema humile* [=*Iridomyrmex humilis*], certain parasitic species suppress populations of lecaniine scale insects. On the other hand, Bess (1958) reported that green scales (*Coccus viridis* [Green]) could not survive without large red ants (*Oecophylla smaragdina* [Fabricius]), but the ants did not reduce parasite and predator attacks on them.

Illingworth (1931) noted that gray pineapple mealybugs did not thrive on pineapples under laboratory conditions without big-headed ants. He also frequently encountered two species of predacious lady bird beetles (coccinellids) on ant-free pineapples. On ant-infested pineapples, the mealybugs were covered with "bits of trash." Illingworth believed that *P. megacephala* covered the mealybugs to protect them from predators; however, he did not test his hypothesis. The reason that ants cover mealybugs remains unknown.

Su (1977) compared big-headed ant-free and big-headed ant-attended gray pineapple mealybug populations on live potted pineapple plants. He found that the mean number of gray pineapple mealybugs on plants with ants was higher than on plants without ants. He did not analyze his data statistically, since he was trying to determine *if* ants have a positive effect on gray pineapple mealybug population growth, instead of trying to determine *how* ants benefit mealybugs. Because his experiment did not exclude natural enemies, their presence might have suppressed the gray pineapple mealybug population on ant-free plants.

Macion (1978) conducted an experiment on the effect of big-headed ants on gray pineapple mealybug population growth on sisal and found that after three months there were significantly more gray pineapple mealybugs on sisal with big-headed ants. He did not speculate on the cause of this phenomenon. This experiment was conducted in a greenhouse and he made no mention of the presence or absence of natural enemies. Natural enemies might have reduced gray pineapple mealybug populations in the absence of ants.

To determine if gray pineapple mealybugs derive any benefit from big-headed ants other than protection from natural enemies Jahn (1992) repeated the Illingworth (1931) study, but in the absence of natural enemies. He used pineapple plants and pineapple fruits, and found no significant difference in the number of gray pineapple mealybugs produced with and without ants. This suggests that protection from natural enemies is the primary reason that gray pineapple mealybugs thrive when associated with ants in pineapple fields.

To investigate the effect of big-headed ants on gray pineapple mealybugs under field conditions, Jahn (1992) conducted two field experiments. One experiment indicated that big-headed ants aided the establishment of gray pineapple mealybugs on pineapple in the presence of natural enemies. The other experiment indicated that big-headed ants suppress populations of coccinellids that feed on gray pineapple mealybugs. When big-headed ants are eradicated, the coccinellid population increases, resulting in a decline in the gray pineapple mealybug population.

Do big-headed ants "protect" gray pineapple mealybugs from natural enemies?

These prior findings are consistent with the hypothesis that gray pineapple mealybug survival is greater in the presence of the big-headed ants because ants attack the natural enemies of the mealybugs. Throughout this chapter we have loosely referred to this as "protection of mealybugs from natural enemies." However, the word "protection" suggests that ants are attacking the natural enemies of gray pineapple mealybugs for the purpose of keeping the mealybugs from harm. In other words "protect" implies that the behavior of the ants is affected by the presence of the mealybugs. Ants do not need to consciously think about saving a mealybug from a predator or parasite to protect it. All that is required is that colonies of ants have been selected over time to protect specific resources. Since big-headed ants utilize mealybugs as a source of honeydew, the protection of mealybugs may not be much different (in an evolutionary sense) from the protection of any other food resource.

Big-headed ants attacking the natural enemies of gray pineapple mealybugs could result from any of the following behaviors:

1. Big-headed ants hunt specific prey which by chance includes some natural enemies of gray pineapple mealybugs;
2. Big-headed ants indiscriminately attack all arthropods in *Pheidole* territory, except gray pineapple mealybugs and other big-headed ants;
3. Big-headed ants specifically protect gray pineapple mealybugs by attacking their natural enemies;

4. Big-headed ants specifically protect gray pineapple mealybugs by indiscriminately attacking all arthropods other than big-headed ants and gray pineapple mealybugs.

In the first two cases, the behavior of the ants is the same whether gray pineapple mealybugs are present or not. In the last two cases, the ants behave in a certain manner to defend gray pineapple mealybugs as a resource. Even if big-headed ants protect gray pineapple mealybugs, the ants do not necessarily discriminate prey to do so.

The differences between the behavioral options are subtle, but an analogy may clarify the matter. A sheep rancher may invest time and money to kill wolves that enter his property. Of course the rancher would not kill wolves if he were raising fish instead of sheep. On the other hand, a fellow who hunts wolves for a living is undoubtedly aiding the local sheep population. The bounty hunter may even benefit from this situation by getting his wool sweaters cheaper than if he did not hunt. But if all the sheep were to be replaced with fish, this would not affect his decision to continue hunting wolves. In this analogy we could say that the sheep and the rancher have a true mutualistic relationship. The hunter however benefits the sheep by happenstance.

Has there actually been selection for big-headed ants to protect gray pineapple mealybugs as a resource? If so, in our analogy the ants are ranchers and the mealybugs are sheep. Or do the gray pineapple mealybugs simply benefit from the usual behavior of big-headed ants, as the sheep benefit from the bounty hunter? To expand on the analogy, we can imagine two types of ranchers and two types of hunters. One sheep rancher puts up a high voltage electrified fence (around his non-electrified pasture fence) to kill wolves. He gets the job done, but in the process kills many other animals which are harmless to sheep. This rancher is like the indiscriminate ants protecting mealybugs in case no. 4. Another rancher patrols the ranch and only shoots wolves. He is like the ants in case no. 3. In a similar fashion one can imagine a bounty hunter that shoots everything that moves to increase his chances of bagging the desired prey, though he makes an effort to avoid killing humans and sheep. He is like the ants in case no. 2. Finally, there is a bounty hunter that only shoots wolves and beavers. He is like the ants in case no.1.

If ant/mealybug symbiosis is the result of co-evolution, then we could expect to find evidence for the rancher/sheep analogy. Big-headed ants are of Ethiopian origin, while gray pineapple mealybugs are of Neotropical origin. This does not preclude the possibility that co-evolution occurred. It is not known how long the two species have been together. The two species do have a mutualistic relationship on pineapple outside of Hawaii (e.g., in South Africa). Perhaps the two species co-evolved in Africa or elsewhere, and were later brought to Hawaii. Another possibility is that big-headed ants co-evolved with an Ethiopian mealybug, while gray pineapple mealybugs co-evolved with a Neotropical ant. By coincidence, the mutualistic relationship was then transferred to another species.

Keeping with the analogy, to determine if big-headed ants are ranchers or bounty hunters we must answer the question: does the predatory behavior of big-headed ants change in the presence of gray pineapple mealybugs? This could be tested in a laboratory experiment by giving big-headed ants the opportunity to attack a wide range of arthropods (including enemies and non-enemies of gray pineapple mealybugs) in the presence versus the absence of gray pineapple mealybugs. Ants with gray pineapple mealybugs must not be given an alternate source of carbohydrate to mealybug honeydew, since we want to see if they are protecting a resource.

There are four possible outcomes to such an experiment, corresponding to the four hypothetical cases described above:

1. Big-headed ants specialize in attacking certain arthropods. Some of these are natural enemies of gray pineapple mealybugs. The time spent attacking these arthropods is not affected by the presence of gray pineapple mealybugs. If so, then big-headed ants are "wolf and beaver hunters" and co-evolution has not occurred.
2. Whether gray pineapple mealybugs are present or not, big-headed ants will attack any arthropod in its foraging territory except gray pineapple mealybugs and big-headed ants. If so, then big-headed ants are "shoot-if-it-moves hunters" and co-evolution has not occurred.
3. Big-headed ants will attack the natural enemies of gray pineapple mealybugs with greater frequency than it attacks other arthropods, but only in the presence of gray pineapple mealybugs (when honeydew is big-headed ant's carbohydrate source). If so, then big-headed ants are "shoot-the-wolf sheep ranchers" that truly protect gray pineapple mealybugs. This is difficult to explain unless there has been selection for big-headed ants to specifically defend gray pineapple mealybugs, i.e., unless co-evolution has occurred.
4. The frequency of big-headed ant attacks on arthropods in general increases in the presence of gray pineapple mealybugs (being utilized for honeydew) when big-headed ants have ample protein in their diet. If so, then big-headed ants are "electrified-fence sheep ranchers" that truly protect gray pineapple mealybugs without discriminating between natural enemies and other arthropods.

This result suggests that co-evolution has occurred, though not necessarily with this particular species of mealybug.

The factors being measured in such an experiment can be expressed as a model to describe choice of prey (Krebs and Davies 1981). Ants search for T_S seconds and encounter two types of prey at the rates r_1 and r_2. The two kinds of prey contain E_1 and E_2 calories. The two types of prey take h_1 and h_2 seconds to handle. They therefore yield the short term profits of E_1 / h_1 and E_2/h_2. There is the additional profit of preventing the prey from destroying mealybugs: P_1 and P_2 represent the number of mealybugs which would have been destroyed if the prey had not been killed. There is also the potential for ants to be injured when they attack their prey. Wasps, for example, represent a greater risk than flies for attacking ants. The risk of injury when handling the prey is represented by i_1 and i_2, where i is the number of ants per 100 that die attacking this species, i.e., the probability that the attack will result in the death of an ant. Thus the relative profits of hunting prey types 1 and 2 are: $(E_1/h_1) + P_1 - i_1$ and $(E_2/h_2) + P_2 - i_2$.

If ants eat both kinds of prey they obtain the following in T_S seconds:
$E = T_S\,(r_1(E_1 + P_1 - i_1) + r_2(E_2 + P_2 - i_2))$
and this will take a total time: $T = T_S + T_S(r_1h_1 + r_2h_2)$.
Therefore the rate of intake is
$E/T = [T_S(r_1(E_1{+}P_1{-}i_1){+}r_2(E_2{+}P_2{-}i_2))] / [T_S{+}T_S(r_1h_1{+}r_2h_2)]$.
Dividing by T_S gives
$E/T = [r_1(E_1{+}P_1{-}i_1) + r_2(E_2{+}P_2{-}i_2)] / [1 + r_1h_1 + r_2h_2]$.
Let $E_1 = E_2$, $h_1 = h_2$, $r_1 = r_2$, *and* $i_1 = i_2$.

If ants are protecting mealybugs as a resource then to maximize E/T ants should specialize on type 1 prey if $P_1 > P_2$. But if ants are not protecting mealybugs as a resource then they should eat types 1 and 2 with equal frequency.

Acknowledgments

This is Journal Series No. 3700 of the Hawaii Institute of Tropical Agriculture and Human Resources.

Resumen

En este capítulo se revisa la evidencia de que la hormiga cabezona, *Pheidole megacephala,* interfiere con el control biologico de los piojos harinosos de la piña, *Dysmicoccus neobrevipes* en Hawaii. Se discuten los aspectos evolucionarios y de comportamiento de las hormigas que protegen los piojos harinosos de sus enemigos naturales, asi como también el uso de una prueba para determinar si ha ocurrido el proceso de co-evolución, y si una descripción matematica simple de estos factores pudiera ser medida mediante dicha prueba.

References

Bartlett, B. R. 1961. The influence of ants upon parasites, predators, and scale insects. *Ann. Entomol. Soc. Am.* 54: 543-551.

Beardsley, J. W., T. H. Su, F. L. McEwen, and D. Gerling. 1982. Field investigations on the interrelationships of the big-headed ant, the gray pineapple mealybug, and the pineapple mealybug wilt disease in Hawaii. *Proc. Hawaiian. Entomol. Soc.* 24: 51-67.

Bess, H. A. 1958. The green scale, *Coccus viridis* (Green) (Homoptera: Coccidae), and ants. *Proc. Hawaiian Entomol. Soc.* 16: 349-355.

Blackburn, T., and W. F. Kirby. 1880. Notes on species of Aculeate Hymenoptera occurring in the Hawaiian Islands. *Entomol. Monthly. Mag.* 17: 85-89.

Gray, R. A. 1952. Composition of honeydew excreted by pineapple mealybugs. *Science* 115: 129-133.

Illingworth, J. F. 1917. Economic aspects of our predaceous ant. *Proc. Hawaiian Ent. Soc.* 3: 349-368.

_____ . 1926a. A study of ants in their relation to the growing of pineapples in Hawaii. *Exp. Stat. Assoc. Hawaiian Pineapple Canners Bull.* 7: 2-16.

_____ . 1926b. Pineapple Insects and some related pests. *Exp. Stat. Assoc. Hawaiian Pineapple Canners Bull.* 9: 19-22.

_____ . 1931. Preliminary report on evidence that mealy bugs are an important factor in pineapple wilt. *J. Econ. Entomol.* 24: 877-889.

Jahn, G. C. 1992. The ecological significance of the big-headed ant in mealybug wilt disease of pineapple. *Ph.D. Dissertation*, Univ. Hawaii, Honolulu.

Krebs, J.R., and N.B. Davies. 1981. *An Introduction to Behavioural Ecology*. Sinauer Associates, Inc., Sunderland, Massachusetts. p. 55.

Macion, E. A. 1978. Ecological, pathological and control aspects of the mealybug-ant-wilt complex in pineapple. *Ph.D. Dissertation*, Univ. Hawaii, Honolulu.

Reimer, N. J., J. W. Beardsley, and G. C. Jahn. 1990. Pp.40-50. Pest ants in the Hawaiian Islands. In: R.K. Vander Meer, K. Jaffe and A. Cedeno [eds.]. *Applied Myrmecology: a World Perspective.* Westview Press, Oxford.

Rohrbach, K. G., J. W. Beardsley, T. L. German, N. Reimer, and W. G. Sanford. 1988. Mealybug wilt, mealybugs and ants on pineapple. *Plant Disease* 72: 558-565.

Su, T. H. 1977. The distribution of ants and mealybugs in a new planting of pineapple. *M.S. Thesis*, Univ. Hawaii, Honolulu.

18

The Impact and Control of the Crazy Ant, *Anoplolepis Longipes* (Jerd.), in the Seychelles

I. H. Haines, J. B. Haines and J. M. Cherrett

Introduction

The crazy ant, *Anoplolepis longipes* (Jerd.), was first reported in Mahe in the Seychelles in 1972 (Lewis et al. 1976). From 1974 to 1977, the former British Ministry of Overseas Development in cooperation with the Department and Ministry of Agriculture in the Seychelles, conducted an intense research and control campaign aimed at limiting the spread of the ant in the Seychelles, assessing its impact and pest status, and devising a long-term strategy to minimize any harmful effects. This work was fully reported at the time (Lewis et al. 1976; Haines and Haines 1978a and b, 1979a, b and c). The ant control team of the Seychelles Ministry of Agriculture and Fisheries, set up in the early '70s, has continued operating a control program up to the present (W. G. Dogley, pers. com.).

The purpose of this account therefore is to summarize the work that was done in the early '70s, to report on the successes and failures of the control program to date, and to draw some general conclusions about the impact and control of exotic ants from this case history.

Biology of *Anoplolepis Longipes*

Anoplolepis longipes, a formicine ant belonging to the tribe *Plagiolepidini* (Holldobler and Wilson 1990), is often called the crazy ant because of its rapid movements and frequent changes of direction. In

Mahe, it nests mainly in crevices in the ground, and in moist places under stones or coconut husks (Haines and Haines 1978a). It also nests arboreally, especially in the crowns of coconut palms. Census counts of the contents of 75 nests collected over 21 months showed that they were multiqueened (mean 39, standard error 6.5, range 1-320), and possessed on average 3790 workers (645 standard error, range 2500-36,000). In Mahe, colonies were ill-defined, and individual workers from different nests exhibited no aggression towards each other. In the infested study sites, nest densities ranged from 92 to 671 per ha suggesting maximum ant populations of 5-10 million per ha.

Both males and queens are able to fly, but in two years of regular light trapping using a Rothamsted trap only 131 males and 11 queens were caught, and no mating flights were observed. It appears therefore that the usual method of colony reproduction is budding. Over a 13-month period, 60 queens have been seen walking in the open, 44 being alates accompanied by some workers.

Like all ants, *A. longipes* requires protein-rich foods for egg production and growth of the larvae. They need carbohydrate-rich foods to provide energy for foraging and colony activity. The former is usually collected in solid form, and consists in the main of dead or dying invertebrates, especially insects. In culture, the ants attack, kill and dismember arthropods of considerable size such as cockroaches and centipedes. In the field they have been observed bringing in ants of other species such as *Camponotus maculatus* Forel, *C. grandidieri* Forel, *Odontomachus troglodytes* Sant., *Technomyrmex albipes* (Smith), *Paratrechina vividula* (Nyl.) and *Plagiolepis madecassae* Forel, (Haines and Haines 1978a), as well as a wide range of other insects, isopods, myriapods, molluscs and arachnids. They have also been observed occasionally carrying in small reptiles and other vertebrates. For carbohydrates, the ants drink juice from fruit, plant exudates, and especially the honeydew produced by homoptera. They are particularly associated with the plant-feeding coccids *Ceratplastes rubens* Maskell and *Coccus viridis* (Green). Workers descending trees after tending these species often have greatly distended, almost translucent gasters. The quantity of honeydew has been estimated to constitute 50% of the body weight of a 2.5mg worker.

In the field, *A. longipes* forages during both day and night, although it is inhibited by heavy rain and strong winds. Foraging continues throughout the year, as there is a continuous production of worker larvae, pupae and eggs without any clear seasonality. Sexuals are also found throughout the year, most being produced shortly before or shortly after the rainy season which is from December to February.

Introduction and Spread in the Seychelles

Anoplolepis longipes is probably a native of China (Wheeler 1910) and tropical Africa (Wilson and Taylor 1967), but it has now become a `tramp ant', spread widely by mankind to many tropical countries. The Seychelles are a group of 92 tropical oceanic islands in the Indian Ocean comprising in all, about 250km^2. They are within 5° south of the equator, on the Mascaren Ridge, and many represent the tips of ancient submerged mountains. The largest island, containing the capitol, is Mahe (1443km^2), which rises to 911m. It is on this island that *A. longipes* was first recorded by Government Entomologist P. L. Mathias in 1969 (Lewis et al. 1976). How the ant was first introduced is not known. It is thought to have occurred about 1962, and as the initial infestation was centered on the remote village of Maldive at the end of a track in the center of the northern peninsula; it seems likely that the ants were brought in with some package from a passing ship. In recognition of its supposed place of origin, its local name is "Fourmi Maldive".

By 1970, it had spread to an area of about 220ha, centered in the village, but by 1973 some 1000ha of the northern peninsula were infested, and it had broken out of this area to set up at least 7 other centers totalling another 100ha around the island. By 1976, it was in 19 centers totalling 1417ha (Haines and Haines 1978b). A subsequent survey in 1989 (W.G. Dogley, personal communication) showed spread to most of the island. The only other island currently infested is Felicite, where the ants are distributed over about 60% of the land, reputedly after a deliberate introduction following a dispute. Studies of the rate of natural spread gave estimates which varied from 0.1m to 1.1m per day (Haines and Haines 1978a). Establishment in new areas is almost certainly due to the accidental transfer of colonies in vehicles and in produce, such as coconut husks, vegetables and building materials.

Impact

Anoplolepis longipes is sometimes regarded as a beneficial insect, acting as a natural enemy of some pests such as the beetles, *Melittomma insulare,* Fairm. and *Oryctes monoceros* (O.) and the coreid bug, *Amblypelta cocophaga* China. All of these insects damage coconut palms (Brown 1959, Greenslade 1971, Room and Smith 1975). Even in Mahe, a few people welcomed the ants on the grounds that they reduce rat and cockroach problems in houses. However, a questionnaire given to 246 people requesting poison bait for ant control showed that they regarded it primarily as a pest (Table 18.1), most commenting on its general nuisance value in the home.

Anoplolepis longipes greatly enhanced coccid populations, and their production of honeydew, which was considerably in excess of what the ants could consume. As a result, the honeydew contaminated foliage induced sooty mold to grow on it. When a site at Maldive, where the ants had been living for 15 years, was compared with one at Glacis:Sunset which was still free of ants, 93% of cinnamon leaves had coccids, compared with 23.6%. The number of coccids per leaf was 33.4 compared with 2.1, whilst in general, 98% of the plants had sooty mold versus 41% in the ant-free site, 60% of the canopy was covered compared with 18%. Indeed, sooty mold growth was so marked that it was possible to drive around the island mapping the extent of the *A. longipes* infestation by noting the presence or absence of black (sooty mold contaminated) trees and bushes.

Considerable concern was expressed over the effects of the ants on other animals. There were records of newly-hatched chickens and newly born domestic animals such as pigs, dogs, cats and rabbits being killed, whilst older animals were irritated and forced to move away. The effects on wild animals were less easily observed, but none-the-less of great potential importance. The Seychelles, like the Galapagos and Hawaii, are noted for the large number of endemic plant and animal species they possess which are of great conservational value. It was noted, for example, that none of the cinnamon bushes at Maldive were occupied by the native ant, *Technomyrmex albipes,* whereas at Glacis: Sunset, where *A. longipes* was still absent, they were all occupied. Similarly, a one hour search at Glacis revealed over 100 individuals of the local skink, *Mabuya sechellensis,* whereas around Maldive, a two-hour search produced less than 10. However the effect of *A. longipes* on local communities is complex as indicated by pitfall trapping carried out in two sites, Maldive and La Louise, an area which *A. longipes* was just beginning to invade.

TABLE 18.1. The pest status of *A. longipes* as revealed by 246 answers to a questionnaire given to people requesting toxic bait for ant control. (From Haines and Haines 1978b.)

Effect of ant	*Observed effect (%)*	*Details of effect*
General nuisance	75	Invaded homes, crawled over people & food.
Medical problem	8	Caused acute distress by entering ears, nose, eyes & open wounds, especially in young and old.
On plants	17	Removed soil from roots, induced sooty mold on fruit & foliage.
On mammals	7	Crawled over & irritated dogs, pigs & rats.
	4	Killed newborn pigs, dogs, cats, rabbits and rats.
On birds	11	Crawled over & irritated chickens at all ages, but mostly when brooding, often forcing bird to move away.
	13	Killed chickens, mostly new chicks when moist from the egg.
On reptiles	2	Drove away snakes & lizards.
	2	Killed snakes & lizard.
On invertebrates	2	Drove away centipedes & cockroaches.
	4	Killed insects, especially cockroaches & centipedes. One report of bees being killed.

Table 18.2 shows the frequency of occurrence of various invertebrate groups from 45 pitfall traps at each site operated for 24h. It appears that *A. longipes* depressed other ants, beetles, diptera, spiders and collembola, but benefitted the smaller cockroaches and isopods. It is probable that these differential group effects would also apply to different species within any one group. For some rare species therefore, the arrival of *A. longipes* may have been their salvation, but for many others it could prove the final disaster.

TABLE 18.2. The percentage of 24 pitfall traps containing various groups of animals, at two sites: Maldive, where *A. longipes* was abundant, and La Louise where it was recently introduced and rare. (From Haines and Haines 1978b.)

Fauna	*Maldive*	*La Louise*
Insecta		
Formicidae (not *A. longipes)*	7	24
Dictyoptera	27	7
Hemiptera	13	4
Coleoptera	11	27
Diptera	9	33
Lepidoptera	16	7
Dermaptera	0	2
Orthoptera	27	24
Unidentified insects	0	2
Isopoda	52	18
Chilopoda	0	2
Diplopoda	20	29
Mollusca	9	9
Araneae	9	44
Acarina	0	2
Collembola	4	31

If species survive, they may recolonize if *A. longipes* populations subsequently decline. This was well illustrated at the Union Vale site, where the number of ants coming to sugar baits was monitored regularly for 2 years. Initially, in June 1974, only *A. longipes* was recorded, 500 individuals being counted. By June 1975, fewer than 50 *A. longipes* were seen at bait stations, and their place had been taken by about 250 *Plagiolepis madecassa*; in June 1976, no *A. longipes* or *P. madecassa* were found, but 600 *Pheidole punctulata* were recorded. This coincided with the reappearance of centipedes, absent when *A. longipes* was dominant.

Control Techniques

In view of the very large initial populations of *A. longipes*, its pest status, particularly as a domestic nuisance, and the danger that it might spread to other islands in the Seychelles group, it was decided that control techniques needed to be developed. In the time available, and in view of the potential dangers of biological control introducing yet more species into a fragile and vulnerable ecosystem, it was decided to concentrate on chemical control. Two approaches were attempted, the development of a toxic bait, using low concentrations of active ingredient for large-scale field control of colonies, and also a suitable contact spray for the direct treatment of domestic premises, shops, hospitals and hotels aimed at killing the foraging workers.

Ant preferences for the type of bait matrix to be used were assessed by presenting solids or liquids on glass sheets in a Latin square design, and recording the number of each type of bait taken. For liquids, the number of workers seen drinking at a particular drop was recorded. In this way, a wide range of potential food stuffs were screened for palatability both individually and in mixtures, to produce the most acceptable bait. The formulations of the best solid and liquid baits developed are given in Table 18.3.

Coir, the waste from the coconut fiber industry, was sieved to exclude both large particles and fine dust. This provided the basic absorbant structural carrier for the solid baits. Salt, sugar and yeast extract provided the basic attractants and phagostimulants. Animal fat acted as the solvent for the toxicant and also reduced leaching and extended shelf life. In practice, although not quite so effective, coconut oil was used instead of animal fat for cheapness and convenience. After testing a short list of more than 20 toxicants, Haines and Haines (1979c) concluded that of those readily available to them at that time, none was as effective as aldrin, despite its unwelcome property of environmental persistence. Ten kg per ha of bait, containing 1% active ingredient of aldrin, gave good control (Haines and Haines 1979b). Typically, ant abundance decreased by more than 90% within the first few days after baiting, but thereafter populations slowly recovered with speed of recovery depending on the rate of reinvasion from outside the treated area. Consequently, the area baited was a key factor in the length of time that relief from the ant nuisance was obtained. Thus a 2.5% aldrin bait applied to an area 100m^2 around each nest allowed 50% recovery in

TABLE 18.3. Composition of the most acceptable solid and liquid baits for *A. longipes* control. (After Lewis et al. 1976 and Haines and Haines 1979a.)

Solid bait Ingredient	*%W/W in bait*	*Liquid bait Ingredient*	*% W/W in bait*
Sieved coir waste	41	Sugar	83
Salt	4	Tween 80	2
Sugar	2	Water	15
Yeast extract	23		
Water	23		
Animal fat	7		
Proprionic acid	0.25		

9 days, whereas a treated area of 1ha gave 35 days relief (Lewis et al. 1976). In contrast, a 1% toxicant bait gave 35 days reduction when 2ha were treated, and >300 days when 50ha were baited (Haines and Haines 1979b).

The liquid bait was prepared by adding 2.5% aldrin to a sugar solution using 2% Tween 80 as an emulsifier. This was sprayed using a Polypak knapsack sprayer at the rate of 16 liters per ha onto ground vegetation, rock surfaces and the bases of tree trunks. One ha treated in this way suppressed ant activity below 50% for about 60 days. As the total weight of aldrin applied per ha was about 1.6 times that used in the comparable solid bait treatment, it was thought that the effectiveness of the two techniques was similar. However, for general use, Haines and Haines (1979a) concluded that in relation to cost, safety, convenience of marketing and ease of application, solid bait formulations were preferred.

Finally, a rapid-acting insecticidal spray was required for use in premises, hotels, hospitals and ships. Sixteen insecticides were screened by applying them as sprays to outdoor walls to determine how long it took ant activity to build up again to 50% of the pre-treatment score. Bendiocarb proved the most effective, and when applied at $67ml/m^2$ with as little as 0.15% a.i., it gave outdoor protection from heavy infestations for 4-9 days in wet weather, and for some 50 days when used indoors (Lewis et al. 1976).

Results of the Control Program

When the control program was conceived in 1973, it was still hoped that the infestation might be contained in the northern peninsula, and then slowly rolled back using toxic baits. In this event, the effects on non-target organisms of intensive pesticide use in a restricted area might have been acceptable. However, the 1973 infestation survey soon showed that this was impractical, as at least 7 secondary outbreaks had occurred in the main part of the island. This view proved correct when, in spite of the control program, 12 other outbreak centers were established by 1976, and most of the island was infested by 1989.

More limited objectives were therefore set, (1) to provide a cheap but effective toxic bait for sale to local people so that they could control ants in and around their houses; (2) to provide a spray service for badly infested houses and public buildings and hospitals, and (3) to try to prevent the establishment of *A. longipes* on other islands in the Seychelles group, some of which like Praslin and Aldabra are of great conservation importance. Accordingly, the Ant Control Unit was set up in the Ministry of Agriculture. The unit manufactures the bait locally, and supplies it to individuals, usually in 1kg quantities. In the first five years, the total quantities sold (kg) were >1500, 620, 490, 320 and 272 respectively, and sales still continue. The annual number of bendiocarb spray treatments did not exceed 14 in the first five years.

In an attempt to prevent spread to other islands, managers of these islands were contacted, told of the dangers of ant establishment, and were sent preserved specimens so that they might recognize them. A publicity film was made and widely shown, and all managers were asked to report any infestations encountered. This policy has had some success. In February 1975, *A. longipes* was accidentally introduced into Praslin in a shipment of prefabricated building panels, and prompt action with toxic baits eradicated the infestation (Haines and Haines 1978b). Subsequently, there has been almost constant reinfestation at Amitie close to the the airstrip, indicating that ants are being regularly flown in from Mahe. To date, eradication measures have continued to be successful. A possible deliberate introduction of *A. longipes* into the island of Felicite resulted in the ants becoming established over more than two-thirds of the island, and the population density is now relatively high (W.G. Dogley, personal communication). Eradication attempts failed because most of the island is rocky and mountainous and because so few people live there that there is a manpower shortage for large-scale baiting. To date, introductions to the island of La Digue have also been eradicated, but the regular shipments of coconuts and animal fodder from nearby Felicite give grounds for concern.

Simmonds and Greathead (1977) discussing the problems of introduced pests and weeds noted that after introduction, there was often a period of slow increase in numbers, perhaps whilst genetic adaptation took place, followed by a period of explosive growth in numbers which leveled out at a high density. Then in some cases, there is a decline in the density of the introduced species, to the point where it is no longer considered a pest. This seems to have been the case with *A. longipes* in some areas in Mahe. Lewis et al, (1976) reported that as early as 1974, some reduction in ant activity had occurred in the original outbreak area, and they noted that on two other islands in the Indian Ocean, Rodriguez and Agalega, *A. longipes* populations had also declined 10-15 years after introduction.

This phenomenon was confirmed by counting the number of workers coming to sucrose-impregnated filter paper discs at 3 sites in Mahe in 1975 and again in 1980 during a short visit by I. H. Haines. The results (Table 18.4) demonstrate that although numbers were maintained in Les Mamelles, they were greatly reduced at Maldive, whilst the ants seemed to have disappeared from Union Vale.

TABLE 18.4. Changes in population density of *A. longipes* at 3 sites between 1975 and 1980.

Year	Sites: Maldive	Les Mamelles	Union Vale
(a) No. of *A. longipes* workers coming to 5 sucrose-impregnated filter paper discs			
1975	49	37	22
1980	9	37	0
(b) Percentage of cinnamon bushes occupied by			
A. longipes			
1975	100	100	75
1980	8	56	0
T. albipes			
1975	0	0	19
1980	80	48	94

These changes in abundance were reflected in the ant occupation of cinnamon bushes, where a marked recovery of another ant, *T. albipes,* occurred.

The perceived pest status of *A. longipes* also changed. In 1974, the ant was considered to be such a serious pest that the control program was set up as a matter of urgency, whilst in 1991 it is "...now not regarded as a serious problem..." although "...it can still be considered a pest mainly because of the nuisance it can cause to humans and animals" (W.G. Dogley personal communication).

Discussion

The accidental introduction of *A. longipes* into Mahe provides a good example of the dangers of introducing exotic animals into remote islands. It became a domestic nuisance, especially during the period of explosive population growth, and when people were unfamiliar with it. Increasing familiarity and its population decline has helped to reduce its pest status. As an agricultural pest, *A. longipes* has some impact on domestic animals, but a greater one on shrub and tree crops where, by enhancing coccid populations, it increases sap removal, enhances the dangers of disease transmission, and greatly increases honeydew and sooty mold contamination of leaves and fruit. Perhaps least understood is the impact that it has on other wildlife and on biodiversity in general. A reduction in the local skink population and in populations of invertebrates including other ants, has been recorded. Greenslade (1971) noted that in coconut plantations in the Solomon Islands, species diversity fell whenever *A. longipes* populations flourished, but its true impact in Mahe may never be appreciated as knowledge of the local fauna before its introduction is incomplete.

Controlling *A. longipes* as a domestic, or even as an agricultural pest, is relatively simple compared with trying to control it as a conservation pest threatening rare species and habitats. At present we do not possess pest control techniques which will allow us to remove a single species of ant from a complex of other, less dominant indigenous ones. The development of such techniques is currently one of the great challenges in pest control. Classical biological control would involve trying to reduce the damage from one introduced species by introducing yet more, as predators, parasites or disease-causing organisms. In a remote island, this is potentially a dangerous technique compared with toxicant use, as the latter can always be discontinued if unwelcome side effects appear. At present, baiting seems to offer the most practical solution as it uses small quantities of toxicant formulated to produce as selective a kill as possible. Trying to prevent further introductions of exotic species seems the most sensible policy, but it requires public support, continued vigilance, education about the most dangerous introductions likely to occur, and a core of knowledgeable scientists likely to spot new species at an early stage in their spread. Once an introduction has been found, attempts should be made to eradicate it. In such an emergency, it may be better from a conservation point of view to use an environmentally less acceptable, but effective chemical on a strictly limited scale, than to let the introduced species become established. Those who regulate permitted chemicals have, therefore, a responsibility to ensure that effective toxicants can be made available at short notice. Particularly because, despite vigilance, absolute quarantine is never possible; a single failure to prevent new introductions is all that is required to lead to problems.

Of particular interest is why some pest species subsequently decline in density, because it may be possible to exploit this phenomenon in applied control. Detailed research is needed to produce and compare a series of life tables during the increase and decrease phases in the pest population cycle. One possible explanation of pest decline is increased predatory, parasitic or disease action, and if so, an early introduction of these agents might prevent pest establishment. Of more concern is the possibility that some resource may have become depleted as a result of the initial surge in pest numbers. The danger to conservation here is that the declining resource might itself be an interesting and perhaps rare species, or alternatively, that an interesting and rare species might have been competing with *A. longipes* for this resource. In either case, the outlook for the rare species is poor.

Acknowledgments

We are grateful to the pest and disease control section of the Ministry of Agriculture and Fisheries of the Republic of the Seychelles, and particularly to Mr. W. G. Dogley for allowing us to quote unpublished information. The original studies were funded by the former British Ministry of Overseas Development.

Resumen

La hormiga loca, *A. longipes*, fue introducida accidentalmente en la isla de Mahe (Seychelles) hacia 1962, y desde entonces se ha extendido por toda la isla. Entre 1974 y 1977 se llevo a cabo un intenso programma de investigacion a fin de estudiar su biologia y control. Esta especie esta considerada una plaga por las molestias que causa en casas, hospitales y lugares de trabajo; porque ataca a los animales domesticos, pudiendo matar a sus crias; y porque aumenta en gran proporcion las poblaciones de homopteros chupadores de savia, a los que mantiene para aprovechar sus melazas. Tambien se alimenta de especies locales de invertebrados, y puede interferir en la reproduccion de algunos vertebrados nativos, representando una amenaza para la supervivencia de ciertas especie indigenas. En lugares cerrados la plaga se puede controlar pulverizando las superficies con Bendiocarb, pero en el campo el sistema mas efectivo resulto ser la utilizacion de cebos envenenados, a base de residuos de coco, sal, azucar, extracto de levaduras, agua y grasas animales. En 1974 el Ministerio de Agricultura y Pesca establecio una unidad para el control de esta hormiga, que ha estado funcionando hasta la fecha. Esta unidad ha ayudado eficazmente a los agricultores y habitantes de la isla a mantener las poblaciones de hormiga loca a niveles aceptables en granjas y viviendas, y ha evitado la propagacion de la plaga a otras islas, por medio de cuarentenas, inspecciones y campanas de eradicacion. La unica excepcion es la isla de Felicite, que ha sido recientemente afectada. En la propia isla de Mahe, las poblaciones de *A. longipes* han disminuido en algunas areas, e incluso han desaparecido completamente de zonas previamente infectadas. Este hecho ha ido acompanado de una cierta recuperacion de la fauna nativa. De momento no se conocen exactamente las causas de esta disminucion.

References

Brown, E. S. 1959. Immature nutfall of coconuts in the Solomon Islands. I. Distribution of nutfall in relation to that of *Amblypelta* and of certain species of ants. *Bull. Entomol. Res.* 50: 97-133.

Greenslade, P. J. M. 1971. Interspecific competition and frequency changes among ants in Solomon Islands coconut plantations. *J. App. Ecol.* 8: 323-352.

Haines, I. H., and J. B. Haines. 1978a. Colony structure, seasonality and food requirements of the crazy ant, *Anoplolepis longipes* (Jerd.) in the Seychelles. *Ecol. Entomol.* 3: 109-118.

_____ . 1978b. Pest status of the crazy ant, *Anoplolepis longipes* (Jerdon) (Hymenoptera: Formicidae), in the Seychelles. *Bull. Entomol. Res.* 68: 627-638.

_____ . 1979a. Toxic bait for the control of *Anoplolepis longipes* (Jerdon) (Hymenoptera: Formicidae) in the Seychelles. I. The basic attractant carrier, its production and weathering properties. *Bull. Entomol. Res.* 69: 65-75.

_____ . 1979b. Toxic bait for the control of *Anoplolepis longipes* (Jerdon) (Hymenoptera: Formicidae) in the Seychelles. II. Effectiveness, specificity and cost of baiting in field applications. *Bull. Entomol. Res.* 69: 77-85.

_____ . 1979c. Toxic bait for thecontrol of *Anoplolepis longipes* (Jerdon) (Hymenoptera: Formicidae) in the Seychelles. III. Selection of toxicants. *Bull. Entomol. Res.* 69: 203-211.

Hölldobler, B., and E. O. Wilson. 1990. *The Ants*. Springer-Verlag, Berlin.

Lewis, T., J. M. Cherrett, I. Haines, J. B. Haines and P.L Mathias. 1976. The crazy ant *Anoplolepis longipes* (Jerd.) (Hymenoptera: Formicidae) in the Seychelles, and its chemical control. *Bull. Entomol. Res.* 66: 97-111.

Room, P. M., and E. S. C. Smith. 1975. Relative abundance and distribution of insect pests, ants and other components of the cocoa ecosystem in Papua, New Guinea. *J. App. Ecol.* 12: 31-46.

Simmonds, F. J., and D. J. Greathead. 1977. Introductions and pest and weed problems. Pp. 109-124. In: J. M. Cherrett and G. R. Sagar [eds.]. *Origins of pest, parasite, disease and weed problems.* Symposium of the British Ecological Society, 18. Blackwell Scientific Publications, Oxford.

Wheeler, W. M. 1910. *Ants, their structure, development and behavior.* Columbia Univ. Press, New York.

Wilson, E. O., and R. W. Taylor. 1967. The Ants of Polynesia (Hymenoptera: Formicidae). *Pacific Insects Monograph* 14: 1-109.

19

Control of the Little Fire Ant, *Wasmannia Auropunctata*, on Santa Fe Island in the Galapagos Islands

Sandra Abedrabbo

Introduction

The introduction of exotic flora and fauna onto the Galapagos Islands is a continual problem. One of the more troublesome of these introductions has been the little fire ant, *Wasmannia auropunctata*. This ant was introduced accidentally to Santa Cruz and has reached the other inhabited islands (San Cristobol, Floreana, and southern Isabela), as well as the uninhabited islands of Santiago, Pinzon, and Marchena (Lubin 1984; Abedrabbo, pers. obs.). Since its introduction, it has become a troublesome pest to humans and many of the endemic fauna on several Galapagos islands (Clark et al. 1982). Because of its potential danger to the fragile ecosystems of the Galapagos Islands, attempts have been made to eradicate *W. auropunctata*. These programs have been conducted by the Galapagos National Park Service (GNPS) and the Charles Darwin Research Station (CDRS). This paper reports the studies conducted on the control of *W. auropunctata* at "La Caleta" in the northeast sector of Santa Fe Island along the tourist trail.

W. auropunctata was first detected on Santa Fe in May 1975 (DeVries 1975). An intensive eradication program was started that year and continued into 1976 to prevent the spread of the ants to the rest of the island (Clark 1975 and 1976). The program consisted of removing rocks and cutting or removing herbaceous vegetation. This was followed by burning of the area and application of insecticides (a mixture of pyrethrin + resmethrin [PIX] and DDT). The use of these control techniques was considered justified since it was thought that *W. auropunctata* might spread throughout the island with the arrival of the rainy season. Santa Fe Island has at least seven native ant species and an invertebrate fauna typical of low islands in the Galapagos. The spread of *W. auropunctata* posed a potentially great loss to this fauna.

In spite of the prior control measures, *W. auropunctata* was detected again in November 1983 on the tourist trail on Santa Fe. It is possible that the ant was re-introduced as this is a frequently visited tourist site; however it appears more likely that the previous attempts to eradicate the ant were not successful (Meier 1983). In April 1985, the presence of the ant in the area was verified and the area was declared infested again (Wilson et al. 1985).

GNPS and CDRS personnel began another eradication campaign in December 1985. The vegetation was cut and partially burned and the area was treated with the insecticides Aldrin, Ostation and Basudin. Observations were made prior to these treatments on other invertebrates in the treated area. A check was also made on the vegetation before treatments (Lawesson 1986) and 5 months afterwards (Nowak 1986). J. Lawesson, who was the CDRS botanist at that time, stated that the vegetation in this area of Santa Fe also occurred on other islands and if any species were affected, they could be replaced.

The site was monitored monthly during 1986, with less frequent monitoring in following years. Unfortunately, due to the lack of permanent staff, the follow-up study to assess the environmental impact of the insecticides could not be continued. Instead, the project was limited to checks for the presence of *W. auropunctata* and additional control procedures if any ants were found.

In June 1987, Dr. D. F. Williams of the Medical and Veterinary Research Laboratory, U. S. Department of Agriculture in Gainesville, Florida, visited the CDRS as a specialist on ant control. He initiated experiments with specific chemical baits used for ants. The baits are transported by worker ants directly to the ant's nest. Initial tests were made with two commercial formulations developed for the control of the fire ant, *Solenopsis invicta*: Amdro (a. i.-hydramethylnon) and Logic (a. i.-fenoxycarb). At this time, only Amdro is being evaluated as its effectiveness against *W. auropunctata* has been demonstrated.

Description of the Study Area

The Galapagos Islands are situated on a submarine platform in the Pacific ocean 972 km from continental Ecuador. The Archipelago has an area of 8,006 km^2 (Black 1973), and has 13 main islands and dozens of islets and rocks which protrude from the sea surface (Figure 19.1). The island of Santa Fe has an area of 2,413 ha, with a maximum altitude of 259 m (Figure 19.2). One part of the area is littoral fringe, which is dominated by evergreen bushes of *Cryptocarpus pyriformis* and lesser amounts of *Scutia spicata* and *Maytenus octogona*. Depending on the soil type, perennial and annual plants also occur. The other part of the study area, which has very eroded clay soil, is dominated by shrubs such as *Cordia lutea, Bursera graveolens* and *Croton scouleri*, all of which are drought deciduous. These shrubs, along with *Opuntia echios*, constitute the representative floral elements of the island's interior.

FIGURE 19.1. Galapagos Islands

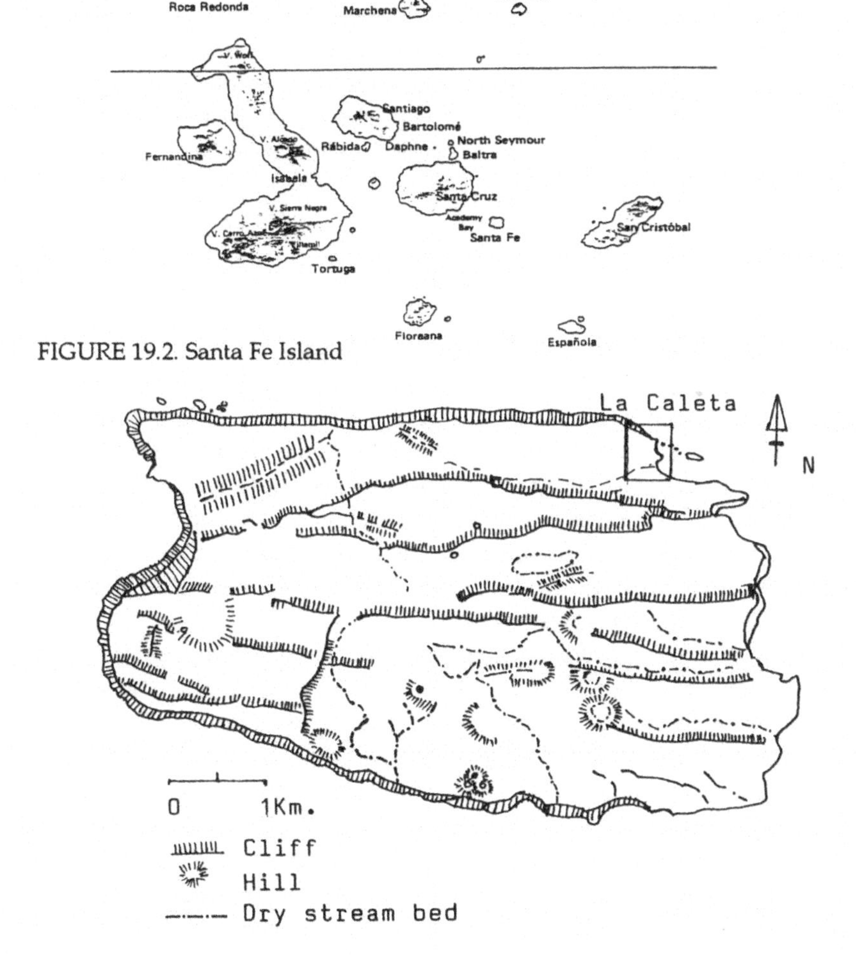

FIGURE 19.2. Santa Fe Island

Methods

The study area was located on the northeast corner of Santa Fe Island, an area that has been monitored for *W. auropunctata* since 1975. The estimated area of infestation in 1985 was 2 ha. Evaluation of Amdro for control was started in 1987 with the visit of Dr. D .F. Williams, mentioned previously. Transects with a grid design of bait stations spaced every 5 to 10 meters were established over an area of 3 ha. This provided sufficient area to determine whether or not the ants might be expanding their territories. Tuna fish or peanut butter baits were used to detect the presence of *W. auropunctata*. These were found to be very attractive to the ants. However, other animals such as the endemic rat (*Oryzomys bauri*), the Galapagos mocking-bird (*Nesomimus parvulus*), and the lava lizard (*Tropidurus albemarlensis*) were also attracted to the baits and consumed them. When *W. auropunctata* was found at a station, a 1-meter grid was established to locate the center of the infestation. This was where the Amdro was applied. The presence of other ant species was also noted.

Results

Beginning in May 1975, the seasonal occurrence of *W. auropunctata* in the northeast sector of Santa Fe was monitored (Table 19.1). General observations have shown that the ants are easiest to locate in wet conditions such as occurred in November 1983, shortly after the El Niño of 1982-83. In dry periods, the ants were found less frequently. For example, in 1988 and 1989 (dry years), it was very difficult to detect *W. auropunctata*. Under these conditions, the ants remain in the soil and in the subterranean parts of shrubs. Because of this, the best time to monitor for *W. auropunctata* is after rains.

The frequency of occurrence of ant species collected from various studies is shown in Tables 19.2-4. In April 1985, prior to our studies, 91 bait stations were placed within the *W. auropunctata* infested area and 34 stations outside the area. *W. auropunctata* was detected at 37% of the bait traps (Table 19.2) in the infested area. *Solenopsis globularia* was observed at 45% of the bait stations in the infested area and 47% in the uninfested areas. *Monomorium floricola* occurred at 33% in the infested and 21% in the uninfested areas. In April 1990, after the Amdro treatments, 230 bait stations revealed only one focus (0.5%) of *W. auropunctata* (Table 19.3). This time *Tetramorium bicarinatum* (listed as *Tetramorium sp.* 2) showed the highest percentage of occurrence: 40% frequency.

TABLE 19.1. Seasonal occurrence [presence (+) or absence (-)] of *W. auropuncta-ta* on Santa Fe Island, Galapagos Islands.

	Jan	*Feb*	*Mar*	*Apr*	*May*	*Jun*	*Jul*	*Aug*	*Sep*	*Oct*	*Nov*	*Dec*
1975					+					+	+	+
1976	+	+	+					-	-			
1977								-				
1983											+	+
1984							+	-				
1985				+							+	+
1986	+	-	-	-	-	+	+	+		-		
1987	-			-		+	+	+				+
1988							-			+		-
1989				-			-			-		
1990				+	-	-					-	
1991			-			-			-			

TABLE 19.2. Frequency of occurrence of ants in the infested area (91 bait stations) and outside the area (34 bait stations) on Santa Fe Island. (Data from Wilson, April 1985.)

	% Frequency	
Genera	*Infested area*	*Uninfested area*
Wasmannia sp.	37	0
Solenopsis sp.	45	47
Monomorium sp.	33	21
Tapinoma sp.	18	18
Paratrechina sp.	1	0
Pheidole sp.	0	12
Dorymyrmex sp.	0	26

TABLE 19.3. Frequency of occurrence of ants in the infested area (230 bait stations) on Santa Fe Island. (Data from Martínez, April 1990.)

Genera	*% Frequency*
Wasmannia sp.	0.5
Solenopsis sp.	15
Monomorium sp.	4
Tapinoma sp.	13
Pheidole sp.	2
Dorymyrmex sp.	4
Tetramorium sp. 1	10
Tetramorium sp. 2	40
Subfamily *Myrmicinae*	4

TABLE 19.4. Frequency of occurrence of ants in the infested area (250 bait stations) on Santa Fe Island. (Data from Abedrabbo, November, 1991.)

Genera	*% Frequency*
Pheidole sp.	47
Monomorium floricola	16
Paratrechina sp.	15
Tapinoma melanocephalum	14
Tetramorium bicarinatum	8.4
Tetramorium simillimum	6
Solenopsis globularia	8
Dorymyrmex pyramicus	0.8
Lost or unvisited station	16

In September 1991, no *W. auropunctata* were found on 250 bait stations (Table 19.4). At this time, a *Pheidole sp.* was the most abundant ant (47%) followed by *M. floricola, Paratrechina sp.* and *Tapinoma melanocephalum.* It is interesting to note that the *Pheidole sp.* was often found on the upper part of the bait papers while *M. floricola* was on the lower part.

We believe at this time that we eliminated *W. auropunctata* from Santa Fe Island using Amdro. This pest ant has not been found since May 1990, even though during a survey in March 1991, the area had three months of heavy rains, was humid and had green vegetation, ideal conditions for heavy populations of *W. auropunctata.*

Several other ant species as well as other invertebrates, were also found. The other ant species found in the area were *T. bicarinatum, T. simillimum, Dorymyrmex sp., T. melanocephalum, Paratrechina sp., M. floricola, Pheidole sp., S. globularia* and one other unidentified species *(Myrmicinae).* The density of these ants increased as *W. auropunctata* decreased. Other common invertebrates found in the area were *Isopoda, Scorpionida (Hadruroides lunatus), Araneae (Neoscona cooksoni, Latrodectus apicallis), Orthoptera (Schistocerca melanocera), Coleoptera: Carabidae (Calosoma* sp.), *Tenebrionidae (Phaleria manicata, Ammophorus* sp.), *Curculionidae (Gerstaeckeria galapagoensis), Chrysomelidae, Coccinellidae (Cycloneda sanguinea),* and *Scaphidiidae.*

An examination by B. Nowak on the vegetation in May 1986, five months after the burning, showed some revegetation. This appeared to be due to the survival and regrowth of woody species and new growth of herbaceous species. Because of the rapid regeneration of the area, it appears a large quantity of seeds survived the fire. The majority of the *Cryptocarpus pyriformis* shrubs returned and the herbaceous species were particularly active in the recolonization. The herbaceous annuals dominant in sandy soils were *Tiquilia galapgoa, Heliotropium angiospermum* and *Kallstroemia adscendens,* while in rocky soil *Mentzella aspera* and *Acalypha parvula* were dominant. Regeneration of trees took longer, while some *Cordia lutea* did not recover.

Discussion

Unfortunately the bait stations did not have permanent markers and so gradually their locations were changed. The transects and the number of bait stations varied over the years depending on the individuals who conducted the monitoring and the amount of time available to them. Because of this lack of continuity, it was difficult to compare the results from the different monitoring periods from the beginning of the program to the present.

The application of residual insecticides such as aldrin and DDT to control *W. auropunctata* only results in killing ants on the surface since the materials never reach the queens which are located in the nests below ground. Because of this, these chemicals were ineffective. *W. auropunctata* is adapted to dry periods by penetrating the soil to depths of 30 cm, then reappearing following rains. The use of chemicals is controversial and not readily accepted, especially in an environment such as the Galapagos Islands. It is evident that DDT and the other chemicals used in the past did not eradicate *W. auropunctata.* However, it is important to note that the density of these ants was higher before the control program and their advance throughout the island was probably halted by the application of these chemicals.

The plant species found in May 1986 following the treatment are the same as the original species of the area. The littoral fringe is permanently exposed to disturbances such as high tides, sea lions and tourism.

Recommendations

The use of fire can be very dangerous as seen by its damage on southern Isabela. Insecticides such as DDT, aldrin, Ostation and Basudin should not be used in fragile ecosystems such as the Galapagos Islands. It is better to use specific poisons, especially baits which control the pest, but do not cause damage to the Galapagos ecosystem. The northeast area of Santa Fe has a large colony of sea lions, as well as land iguanas, endemic rice rats, birds, snakes, and lizards that could be affected by the use of chemicals that have residual effects.

The use of the fire ant bait Amdro seems to be effective for the control of *W. auropunctata*. If we can confirm the success of the eradication trial on the island of Santa Fe, then the next step is to delimit the areas infested with *W. auropunctata* on the islands of Pinzon and Marchena, and evaluate the possibilities of eradicating this ant on these islands with Amdro. It is always very important and necessary to maintain a constant vigil of the reoccurrence of *W. auropunctata* and this should be accomplished with the help of guides and visiting scientists. Plans are to monitor *W. auropunctata* on Santa Fe during the rainy season of 1992 to assure that the control program is still effective. This sector of Santa Fe is in daily use, being visited by many tourists. Because of this, the Galapagos National Park rule prohibiting bringing food onto any of the islands must be maintained. If the necessary precautions are not taken, it is highly probable that *W. auropunctata* will be introduced or reintroduced onto many of the islands.

Resumen

La presencia de hormiga colorada, *Wasmannia auropunctata* en Santa Fe, junto al sendero turístico, fue detectada por primera vez en mayo de 1975, en cuyo año se inició un programa de erradicación con personal del Servicio Parque Nacional Galápos y la Estación Científica Charles Darwin. En noviembre de 1983 se encontró nuevamente *W. auropunctata* en este sector y se prodría decir que fue re-introducida o los intentos anteriores de erradicarla no fueron exitosos. En 1985 se inició un nuevo programa de control y desde 1987 con la visita del Dr. D. F. Williams del Departamento de Agricultura de Gainesville, Florida, EE. UU. se utiliza Amdro (componente activo-hydramethylnon) y Logic (componente activo-fenoxycarb), que son transportados por las obreras al nido.

Actualmente creemos que fue posible erradicar *W. auropunctata* utilizando Amdro en las 2 hectáreas de Santa Fe; pues no se ha encontrado hormiga colorada en este sector desde mayo de 1990, a pesar que en el monitoreo de marzo de 1991, después de tres meses de lluvias fuertes, el área se encontró húmeda, la vegetación verde y se encontró siete otras especies de hormigas y otros invertebrados.

El uso de químicos es controvertido y poco aceptado, especialmente en un ambiente frágil como es Galápagos; por lo tanto es recomendable siempre averiguar venenos específicos que aseguren su control y no causen un impacto al ecosistema galapagueño.

Este sector de Santa Fe es de uso diario, pues es muy visitado por los turistas, por lo cual hay que aplicar las leyes del Parque Nacional Galápagos, concernientes a la prohibición de bajar comida a las islas. Sin tomar las debidas precauciones es altamente probable la introducción de *W. auropunctata* a las otras islas, en las cuales no está presente.

References

Black, J. 1973. *Galápagos Archipiélago del Ecuador.* Imprenta Europa. Quito, Ecuador.

Clark, D. B., C. Guayasamín, O. Pazmiño, C. Donoso, and Y. Páez de Villacís. 1982. The tramp ant *Wasmannia auropunctata*: Autecology and effects on ant diversity and distribution on Santa Cruz Island, Galapagos. *Biotropica* 14: 196-207.

Clark, D. 1975 and 1976. Field Trip Itinerary Report. Charles Darwin Research Station. Unpublished.

DeVries, T. 1975. Field Trip Itinerary Report. Charles Darwin Research Station. Unpublished.

Lawesson, J. 1986. Chequeo de la vegetación en Santa Fe. Informe de campo. Charles Darwin Research Station. Unpublished.

Lubin Y. D. 1984. Changes in the native fauna of the Galapagos Islands following invasion by the little red fire ant, *Wasmannia auropunctata. Biol. J. Linn. Soc.* 21: 229-242.

Meier, R. E. 1983. Coexisting patterns and foraging behaviour of ants within the arid zone of three Galapagos Islands. Annual Report 1983. Charles Darwin Research Station.

Nowak, B. 1986. Chequeo de la vegetación en Santa Fe. Informe de campo. Charles Darwin Research Station. Unpublished.

Wilson, M., A. Wilson, H. Schatz, I. Schatz, and H. Serrano. 1985. Distribución de *Wasmannia auropunctata* en el noreste de Santa Fe. Informe de campo. Charles Darwin Research Station. Unpublished.

20

Foraging of the Pharaoh Ant, *Monomorium Pharaonis*: An Exotic in the Urban Environment

Karen M. Vail and D. F. Williams

History and General Biology

The Pharaoh ant, *Monomorium pharaonis* (L.), is a small species ca. 2 mm long which ranges from yellow to light red. This ant was first identified as *Formica pharaonis* by Linnaeus in 1758 and was given at least six other specific names (Bolton 1987) before receiving its current name (Mayr 1862). It was thought that this species originated in South America (Arnold 1916), or the Afrotropical region (Bernard 1952); however, Bolton (1987) suggests that its origin is India, an idea that agrees with Emery (1922) and Wilson and Taylor (1967). Because it has been introduced to many parts of the world through international trade, it is doubtful its place of origin will ever be determined. Smith (1965) thought it was so common a pest as to state "the Pharaoh ant is probably found in every town or city of commercial importance in the United States". The Pharaoh ant is a major pest ant in Florida (Bieman and Wojcik 1990) and the majority of the contents of this chapter reflect research conducted in this state.

The Pharaoh ant's unusual reproductive strategy undoubtedly contributes to its pest status. Unlike many ants, this species does not require mating flights for sexual reproduction; they reproduce by budding. To start a new colony, workers carry their brood to another location. This movement may be accompanied by queens, but it is not necessary (Peacock et al. 1955). Workers then rear sexual brood and adult reproductives to perpetuate the colony.

Budding makes the Pharaoh ant difficult to control because there are often groups of related nests instead of only one. Factors that influence budding include changes in the environment such as temperature and food and water resources, or overcrowding (Edwards 1986). The presence of insecticides is also suspected to cause budding (Green et al. 1954).

Nest location can also make control difficult. Nests are usually located in inaccessible or difficult to reach areas for the pest control operator, including interior wall voids, areas under or behind window sills, toilets, sinks, switch plates, lights, etc. Edwards (1986) suggests that any warm area with high humidity can serve as a nest site. We have noticed at several hotels, motels and hospitals in North Florida that aluminum window and door frames are a popular nest site. Although there are reports that suggest the Pharaoh ant nests outdoors in South Florida (Creighton 1950), in temperate (nontropical or nonsubtropical) regions it usually nests indoors (Kohn and Vlcek 1986). In North Florida, we observed ants foraging both indoors and along the outdoor periphery of the building. Further south in areas such as Tampa, foraging is seen further from structures with trails often following fences.

Pharaoh ant foraging has been described by Sudd (1957, 1960). To initiate their search for a food source, scouts followed routes along cracks or edges of structures, a phenomenon described by Klotz and Reid (1992) as guideline orientation. As the scouts approach a food source, the trails branch. The return trips to the nest were more direct than the outgoing trips possibly indicating workers use visual orientation as well as odor trails. Workers in the nest were recruited by stimulating them to search for food. This stimulation was accomplished by workers returning to and rushing about the nest without any apparent stimulatory contact with other workers. Food location was identified by following the trail pheromone, faranal, which was laid by the scouts. Foragers also reinforced the trail pheromone. Faranal ($C_{17}H_{30}$), is a terpenoid produced by the Dufour's gland (Ritter et al. 1977).

Pharaoh ant control currently emphasizes the use of toxic baits. For a review and history of control methods, see Edwards (1986). Insecticidal sprays and dusts are often ineffective because they only affect a small percentage of the workers that forage (Williams 1990) and may also cause budding which amplifies the problem. Multiple nests within a relatively small area also reduce the effectiveness of sprays. Residual sprays may be used to reduce foragers in sensitive areas such as surgery and neonatal care units in hospitals, but use of sprays in hospitals is severely restricted in Florida. Unlike sprays which usually do not contact a nest, a toxic bait can be transported to the nest by foragers which feed on the bait and then distribute the toxin throughout the colony by trophallaxis.

Stomach poisons or insect growth regulators (IGRs) are the active ingredients in toxic baits. Stomach poisons used in some commercial baits include Maxforce Pharaoh Ant Killer (hydramethylnon, Clorox Co., Pleasanton, CA), Raid Max Ant Bait (sulfluramid, S. C. Johnson, Racine, WI), Pro-Control (sulfluramid, Micro-Gen Equipment Corp., San Antonio, TX), Drax Ant Kil Gel (orthoboric acid, Waterbury Co. Inc., Waterbury, CT)) and Terro Ant Killer II (sodium tetraborate decahydrate [borax], Senoret Chemical Co., St. Louis, MO). An early recommendation for Pharaoh ant control was the use of borax and sugar dissolved in boiling water and placed on broken crockery (Riley 1889). Although our use of boric acid in laboratory evaluations has given variable results, pest control operators (PCOs) use this option because they can choose the attractant. Foraging workers are attracted to many compounds and the PCOs mix the boric acid with their choice of attractive ingredient or combination of ingredients.

Currently, the only insect growth regulator registered for Pharaoh ant control is methoprene, marketed under the name of Pharorid (Zoecon Corp., Dallas, TX). IGRs are much slower acting than stomach poisons because typically only the queen and brood are affected. Methoprene induces sterility in queens and disrupts the brood stages (Edwards 1975). Workers are unaffected and are still evident for weeks or months after treatment. Because the maximum worker life span at 27^{0}C is estimated at 9-10 weeks (Peacock and Baxter 1950), they may be present long after queens have ceased egg production and the brood has died. The advantage of using insect growth regulators is that they are more likely to be distributed throughout the entire colony because they do not adversely effect workers. Also IGRs are more acceptable to the consumer since they are considered safer compounds. The oral LD_{50} of methoprene to rats is >34,000 mg/kg (Ware 1983).

Stomach poisons work relatively fast compared to IGRs, i.e., worker numbers are reduced in a few days and complete colony elimination can occur in a few days to a few weeks. However, stomach poisons may work too quickly, thereby eliminating the worker force before the insecticide can be distributed to the entire colony. Relatively few workers and brood need to survive to perpetuate the colony. Peacock et al. (1955) have reported just 100 workers and 50 pieces of brood can initiate a successful colony. Vail (unpublished data) has found that 5 workers and 30 eggs, 19 larvae and 3 pupae can start a new colony.

Foraging study

In July 1990, we were asked to control a Pharaoh ant infestation at the Bachelor Officer's Quarters (BOQ), Jacksonville Naval Air Station, Jacksonville, FL. Fenoxycarb (Maag Agrochemicals now part of Ciba-Geigy, Greensboro, N.C.) an insect growth regulator, was evaluated for control of the Pharaoh ants in this approximately 7841 m^2 facility (Williams and Vail 1993b). Fenoxycarb affects Pharaoh ants similarly to methoprene: it decreases brood and egg production and has little effect on Pharaoh ant workers (Williams and Vail 1993a). The east wing (2319 m^2) of the BOQ was treated with Raid Max Ant Bait, the west wing (2040 m^2) with 0.5% fenoxycarb, and the south wing (2787 m^2) with 1% fenoxycarb. All wings, except the north wing, consisted of two floors. The north wing (412 m^2), which was separated from the other wings by a large auditorium, served as a control. Fenoxycarb baits were formulated by weight in peanut butter oil and this oil/IGR solution was then applied to a pregel defatted corn grit at 30% by weight.

Within six weeks after the last treatment, no ants were found in the Raid Max or 0.5% fenoxycarb-treated wings. Ants were still present in the 1% fenoxycarb-treated and control wings. Although ants were found in the Raid Max-treated wing at 12 wk, a re-application of baits at week 12 and 14 killed the remaining colonies because no ants were detected after this treatment. Ants were not found again until week 24 in the 0.5% fenoxycarb-treated wing indicating the IGR gave at least 18 weeks of control.

The presence of ants in the 0.5% fenoxycarb-treatment suggested to us that foragers from the south wing were entering the west wing where a vending machine area was located. To determine if Pharaoh ants could travel that distance for food, a foraging study was initiated. Sudd (1957, 1960) reported on foraging, recruitment and communication of Pharaoh ants, and Haack (1987) studied food flow in small colonies of the Pharaoh ant, but neither observed foraging in a building the size of the BOQ with such a large Pharaoh ant infestation.

Materials and Methods

On 18 March 1991, Pharaoh ants were monitored at the BOQ by placing two to six baits per room (one to two in the bedroom and living room windows, and two on bathroom floors where applicable) in preselected rooms in all wings. The bait consisted of 2 cc's of natural peanut butter applied from a syringe onto an index card. Two h later, the baits were checked for the number of Pharaoh ants and removed. A bait card holding 6 cc's of 2% calco red dye formulated in peanut butter was placed in room 162, the room with the highest Pharaoh ant count in the north wing. A similar bait card with 6 cc's of 2% calco blue in peanut butter was placed in room 146, the room with the highest infestation in the south wing. Laboratory tests indicated that the dyed peanut butter did not cause mortality. Dye mixtures were left in place for 4 d, until 22 March 1991. On this day, we removed the baits containing dyed peanut butter mixtures and baited the BOQ again with plain peanut butter as on 18 March 1991 except that baits were not applied to the east wing because ants were not found there on 18 March 1991. We also placed baits on the ground near the outside perimeter of the building starting from the entrance by the desk continuing around the north, west and south wings in a counter-clockwise direction until we reached the east wing (see Figure 20.1). After 2 h, we collected the plain peanut butter baits. We tapped any ants found at the baits onto white paper, folded the paper over, placed the paper on a clipboard and crushed the ants with a heavy metal roller. If ants contained dye, we recorded the location where they were collected on a data sheet (drawn on a map for outside baits) and on the white paper on which they were crushed. The distance from the dyed bait at which ants containing dye were found was determined and noted. The percent of workers containing dye on each card was determined in the laboratory the next day.

We returned to the BOQ on 29 March 1991 (dyed baits had been removed for 7 days) and repeated the procedures of 22 March 1991 with a few exceptions. Only the lounge, soda mess (vending machine area) and rooms 228 and 229 were baited in the west wing. Also, baits found with ants were picked up and placed in a Ziploc® bag and the location marked on the bag. We placed the bags in a cooler with ice and then placed them in a freezer in the laboratory upon our return. The locations where ants were collected, percent of ants containing dye and foraging distance were determined in the laboratory.

Foraging distance was calculated under several assumptions. We assumed ants foraged along the outside wall because previous work in the BOQ indicated the majority of foragers were found near walls, windows or pipes. When ants were found in rooms on the other side of the building from where the dyed baits were placed, we assumed that they travelled along structural guidelines (Klotz and Reid 1992) following right angles because they would either be following pipes or beams. Sudd (1960) found Pharaoh ants using trails that followed edges of doors and floors. Also, we assumed that the colony was near to the position where the dyed baits were placed, since these positions had the highest infestations in each wing on 18 March 1991. All foraging dis-

FIGURE 20.1. Results of Pharaoh ant baiting in the BOQ on 22 March 1991. Building plan is drawn to show the locations of the wings and the baits, and is not drawn to scale.

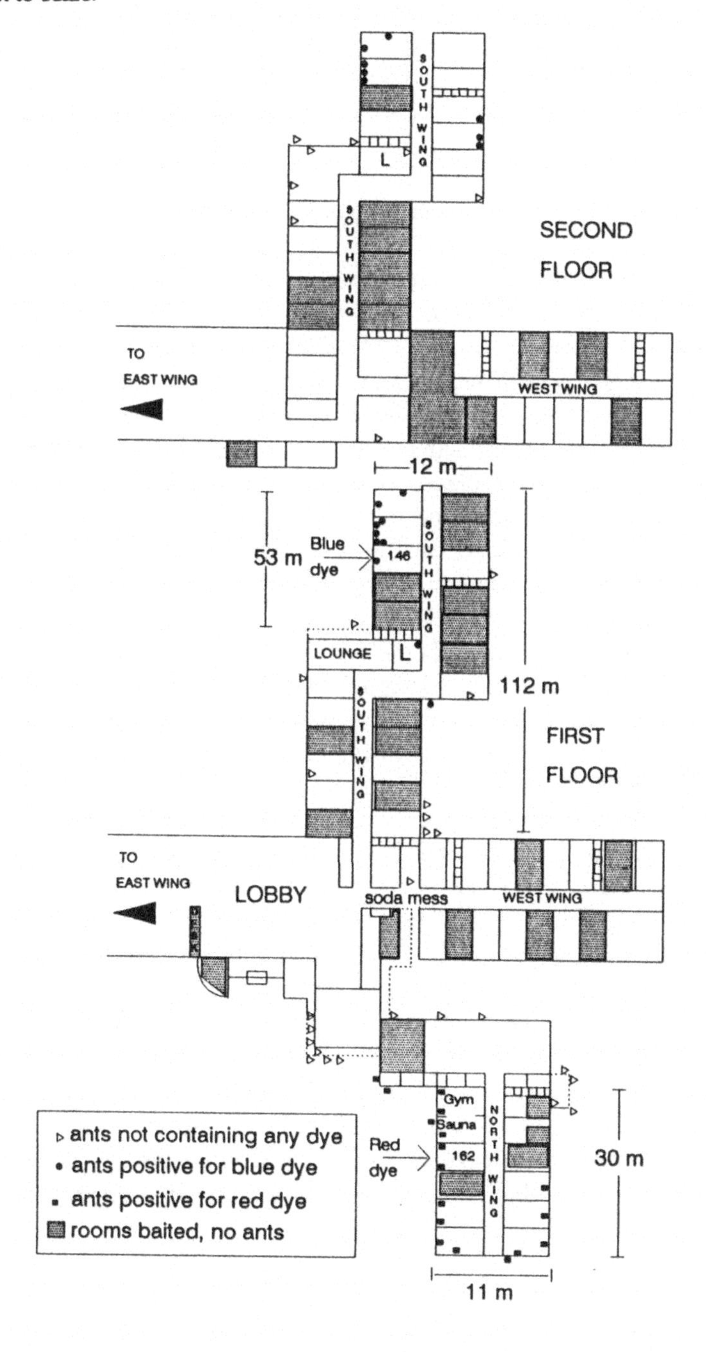

tances were measured from the position of the dyed bait. We verified the foraging distance measurements by using blueprints obtained from the public works department at the Jacksonville Naval Air Station.

We calculated foraging distance and compared foraging (number of ants at bait, percent marked and foraging distance) of ants outside to ants indoors using general linear models (GLM) (SAS Institute 1988). The same variables were compared for the two dates using the same test.

Results and Discussion

Both marked and unmarked ants were found on the plain peanut butter bait cards collected both indoors and outdoors on 22 March (Figure 20.1) and 29 March 1991 (Figure 20.2). In both the north and south wing, Pharaoh ant foraging was extensive. Both horizontal and vertical movement of foragers was detected. Marked foragers were found in rooms on the same side of the hall and across the hall from where the baits were placed (Figures 20.1 and 20.2) as well as on the floor above. The foraging territory (area where dyed ants were detected) was increased on the second monitoring date in the north wing. Ants were found containing dye further south on the west side of the building.

The results presented hereafter shall only refer to baits containing marked ants. The mean foraging distance (±SD) for both dates combined was 16.2 ± 9.6m. The maximum distance a dyed ant was found from the dyed bait card was 45 m. This was greater than that found by Sudd (1960), Peacock and Baxter (1949) and Ritter et al. (1977). The longest Pharaoh ant trail that Sudd (1960) measured in houses in Nigeria was 9.5 m. Peacock and Baxter (1949) measured a 12.2 m trail in Scotland. Ritter et al. (1977) mentions that Pharaoh ant trails are many meters long. The foraging distances recorded at the BOQ indicated the ants did have the potential to travel from one wing to the other.

The mean number of ants per bait card (117.2 ± 96.1) indicates the severity of the infestation. A mean of 49.9 ± 40.8% of the foragers crushed on the paper contained dye. There was a very weak negative relationship between the percent of the ants marked and the distance from the dyed bait (% marked=80.90-12.91*[ln(meters from the bait)], $r^2=0.14$). Percent of workers found containing dye on 29 March was significantly less than those found containing dye on 22 March 1991 (Table 20.1).

FIGURE 20.2. Results of Pharaoh ant baiting in the BOQ on 29 March 1991. Building plan is drawn to show the locations of the wings and the baits, and is not drawn to scale.

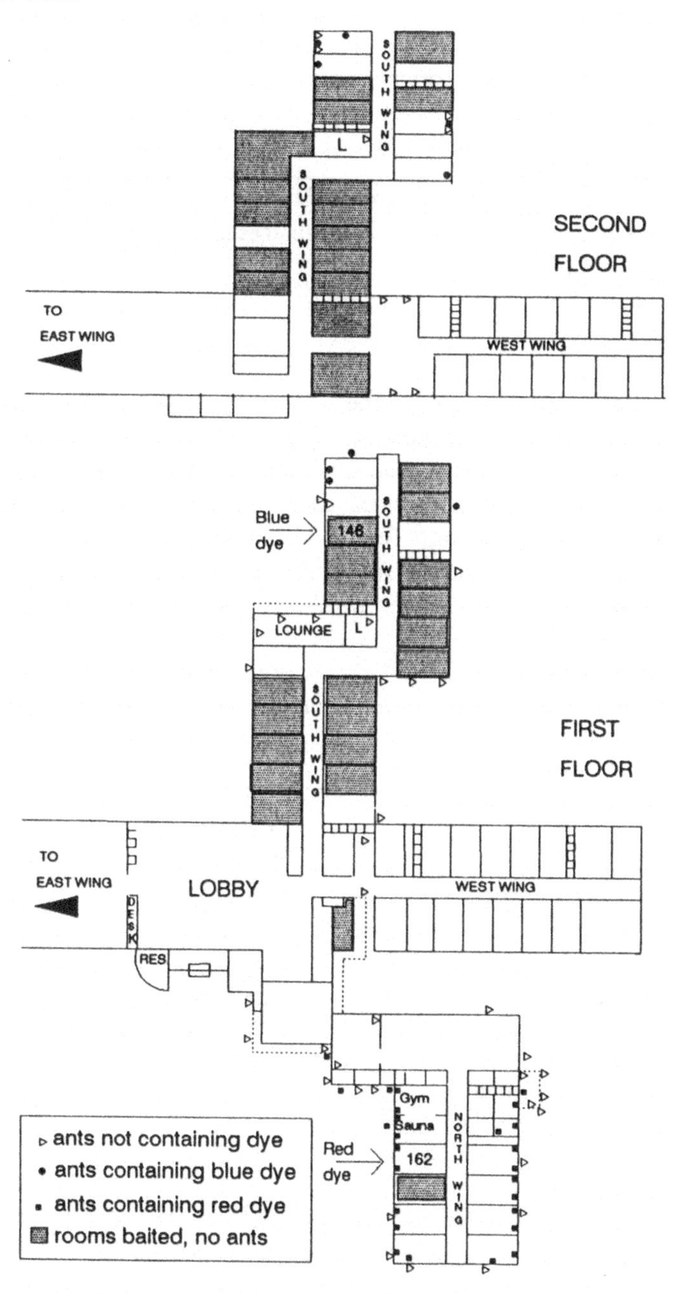

TABLE 20.1. Comparison by sampling date of percent workers containing dye, number of ants per bait card containing dye, and foraging distance

	3/22/91 *Mean (SE)*	*3/29/91* *Mean (SE)*
% workers with dye[a]	74.7(6.0)	26.6(5.5)*
number ants/card[b]	98.8(13.7)	136.1(18.2)NS
foraging distance (m)[c]	15.7(1.7)	16.7(1.6)NS

[a]percentages were arcsine transformed before GLM (F=33.7; df= 1,64; P=0.001) was performed. Untransformed percentages are presented.
[b](F=2.7; df=1,67; P=0.11).
[c](F=0.16; df=1,67; P=0.69)
*significant, NS=not significant.

Foraging distance along the outdoor periphery was significantly longer than that occurring indoors (Table 20.2). However, the number of ants per card was higher indoors than outdoors indicating that more workers were foraging indoors than outdoors. A similar study conducted in quadraplex units (buildings consisting of four apartments) in September and October in north Florida indicated a majority of the Pharaoh ant foraging occurred outdoors (Oi et al., in review).

TABLE 20.2. Comparison by indoor and outdoor foraging environment of percent workers containing dye, number of ants per bait card containing dye, and foraging distance.

	Indoors *Mean (SE)*	*Outdoors* *Mean (SE)*
% workers with dye[a]	49.9 (5.4)	50.4 (14.1)NS
number ants/card[b]	128.4 (13.0)	64.0 (18.4)*
foraging distance (m)[c]	14.8 (1.2)	22.6 (2.7)*

[a]percentages were arcsine transformed before GLM (F=0.01; df=1,64; P=0.94) was performed. Untransformed percentages are presented.
[b](df 1,67 F=4.7, P=0.03).
[c](df 1,67 F=7.11, P=0.01).
*significant, NS=not significant.

Because dyes used could be spread to other colonies through trophallaxis, we are uncertain whether we measured the foraging of one colony, several colonies or the degree of trophallaxis between colonies.

Because of the lack of aggression between colonies, these colonies could be considered one supercolony. A supercolony as defined by Holldobler and Wilson (1990) is a "unicolonial population, in which workers move freely from one nest to another, so that the entire population is a single colony". This study documented the spread of a bait through an area, regardless of whether one colony or several colonies fed on the bait. Less than 6 cc's and 2 cc's of peanut butter was spread throughout the north and south wing, respectively, indicating that the use of baits as a control for Pharaoh ants should be very efficient.

To better quantify foraging of Pharaoh ants, a dye should be used that would mark workers and could be observed without killing the ant yet not be transferred to other individuals through trophallaxis. A dye which marks fat bodies would achieve these results; however, such a dye has yet to be located. Also, offering the baits containing dye for a shorter period of time may ensure that foragers from only the closest colony would have time to find the bait and be marked. Such modification of our techniques may allow for the estimation of colony size using a mark-recapture method similar to that developed by Su and Scheffrahn (1988) for termites. The ability to estimate colony size would enhance greatly the development of more judicious and efficient methods of Pharaoh ant control.

Acknowledgments

We would like to thank CAPT. F. Santana, USN, MSC (now retired), C. Strong, and D. Hall for assisting in the placement of baits at the BOQ. S. D. Porter, J. Klotz, and J. P. Parkman reviewed an earlier version of this manuscript. Their comments were appreciated, but not always heeded.

Resumen

El regulador del crecimiento, Fenoxycarb, fue muy efectivo contra la hormiga faraona, *Monomorium pharaonis* (L.). El fenoxycarb, reduce la producción de abejas en las hormigas reina y las colonias mueren debido a la atrición. Los mejores resultados de experimentos de laboratorio se obtuvieron al usar concentraciones de 0.25, 0.5 y 1% en aceite de mantequilla de maní. Doce semanas después, estas concentraciones redujeron significativamente el numero de obreras y el numero de cría fue reducido a las 5 semanas o antes; sinembargo, se necesitó mas de un cebo para eliminar completamente las colonias. En algunas pruebas, las colonias alimentadas con cebos en concentraciones de 0.1, 0.25 y 0.5%

tuvieron una demora significativa en la producción de individuos alados. Cuando se utilizaron concentraciones menores (0.05, 0.1 y 0.25%) se observó que en las castas intermedias, los individuos eran mas grandes que las obreras, pero mas pequeños que las reinas. Las concentraciones altas de 2.5 y 5% no fueron efectivas quizás por la repelencia de las hormigas al material. Estos resultados demuestran que fenoxycarb es tan efectivo para el control de la hormiga faraona como el cebo comercial, Pharorid (metropeno).

References

Arnold, G. 1916. A monograph of the Formicidae of South Africa. *Ann. South African Mus.* 14: 159-270.

Bernard, F. 1952. La reserve naturelle integrale du Mt Nimba, part 11. Hymenopteres, Formicidae. *Memoires de l'Institute Francais d'Afrique noire.* 19: 165-270.

Bieman, D. and D. Wojcik. 1990. Tracking ants in Florida. PCO April: 11-13

Bolton, B. 1987. A review of the *Solenopsis* genus-group and revision of Afrotropical *Monomorium* Mayr (Hymenoptera:Formicidae). *Bull. British Mus. (Natural History) Entomol. Ser.* 54: 263-452.

Creighton, W. S. 1950. The ants of North America. *Bull. of Mus. Comp. Zool. Harvard Univ.* 104: 1-585.

Edwards, J. P. 1975. The effects of a juvenile hormone analogue on laboratory colonies of pharaoh's ant, *Monomorium pharaonis* (L.). *Bull. Entomol. Res.* 65: 75-80.

Edwards, J. P. 1986. The biology, economic importance, and control of the pharaoh's ant, *Monomorium pharaonis* (L.). Pp. 257-271. In: S. B. Vinson [ed.]. *Economic impact and control of social insects.* Praeger Publishers, New York.

Emery, C. 1922. Hym. fam. Formicidae subfam. Myrmicinae. In: Wytsman, P. [ed.].*Genera Insectorum.* 174B-174C: 95-397.

Green, A. A., M. J. Kane, P. S. Tyler, and D. G. Halstead. 1954. The control of pharaoh's ants in hospitals. *Pest Infest. Res.* (1953): 24.

Haack, K. 1987. Aspects of the food handling behavior of the Pharaoh ant, *Monomorium pharaonis* (L.). Master's thesis, Texas A & M Univ., College Station, TX.

Holldobler, B. and E. O. Wilson. 1990. *The ants.* The Belknap Press of Harvard Univ. Press, Cambridge, Mass.

Klotz, J. and B. Reid. 1992. The use of spatial cues for structural guideline orientation in *Tapinoma sessile* and *Camponotus pennsylvanicus* (Hymenoptera: Formicidae). *J. Insect Behav.* 5: 71-82.

Kohn, M. and M. Vlcek 1986. Outdoor persistence throughout the year of *Monomorium pharaonis* (Hymenoptera: Formicidae). *Entomol. Gener.* 11: 213-215.

Mayr, G. 1862. Myrmecologische Studien. Verhandlungen der k. k. zoologische-botanischen Gesellschaft in Wien. 12: 649-776.

Oi, D., K. Vail, D. Williams and D. Bieman. 1993. Indoor and outdoor foraging locations of Pharaoh ants (Hymenoptera: Formicidae), with implications for control strategies using bait stations. *Florida Entomol.* (submitted).

Peacock, A. D. and A. T. Baxter. 1949. Studies in pharaoh's ant. I. The rearing of artificial colonies. *Entomol. Mon. Mag.* 85: 256-270.

_____ . 1950. Studies in Pharaoh's ant, *Monomorium pharaonis* (L.). 3. Life history. *Entomol. Mon. Mag.* 86: 171-178.

Peacock, A. D., J. H. Sudd and A. T. Baxter. 1955. Studies in Pharaoh's ant, *Monomorium pharaonis* (L.). 11. Colony foundation. *Entomol. Mon. Mag.* 91: 125-129.

Riley, C. V. 1889. The little red ant. *Insect Life* 2: 106-108.

Ritter, F. J., I. Bruggrmann-Rotgans, P. Verweil, E. Talman, F. Stein, J. LaBrinn, and C. J. Persoons. 1977. Faranal trail pheromone from the dufour's gland of the pharaoh's ant, structurally related to juvenile hormone. *Proc. Eighth Int. Cong.* IUSSI Pudoc.

SAS Institute. 1988. SAS/STAT User's Guide release 6.03. SAS Institute. Cary, N.C.

Smith, M. R. 1965. House-infesting ants of the eastern United States: their recognition, biology and economic importance. *USDA, ARS, Tech. Bull.* 1326.

Su, N-Y, and R. H. Scheffrahn. 1988. Foraging population and territory of the formosan subterranean termite (Isoptera: Rhinotermitidae) in an urban environment. *Sociobiology* 14: 353-359.

Sudd, J. H. 1957. Communication and recruitment in pharaoh's ant. *Brit. J. Animal Behaviour* 5: 104-109.

_____ . 1960. The foraging method of pharaoh's ant, *Monomorium pharaonis* (L.) *Animal Behaviour* 8: 67-75.

Ware, G. W. 1983. *Pesticides, theory and application.* W. H. Freeman and Co., N.Y.

Williams, D. F. 1990. Effects of fenoxycarb baits on laboratory colonies of the pharaoh's ant, *Monomorium pharaonis.* Pp. 671-683. In: R. Vander Meer, K. Jaffe and A. Cedeno [eds.]. *Applied myrmecology, a world perspective.* Westview Press, Boulder, CO.

Williams, D. and K. Vail. 1993a. The Pharaoh ant (Hymenoptera: Formicidae): fenoxycarb baits affect colony development. *J. Econ. Entomol.* (in press).

_____ . 1993b. Control of a natural infestation of the Pharaoh ant (Hymenoptera: Formicidae) with a corn grit bait of fenoxycarb. *J. Econ. Entomol.* (in review).

Wilson, E. O. and R. W. Taylor. 1967. The ants of Polynesia. *Pac. Insects Mon.* 14: 1-109.

21

Impact of the Invasion of *Solenopsis Invicta* (Buren) on Native Food Webs

S. Bradleigh Vinson

Introduction

The genus *Solenopsis*, known as fire ants, appears to have evolved and radiated in South America. Their populations are reasonably low in their homeland (Fowler et al. 1990, Porter et al. 1992) where they appear to be integrated into the local food chain and, as such, represent a minor component. However, some *Solenopsis* species have been successful in invading new ecosystems, and in doing so, appear to have left their numerous parasites, pathogens and predators behind (Jouvenaz et al. 1980, 1981; Williams 1980). In these new environments their population has far exceeded those in their native ecosystem and can no longer be considered a minor component. But in order to document and understand the impact that the *Solenopsis* invasion has and is having on native organisms requires an understanding of the invader species involved, their spread, density, and the kinds of impact they can exert.

The Spread and Density of *Solenopsis* in North America

The invasion of the *Solenopsis* complex into North America has occurred in a series of stages and today there are four major species recognized (Hung et al. 1977). Wilson and Brown (1958) and Brown (1961) suggest that *Solenopsis geminata* (F.) was probably introduced into

North America during pre-Columbian times and it is likely that *Solenopsis xyloni* McCook was also introduced during some time in the distant past. However, populations of these so called "native *Solenopsis*" in modern times were relatively low. In Texas prior to the invasion of *Solenopsis invicta,* populations of *S. xyloni* or *S. geminata* were generally less than 8 mounds per hectare. Today both *S. xyloni* and *S. geminata* are absent in areas invaded by the more recent *Solenopsis* introductions (Hung and Vinson 1978, Moody et al. 1981). Two modern introductions have occurred. *Solenopsis saevissima richteri* Forel was first reported in North America in 1918 by Creighton (1930) and is believed to have originated from southern Brazil, Uruguay and northeast Argentina (Buren et al. 1974). This ant is presently known as *Solenopsis richteri* Forel (Buren 1972). The second modern introduction is believed to have occurred some time in the 1930s in the same area, Mobile, Alabama, as *S. richteri,* and is currently known as *Solenopsis invicta* Buren (Buren 1972). It is believed that *S. invicta* originated from the Pantanal in the state of Mato Grosso, Brazil (Lennartz 1973, Whitcomb 1980, Williams et al. 1975).

Populations of *S. invicta* were considered monogynous (having a single queen) and reached densities of 12 to 32 mounds per hectare in the 1950s (Green 1967), although monogyne densities of over 2500 mounds per hectare were reported (Green 1967, Markin et al. 1973). High densities of monogyne colonies are frequently encountered in newly infested land due to a large number of new colonies, most of which do not survive as neighboring colonies mature (Lofgren et al. 1975). Glancey et al. (1973) reported the presence of multiple-inseminated queens in *S. invicta* colonies and by the 1980s polygyne (multiple queen) colonies of *S. invicta* were reported from Texas to Florida (Fletcher et al. 1980, Glancey et al. 1987, Hung et al. 1974, Mirenda and Vinson 1982).

Densities of polygynous colonies reach 250 mounds per hectare (Greenberg et al. 1985) and reports of over 1000 mounds per hectare occur (Lammers 1987). Unlike the temporary high densities of monogyne colonies, the high densities of polygyne are stable over years (Greenberg et al. 1992). While high mound densities are sometimes reported (Eshelman et al. 1989), whether these are due to new small mounds in an area newly invaded by monogyne or due to polygyne colonies cannot be determined on mound density alone.

The origin of the polygynous colony is not clear. Ross et al. (1985) claimed that the two forms are genetically part of a single population; however, there are numerous differences. Polygynous colonies differ from monogynous colonies in having: (1) greater percentage of smaller workers and absence of large major workers (Greenberg et al. 1985), (2) the ability to produce a high percentage of aspermic males (Hung et al. 1974), (3) queens that are less physogastric and lay fewer eggs per queen (Fletcher et al. 1980), (4) workers that are less aggressive (Mirenda and Vinson 1982), (5) greater mound density and movement of workers from one mound to another (Bhatkar and Vinson 1987, Mirenda and Vinson 1982), (6) spread by budding (Porter et al. 1988), (7) differences in venom and hydrocarbon patterns (Greenberg et al. 1990), and (8) differences in isoenzyme patterns (Dunton et al. 1991). Although many of these differences are secondarily due to the biology of the polygynous form, some may be unique. It is not yet clear whether the polygynous form evolved in North America or represents another introduction. Polygynous colonies have been observed in South America (H. Fowler, pers. comm.; K. Ross, pers. comm.). Regardless of the origin, the polygynous form can be considered another invasion, at least in the impact of their increased density.

To understand the impact of *Solenopsis* on food webs and the ecology of the land that the ant has invaded, it is important to recall the density of *Solenopsis* in different areas before and after each of the different invasions. The southeast United States before 1918 was probably infested with both *S. xyloni* and *S. geminata* from the Carolinas through Texas with a low mound density (20-50 mounds/ha). The invasion of *S.*

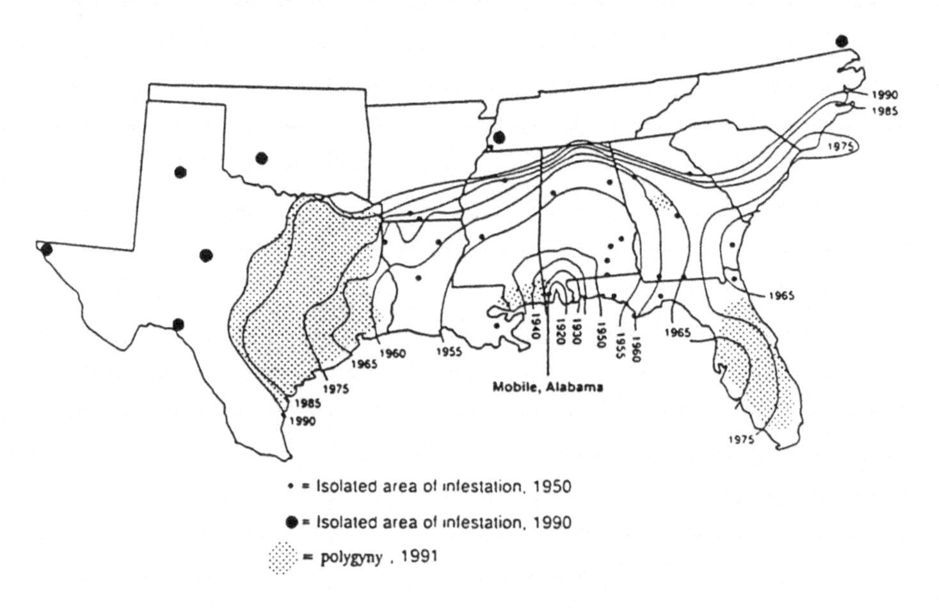

FIGURE 21.1 The spread of the imported fire ant (*Solenopsis richteri* and *Solenopsis invicta*) since their introduction into Mobile, Alabama, in 1918. (Adapted from Vinson 1991a.).

TABLE 21.1. Density of *Solenopsis* species in Brazos Co., Texas, over 15 years.

Date	*Solenopsis*	*Mounds/ha.*	*Reference*
1975	Native	6	Summerlin et al. 1977a
1976	*S. invicta*-SQ	50	Summerlin et al. 1977a
1984	*S. invicta*-SQ	82	Vinson unpub. data
1986	*S. invicta*-MQ	1635	Drees and Visnon. 1990
1990	*S. invicta*-MQ	1057	Greenberg et al. 1992

SQ=monogynous colonies MQ=polygynous colonies ha =hectare

richteri and *S. invicta* began in 1918 and the spread of this complex has been rapid (Figure 21.1). As *S. invicta* spread, *Solenopsis* mound densities increased to 100 to 200 mounds/ha. Although polygynous colonies were reported from several locations in the mid 1970s, the available evidence from published reports and accounts suggest that the spread through 1975 was primarily due to the monogynous form. However, during the 1980s the polygynous form rapidly spread through Texas occupying much of the state by 1990 (Porter et al. 1991). The spread of the polygynous form has been accompanied by an increase in the density of mounds and ants in the already infested regions of North America (Figure 21.1). For example, in College Station, TX there has been a steady increase in the density of *Solenopsis* since 1975 from as few as 6 mounds per hectare consisting of *S. xyloni* and *S. geminata*, increasing to 80 mounds as the native species were replaced by monogyne *S. invicta* to over 1500 today consisting of the polygynous form of *S. invicta* (Table 21.1).

Impact of Invading Species

Organisms can have several different effects as they invade an ecosystem. These effects are at least partly determined by the mechanisms of their invasion (Diamond and Case 1986), which may result from competition, predation or disease. Whether an invader has an impact upon the community it invades also depends on whether it invades a so-called "empty niche," thereby having little impact, or whether it displaces one or many species from occupied niches (Walker and Valentine 1984, Herbold and Moyle 1986). Even a simple addition to an ecosystem may increase some competition for food, space, or other resources, as well as providing another resource for resident organisms to exploit. In fact, a simple replacement of one species for another may have less of an impact. However, many invasions have more serious consequences and it is difficult to predict which species will invade and their effects (Parsons 1983, Ehrlich 1986, Crawley 1987).

What are some attributes of *Solenopsis* that have provided for its rapid expansion in North America? Tschinkel (1986) likened *Solenopsis invicta* to a weed because of its high reproductive potential and rapid development, particularly in disturbed habitats. But considering *Solenopsis invicta* a weed tells very little about its invasive characteristics and effects it may exert on the ecosystem. Characteristics such as their competitive abilities, omnivorous feeding habits, and ability to shift from one resource to another (opportunistic behavior) may be equally important to their high reproductive potential. Furthermore, the apparent absence in areas they invade of effective pathogens and parasites (Jouvenaz et al. 1977), the many biological control agents from the ants native homeland (Jouvenaz et al. 1980) which appear to have failed to follow the ants invasion, along with their wide choice of food (Hays and Hays 1959), and their effective competitiveness (Buren et al. 1978), contribute to their success. Their ability to sting also provides *Solenopsis* with the ability to repel even larger vertebrate competitors from resources.

Competitiveness

Competition between various species of ants is well known (Ward 1987, Savolainen and Vepsäläinen 1988), and competition between *Solenopsis* and other ant species at food resources has been documented (Bhatkar et al. 1972, Banks and Williams 1989). Because *S. invicta* is an efficient forager-recruiter, the workers dominate most available food sources (Horton et al. 1975, Kidd and Apperson 1984, Phillips et al. 1986). Even such tramp species as *Linepithema humile [=Iridomyrmex humilis]* are replaced by *S. invicta* (Long et al. 1987). Some species such as *Monomorium minimum* can compete by emitting a repellent liquid (Baroni-Urbani and Kannowski 1974), although *S. invicta* gradually overcomes the repellent effects due to their large numbers recruited to any substantial food source (Howard and Oliver 1979). Similarly, *Lasius neoniger* can sometimes effectively compete initially with *S. invicta* (Apperson and Powell 1984), but *L. neoniger* is also overcome by large numbers of *S. invicta* (Bhatkar et al. 1972). Whitcomb et al. (1972) reported that *S. invicta* simplified the agroecosystem by displacing other ant species, and as *S. invicta* has spread across the U.S., it has displaced both *S. xyloni,* and *S. geminata* (Wilson and Brown 1958, Roe 1973, Hung and Vinson 1978, Wojcik 1983, Porter et al. 1988). Tschinkel (1986), in referring to *S. invicta* as a weed, suggested that environmental disturbance favors *S. invicta* over *S. geminata* (Tschinkel 1988). In fact, Porter et al. (1988) reported that the replacement of *S. geminata* was at a ratio of 6:1 resulting in a substantial increase in *Solenopsis*. It is important to recall that the Porter et al. (1988) study involved the polygynous form as opposed to the reports of Hung and Vinson (1978) or Roe (1973). Furthermore, *S. invicta* also replaces many species in other ant genera (Wilson 1951, Wilson and Brown 1958, Glancey et al. 1976, Camilo and Phillips 1990, Porter and Savignano 1990).

Yet, there is also little doubt that a number of ant species remain in the ecosystem at least after the invasion of the monogyne form of *S. invicta*. Summerlin et al. (1977a) reported that *Conomyrma insana*, along with *S. invicta*, rapidly reinvaded after a mirex treatment and one year post-treatment only *S. geminata*, *Iridomyrmex pruinosum*, *Pheidole dentata*, and *Monomorium minimum* appeared to have failed to recover to pre-treatment levels. Although several other authors have documented the displacement of other ant genera by *Solenopsis* (Glancey et al. 1976), one of the more revealing studies is provided by Camilo and Phillips (1990). They examined the density of *S. invicta* and correlated this density to the number of coexisting native ant species. They found a negative correlation. The results showed 21 ant species in undisturbed, uninfested plots, and 14 ant species in disturbed uninfested plots; but with the presence of *S. invicta* only 9 species existed in low density *S. invicta* plots and 5 species occurred in high density plots. These results suggest that *S. invicta* is not only capable of replacing ants in similar niches but is able to outcompete many other species. Yet, a few ant species do remain. For example, *C. insana* and *M. minimum* in the Camilo and Phillips (1990) study were positively correlated with increasing *S. invicta* populations, while several species of *Pheidole* were negatively correlated.

Solenopsis invicta is a predator and as such is not only in competition with other ants, but is presumably in competition with other predators. However, *S. invicta*'s ability to eliminate potential predatory competitors by direct predation or resource exclusion has not been thoroughly examined. Sterling et al. (1979) did not detect any effect of *Solenopsis invicta* on spider or entomophagous adult insect abundance. Such data suggests that these insects can escape predation. Whether their reproductive fitness is reduced and whether there is any long-term impact on future generations through reduced prey abundance is difficult to demonstrate.

Food resources

Ants (*Formicidae*) are, in general, omnivorous organisms that feed on any available material in the environment, whether animal or vegetable

(Tevis 1958, Went et al. 1972). There are exceptions and some ant species are restricted in their food choice. For example, Army ants *(Ecitoninae* and *Dorylinae)* are generalized predators (Gotwald 1978, Rettenmeyer et al. 1983) while dacetine ants tend to be more specific predators (Brown and Wilson 1959). A number of species are seed predators and even specialize in seeds (Beattie 1985) or are restricted to feeding on certain plants or plant parts (Janzen 1965, Cherrett 1986). However, most of the more serious ant invader species such as *Linepithemia humile* (Markin 1970), *Wasmannia auropunctata* (Lubin 1984, Ulloa-Chacon and Cherix 1990) or *Paratrechina fulva* (Zenner-Polania 1990) are generalists in their food choices. *Solenopsis* are particularly aggressive generalist (Lofgren 1986) and along with their high densities, they can dominate most potential food sources. These include active predation on invertebrates, vertebrates, and plants; and they can be aggressive scavengers.

Predation (invertebrates). *Solenopsis* are effective invertebrate predators (Lofgren et al. 1975, Risch and Carroll 1982) and have attracted attention because they have been found to reduce many pest insect species. These include the sugarcane borer, *Diatraea saccharalis* (Long et al. 1958, Reagen et al. 1972), cotton pests such as *Heliothis virescens* and *Anthonomous grandis* (Sterling 1978, Sterling et al. 1979, McDaniel and Sterling 1979, 1982), ticks (Harris and Burns 1972, Burns and Melancon 1977), and several species of pest Diptera (Summerlin et al. 1977, 1977b, Howard and Oliver 1978, Summerlin and Kunz 1978, Combs 1982).

In addition to the many pest species attacked (see Lofgren 1986), Wilson and Oliver (1969) reported that *S. invicta* colonies in a Louisiana pasture and nearby forest collected a wide range of food items. Unidentified fragments made up 44-55% of the material collected, with the rest consisting largely of arthropod material although annelids made up from 4.6-8.4% and plant seed made up generally less than 2%. In one area lepidopterous larvae represented nearly 5% of the food collected while in another area they represented less than 1%. These differences may represent both availability and the ability of the ants to capture the prey. Sterling et al. (1979) considered *S. invicta* to be an efficient predator of pest species, without having as much effect on beneficials. This may reflect the low density of beneficial insects, or the increased difficulties in capturing the generally more agile adults over other stages. Some adult beneficials are able to escape the activities of *Solenopsis* (Vinson and Scarborough 1989), but this does not mean that all beneficials are more resistant to fire ant predation. Fire ant predation on immatures may or may not affect adult populations. For example, *Solenopsis* impact the eggs of *Heliothis* (McDaniel and Sterling 1979) but whether they affect adult moth abundance is not known. The fire ants may have just replaced other natural mortality factors. Ricks and

Vinson (1970) showed that, given an equal opportunity, fire ants consumed most insects presented to them, although some were preferred over others. The results support the view that *Solenopsis invicta* will attack almost any invertebrate they can subdue or scavenge. Thus, while *Solenopsis* may not impact agile adult beneficials or other agile adults, they may have an impact on other stages. Lopez (1982) reported that the cocoons of *Cardiochiles nigriceps,* a parasitoid of the cotton pest *Heliothis virescens* were destroyed by *S. invicta* and Nordlund (1988) reported that *S. invicta* reduced the effectiveness of *Trichogramma* released for pest control because *Solenopsis* apparently consumed the parasitized eggs. Vinson and Scarborough (1991) reported that fire ants preferred parasitized aphids over unparasitized ones as food. Such preferences could have a profound impact on biological control agents.

Tedders et al. (1990) found that *S. invicta* would forage to 9m heights in pecan trees, *Carya illinoensis,* where it tended the blackmargined aphid, *Monellia caryella,* deriving honeydew as an important food source. Although *S. invicta* predated on eggs, larvae and pupae of the aphid predators *Chrysoperla fuflabris,* a green lacewing, and pupae of the syrphid, *Allogropta obligua,* outbreaks of the blackmargined aphid on pecans could not be correlated to *Solenopsis* levels in the trees. Ants also did not affect adults or eggs of the aphid predator *Hippodamia convergens* (Tedders et al. 1990), and Vinson and Scarborough (1989) provided evidence that many predators of aphids have defenses against ants like *Solenopsis.* While *S. invicta* did not promote the pest status of the blackmargined aphid on pecans, the fire ant did promote the pest status of another honeydew providing pest, *Dysmicoccus morrisoni* (Tedders et al. 1990). The overall result of the presence of *Solenopsis* in an ecosystem is difficult to predict because there is a decline in predator (Howard and Oliver 1978) and parasite populations (Tedders et al. 1990, Vinson and Scarborough 1991) in addition to pests (Dutcher and Sheppard 1981). But the decline probably extends to many species of arthropods, many of which serve as food for many other animals.

Predation (vertebrates). Mount et al. (1981) provided experimental evidence that *S. invicta* attacked eggs of at least one species of lizard. They also reported predation on eggs of two turtle species and speculated that local declines in some lizard, snake and ground nesting bird populations were attributable to invasion by *S. invicta.* Sikes and Arnold (1986) examined cliff swallows, *Hirundo pyrrhonota,* nesting in drainage culverts in central-east Texas and noticed no effects from 1974 to 1981, although *S. invicta* was present in the region (Summerlin and Green 1977). However, as the density of *S. invicta* increased in the immediate area during the period from 1982 to 1983, Sikes and Arnold (1986) reported a 34% decrease in swallow nesting success. Ridlehuber (1982) reported that *S. invicta* preyed on wood duck, Aix *sponsa,* nestlings and speculated that *S. invicta* were also excluding wood ducks from natural-cavity nest sites. Similar reduction in vertebrate abundance has frequently been attributed to *Solenopsis* (Travis 1938, Kroll et al. 1973, Parker 1977).

The suggestion that the impact of *S. invicta* may be manifested by exclusion of organisms from their normal habitat utilization was suggested by Mount (1981) and supported by work of Smith (1988) and Killion et al. (1991). Both authors present evidence that small mammals avoid areas inhabited by *S. invicta.* Killion et al. (1990) further observed that burrow entrances used by small mammals were less likely to attract and recruit foraging ants as opposed to other areas and suggested that the small mammals can detect and utilize areas of lower ant activity. If mammals avoid high density *S. invicta* populations, then the impact of *Solenopsis* would go beyond resource competition, forcing the mammal onto a smaller resource base while leaving these same resources available in the ant dominated areas for exploitation by other species.

There are a number of studies where *Solenopsis* spp. were not considered to have an impact on native fauna (Emlen 1938, Johnson 1961, Kroll et al. 1973, Delnicki and Bolen 1977). These studies often involved very low *Solenopsis* densities which may explain, in part, the failure to obtain detrimental effects. In low density situations the organism under attack can move and locate a temporary refuge. The importance of this is illustrated by the reports of live trap deaths of small mammals due to *Solenopsis* because the small mammals were unable to escape the ants in the trap (Chabreck et al. 1986, Mosser and Grant 1986, Flickinger 1989). Further, *Solenopsis* may indirectly affect vertebrates by reducing abundance of food (Wilson 1951, Harris and Burns 1972, Porter and Savignano 1990), thus reducing the carrying capacity of the area for the organisms in question, a process that, as suggested by Mount (1981), may take 10-20 years to be manifest.

Plants. *Solenopsis,* particularly members of the *S. geminata* complex, are considered effective seed predators (Carroll and Risch 1984). Early reports of problems associated with the invasion of *S. invicta* indicated that this species was also damaging crops (Wilson and Eads 1949, USDA 1958). However, during the 1960s and 1970s, as *S. invicta* was rapidly expanding its range, several scientific committees evaluating the problem all concluded that *S. invicta* was not a serious row or cultivated crop problem (Bellinger et al. 1965, Mills 1967, Anonymous 1976). This view has changed (Lofgren 1986); and today *S. invicta* is considered to be a serious seed predator (Drees et al. 1991), fire ants attack and tunnel through roots and tubers of potatoes, sunflowers, cucumbers (Boock and Lordello 1952; Adams et al. 1983, 1988; Stewart and Vinson 1991), feed on plants and fruit of soybeans, okra, and eggplants (Lofgren et al. 1975), and girdle young citrus trees (Brown 1982).

Although the impact of fire ants on plants can be considered an increasing problem in agriculture, their impact on plant assemblages in natural ecosystems may be more serious. The ants habit of moving and predating on seeds (Vinson 1972, Drees et al. 1991), reduces the availability of seeds to other seed predators; alters the ratios of the various seeds, as well as alters the distribution of seeds available to develop. Such changes can have a major impact on the ecosystem, but changes in plant assemblages have not been documented or the effects evaluated.

The Effect of *Solenopsis* on Ecological Systems

In order to really understand the effects that a generalist invader has on an ecosystem, we must do more than determine the invader's effect on one or two different species. Although we know that the polygynous form of *S. invicta* is able to replace native ants and monopolize the food utilized by an entire guild (Porter and Savignano 1990) such data does not provide a picture of the total impact that such an invasion may cause. Based on the biomass of replaced ants and other food items, the 20-30 fold increase in *Solenopsis* biomass could not be accounted for, and Porter and Savignano (1990) speculate the *Solenopsis* may divert food from scavengers or find unused food resources. Vinson (1991b) examined the effects of *S. invicta* on a plant decomposing arthropod community. The results demonstrated that fire ants not only prevented colonization of the resource (rotting fruit) by the typical decomposing arthropod community and its guild of parasites and predators, but if the decomposing community was successful in establishing before the arrival of the *Solenopsis*, the *Solenopsis* were effective in utilizing the members of the decomposing community for food, as well as the decomposing resource itself. Summerlin et al. (1984) reported that *Solenopsis* reduced oviposition by adults and predated on larvae of the horn fly, *Haematobia irtitans* (L.), occurring in bovine dung, but coprophagous scarabs were not affected. Whether the *Solenopsis* also consumed components of the dung in competition with the coprophagous scarabs or other dung decomposing organisms was not determined. There is little doubt that *Solenopsis* impacts the community in many ways. For example, Hooper (1976) reported that isopod density in an *S. invicta* infested area was reduced while crickets, *Gryllus* spp., were not affected. It is clear that the relationships within a particular ecosystem can be complex.

Based on the high densities of ants, estimated in excess of 25-50 million per hectare, the data suggest that S. *invicta* may replace many invertebrate predators, scavengers and decomposers, and impact the assemblages of plants through seed predation and movement. The ant's ability to sting may result in the exclusion of many larger organisms (reptiles, birds, and mammals) from vital resources. Such changes would not only result in simplification of the environment, with *Solenopsis* becoming one of the dominant organisms within the system, but the invading ant becomes a keystone species, possibly the dominant population regulator. What effects these changes may have on the long-term evolution of the system, on the system's susceptibility to further invasion, or the susceptibility of the system to collapse due to its domination by a single species, may require years to determine.

Equally serious is that a species such as the polygynous form of S. *invicta* may have already altered the ecological relationships and balance to an extent that its elimination could lead to serious outbreaks of species previously suppressed by a complex of organisms which *Solenopsis* has surplanted. Although we presently appear powerless to deal such a blow, the answer we should be seeking may be in a slower replacement of *Solenopsis* with other ant species as proposed by Buren et al. (1978) in conjunction with biological control agents. Until these approaches are found and developed, we will continue to be forced into utilizing the various pesticides available (Drees and Vinson 1989, 1991) in efforts to manage the problem where it is serious enough to warrant such tactics.

Acknowledgments

Approved as TA 26048 by the Director of the Texas Agricultural Experiment Station. Supported in part by legislative funds for imported fire ant research for the State of Texas. Appreciation is extended to Dr. L. Greenberg for editorial comments and to D. Sennett for secretarial skills.

Resumen

Las especies de *Solenopsis* han invadido Norte America varias veces. Se presume que muchas de estas invasiones ocurrieron antes del descubrimiento de America. Las introducciones mas recientes, desde 1900, han dado como resultado un incremento 50 a 75 veces mayor en la densidad de *Solenopsis*, las cuales han ocurrido en diversos estados. Se revisa el impacto que ha tenido esta invasión en la flora y fauna. Los datos obtenidos sugieren que las especies de hormigas vagabundas invasoras han tenido un efecto devastador en el ecosistema el cual requeriría varios años antes de ser observado completamente. Mas aún, una vez que la hormiga haya invadido y se haya convertido en una fuerza dominante del ecosystema, su eliminación puede conducir a otras invasiones o a una explosión de la población.

References

Adams, C. T., W. A. Banks, and C. S. Lofgren. 1988. Red imported fire ant (Hymenoptera: Formicidae) correlation of ant density with damage to two cultivars of potatoes (*Solanum tuberosum* L.). *J. Econ. Entomol.* 81: 905-909.

Adams, C. T., W. A. Banks, C. S. Lofgren, B. J. Smittle, and D. P. Harlan. 1983. Impact of the red imported fire ant, *Solenopsis invicta* (Hymenoptera: Formicidae), on the growth and yield of soybeans. *J. Econ. Entomol.* 76: 1129-1132.

Anonymous. 1976. Fire ant control. Council for Agricultural Sciences and Technology. CAST Report 62, 2nd Ed. Rep. No. 65. Iowa State Univ., Ames.

Apperson, C. S., and E. E. Powell. 1984. Foraging activity of ants (Hymenoptera: Formicidae) in a pasture inhabited by the red imported fire ant. *Florida Entomol.* 67: 383-393.

Banks, W. A., and D. F. Williams. 1989. Competitive displacement of *Paratrechina longicornis* (Latreille) (Hymenoptera: Formicidae) from baits by fire ants in Mato Grosso, Brazil. *J. Entomol. Sci.* 24: 381-391.

Baroni-Urbani, C., and P. B. Kannowski. 1974. Patterns in the red imported fire ant settlement of a Louisiana pasture: some demographic parameters, interspecific competition and food sharing. *Environ. Entomol.* 3: 755-760.

Beattie, A. J. 1985. *The evolutionary ecology of ant-plant mutualisms.* Cambridge Univ. Press, England.

Bellinger, F., R. E. Dyer, R. King, and R. B. Platt. 1965. A review of the problem of the imported fire ant. *Bull. Georgia Acad. Sci.* 23: 1-22.

Bhatkar, A. P., and S. B. Vinson. 1987. Colony limits in *Solenopsis invicta* Buren. Pp. 599-600. In: J. Eder and H. Rembold [eds.]. *Chemistry and Biology of Social. Insects.* Verlang J. Peperny, München.

Bhatkar, A., W. H. Whitcomb, W. F. Buren, P. Callahan, and T. Carlysle. 1972. Confrontation behavior between *Lasius neoniger* (Hymenoptera: Formicidae) and the imported fire ant. *Environ. Entomol.* 1: 274-279.

Boock, O. J., and L. G. E. Lordello. 1952. Formiga "Lava-pe" praga de batatinha-*Solanum tuberosum* L. *Rev. Agric. (Piracicaba)* 27: 377-379.

Brown, R. 1982. IFA damage to young citrus trees. Paper read at Imported Fire Ant Conference, March 25-26, 1982, Austin, TX.

Brown, W. L., Jr. 1961. Mass insect control programs: Four case histories. *Psyche* 68: 75-109.

Brown, W. L., Jr., and E. O. Wilson. 1959. The evolution of Dacetine ants. *Quart. Review Biol.* 34: 278-294.

Buren, W. F. 1972. Revisionary studies on the taxonomy of the imported fire ants. *J. Georgia Entomol. Soc.* 7: 1-26.

Buren, W. F., G. E. Allen, W. H. Whitcomb, F. E. Lennartz, and R. N. Williams. 1974. Zoogeography of the imported fire ants. *J. New York Entomol. Soc.* 82: 113-124.

Buren, W. F., G. F. Allen, and R. N. Williams. 1978. Approaches toward possible pest management of the imported fire ants. *Bull. Entomol. Soc. Am.* 24: 418-421.

Burns, E. C., and D. G. Melancon. 1977. Effect of imported fire ant (Hymenoptera: Formicidae) invasion on lone star tick (Acarina: Ixodidae) populations. *J. Med. Entomol.* 14: 247-249.

Camilo, G. R., and S. A. Phillips, Jr. 1990. Evolution of ant communities in response to invasion by the fire ant *Solenopsis invicta*. Pp. 190-198. In: R.K. Vander Meer, K. Jaffe and A. Cedeno [eds.]. *Applied Myrmecology: a World Perspective.* Westview Press, Boulder, CO.

Carroll, C. R., and S. J. Risch. 1984. The dynamics of seed harvesting in early successional communities by a tropical ant, *Solenopsis geminata. Oecologia* 61: 388-392.

Chabreck, R. H., B. U. Constanin, and R. B. Hamilton. 1986. Use of chemical ant repellents during small mammal trapping. *Southwest. Natur.* 31: 109-110.

Cherrett, J. M. 1986. The economic importance and control of leaf-cutting ants. Pp. 165-192. In: S. B. Vinson [ed.]. *Economic impact and control of social insects.* Praeger Press, New York.

Combs, R. L., Jr. 1982. The black imported fire ant, a predator of the face fly in Northeast Mississippi. *J. Georgia Entomol. Soc.* 17: 496-501.

Crawley, M. J. 1987. What makes a community invisible? Pp. 429-453. In: A. J. Gray, M. J. Crawley, and P. J. Edwards [eds.]. *Colonization, Succession, and Stability.* Blackwell Publ., Oxford.

Creighton, W. S. 1930. The New World species of the genus *Solenopsis. Proc. Am. Acad. Arts Sci.* 66: 39-151, plates 1-8.

Delnicki, D. E., and E. G. Bolen. 1977. Use of black-bellied whistling duck nest sites by other species. *Southwest. Natur.* 22: 275-277.

Diamond, J., and T. J. Case. 1986. Overview: Introductions, extinctions, exterminations, and invasions. Pp. 65-79. In: J. Diamond and T. J. Case [eds.]. *Community Ecology.* Harper & Row, NY.

Drees, B. M., L. A. Berger, R. Cavazos, and S. B. Vinson. 1991. Factors affecting sorghum and corn seed predation by foraging red imported fire ants (Hymenoptera: Formicidae). *J. Econ. Entomol.* 84: 285-289.

Drees, B. M., and S. B. Vinson. 1989. Fire ants and their control. *Texas Agric. Extension Ser. B-1536, 20M-3-89-revised.* Texas A&M Univ., College Station, TX.

______ . 1990. Comparison of the control of monogynous and polygynous forms of the red imported fire ant (Hymenoptera: Formicidae) with a chlorpyrifos mound drench. *J. Entomol. Sci.* 25: 317-324.

______ . 1991. Fire ants and their management. *Texas Agric. Extension Ser. B-1536, 20-M-3-91-revised.* Texas A&M Univ., College Station, TX.

Dunton, R. F., S. B. Vinson, and J. S. Johnston. 1991. Unique isozyme electromorphs in polygynous red imported fire ant populations. *Biochem. System. Ecol.* 19: 453-460.

Dutcher, J. D., and D. C. Sheppard. 1981. Predation of pecan weevil larvae by red imported fire ants. *J. Georgia Entomol. Soc.* 16: 210-213.

Emlen, J. T. 1938. Fire ants attacking California quail chicks. *Condor* 40: 85-86.

Erhlich, P. R. 1986. Which animals will invade? Pp. 79-95. In: H. A. Mooney and J. A. Drake [eds.]. *Ecology of biological invasions of North America and Hawaii.* Springer-Verlag, NY.

Eshelman, B. D., G. N. Cameron, and E. N. N. Huynh. 1989. Density, dispersion, and persistence of fire ant mounds on the Texas coastal prairie. Pp. 14-18. In: *Proc. 1989 Fire Ant Research Conf.*, Biloxi, MS.

Fletcher, D. J. C., M. S. Blum, T. V. Whitt, and N. Temple. 1980. Monogamy and polygamy in the fire ant. *Ann. Entomol. Soc. Am.* 73: 658-661.

Flickinger, E. L. 1989. Observations of predation by red imported fire ants on live-trapped wild cotton rats. *Texas J. Sci.* 41: 223-224.

Fowler, H. G., J. V. E. Bernardi, and L. F. T. di Romagnano. 1990. Community structure and Solenopsis invicta in Sao Paulo. Pp. 199-209. In: R.K. Vander Meer, K. Jaffe and A. Cedeno [eds.]. *Applied Myrmecology: a World Perspective.* Westview Press, Boulder, CO.

Glancey, B. M., C. H. Craig, C. E. Stringer, and P. M. Bishop. 1973. Multiple fertile queens in colonies of the imported fire ant, *Solenopsis invicta. J. Georgia Entomol. Soc.* 8: 237-238.

Glancey, B. M., J. C. E. Nickerson, D. Wojcik, J. Trager, W. A. Banks, and C. T. Adams. 1987. The increasing incidence of the polygynous form of the red imported fire ant, *Solenopsis invicta* (Hymenoptera: Formicidae) in Florida. *Florida Entomol.* 70: 400-402.

Glancey, B. M, D. P. Wojcik, C. H. Craig and J. A. Mitchell. 1976. Ants of Mobile county, as monitored by bait transects. *J. Georgia Entomol. Soc.* 11: 191-197.

Gotwald, W. H. 1978. Trophic ecology and adaptation in tropical old world ants of the subfamily *Dorylinae* (Hymenoptera: Formicidae). *Biotropica* 10: 161-169.

Green, H. B. 1967. The imported fire ant in Mississippi. *Mississippi State Univ. Exp. Sta. Bull.* 737: 1-23.

Greenberg, L., D. J. C. Fletcher, and S. B. Vinson. 1985. Differences in worker size and mound distribution in monogynous and polygynous colonies of the fire ant, *Solenopsis invicta* Buren. *J. Kansas Entomol. Soc.* 58: 9-18.

Greenberg, L., S. B. Vinson, and S. Ellison. 1992. A 9-year study of a field containing both monogyne and polygyne red imported fire ants, *Solenopsis invicta* (Hymenoptera: Formicidae). *Ann. Entomol. Soc. Am.* 85: 686-695.

Greenberg, L., H. J. Williams, and S. B. Vinson. 1990. A comparison of venom and hydrocarbon profiles from alates in Texas monogyne and polygyne fire ants, *Solenopsis invicta.* Pp. 95-101. In: R.K. Vander Meer, K. Jaffe and A. Cedeno [eds.]. *Applied Myrmecology: a World Perspective.* Westview Press, Boulder, CO.

Harris, W. G., and E. C. Burns. 1972. Predation on the lone star tick by the imported fire ant. *Environ. Entomol.* 1: 362-365.

Hays, S. B., and K. I. Hays. 1959. Food habits of *Solenopsis saevissima richteri* Forel. *J. Econ. Entomol.* 52: 455-457.

Herbold, B., and P. B. Moyle. 1986. Introduced species and vacant niches. *Am. Natur.* 128: 751-760.

Hooper, M. W. 1976. The effects of the imported fire ant, *Solenopsis invicta,* on the East Texas arthropod community. Master Thesis, Univ. Texas, Austin.

Horton, P.M., S. B. Hays, and J. R. Holman. 1975. Food carrying ability and recruitment time of the red imported fire ant. *J. Georgia Entomol. Soc.* 10: 207-213.

Howard, F. W., and A. D. Oliver. 1978. Anthropod populations in permanent pastures treated and untreated with mirex for red imported fire ant control. *Environ. Entomol.* 7: 901-903.

Howard, F. W., and A. D. Oliver. 1979. Field observations of ants (Hymenoptera: Formicidae) associated with red imported fire ants, *Solenopsis invicta* Buren, in Louisiana pastures. *J. Georgia Entomol. Soc.* 14: 259-263.

Hung, A. C. F., M. S. Barlin, and S. B. Vinson. 1977. Identification, distribution and ecology of the fire ants of Texas. *Bull. Texas Agric. Exptl. Sta. B-1185.*

Hung, A. C. F., and S. B. Vinson. 1978. Factors affecting the distribution of fire ants in Texas (Hymenoptera: Formicidae). *Southwest. Natur.* 23: 205-213.

Hung, A. C. F., S. B. Vinson, and J. W. Summerlin. 1974. Male sterility in the red imported fire ant, *Solenopsis invicta. Ann. Entomol. Soc. Am.* 67: 909-912.

Janzen, D. H. 1965. Coevolution of mutualism between ants and acacias in central America. *Evolution* 20: 249-275.

Johnson, A. S. 1961. Antagonistic relationships between ants and wildlife with special reference to imported fire ants and bobwhite quail in the southeast. *Proc. Ann. Con. Southeast. Assoc. Game Fish Com.* 15: 88-107.

Jouvenaz, D. P., G. E. Allen, W. A. Banks, and D. P. Wojcik. 1977. A survey for pathogens of fire ants, Solenopsis spp., in the southeastern United States. *Florida Entomol.* 60: 275-279.

Jouvenaz, D. P., W. A. Banks, and J. D. Atwood. 1980. Incidence of pathogens in fire ants, *Solenopsis* spp., in Brazil. *Florida Entomol.* 63: 345-346.

Jouvenaz, D. P., C. S. Lofgren, and W. A. Banks. 1981. Biological control of imported fire ants: A review of current knowledge. *Bull. Entomol. Soc. Am.* 27: 203-208.

Kidd, K. A., and C. S. Apperson. 1984. Environmental factors affecting relative distribution of foraging fire ants in a soybean field on soil and plants. *J. Agric. Entomol.* 1: 212-218.

Killion, M. J., W. E. Grant, and S. B. Vinson. 1990. Influence of red imported fire ants *Solenopsis invicta* on small mammal habitat utilization. Pp. 43-44. In: M. E. Mispagel [ed.]. *Proc. 1990 Imported Fire Ant Conference.* College Station, TX.

Kroll, J. C., K. A. Arnold, and R. F. Gotic. 1973. An observation of predation by native fire ants on nestling barn swallows. *Wilson Bull.* 85: 478-479.

Lammers, J. M. 1987. Mortality factors associated with the founding queens of *Solenopsis invicta* Buren, the red imported fire ant: a study of the native ant community in central Texas. Ph. D. Dissert., Texas A&M Univ.

Lennartz, F. E. 1973. Modes of dispersal of *Solenopsis invicta* from Brazil into the continental United States: A study in spatial diffusion. M. S. Thesis. Univ. Florida, Gainesville, FL.

Lofgren, C. S. 1986. The economic importance and control of imported fire ants in the United States. Pp. 227-256. In: S. B. Vinson [ed.]. *Economic impact and control of social insects.* Praeger Press, New York.

Lofgren, C. S., W. A. Banks, and B. M. Glancey. 1975. Biology and control of imported fire ants. *Ann. Rev. Entomol.* 20: 1-30.

Long, W. H., E. A. Cancienne, E. J. Concienne, R. N. Dobson, and L. D. Newsom. 1958. Fire-ant eradication program increases damage by the sugarcane borer. *Sugar Bull.* 37: 62-63.

Long, W., L. Nelson, P. Templet, and C. P. Viator. 1987. Abundance of foraging ant predators of the sugarcane borer in relation to soil and other factors. *J. Am. Soc. Sugar Cane Tech.* 7: 5-14.

Lopez, J. D. 1982. Emergence pattern of an overwintering population of *Cardiochiles nigriceps* in central Texas. *Environ. Entomol.* 11: 838-842.

Lubin Y. D. 1984. Changes in the native fauna of the Galápagos Islands following invasion by the little red fire ant, *Wasmannia auropunctata. Biol. J. Linn. Soc.* 21: 229-242.

Markin, G. P. 1970. Foraging behavior of the Argentine ant in a California citrus grove. *J. Econ. Entomol.* 63: 740-742.

Markin, G. P., J. H. Dillier, and H. C. Collins. 1973. Growth and development of colonies of the red imported fire ant, *Solenopsis invicta. Ann. Entomol. Soc. Am.* 66: 803-808.

Masser, M. P., and W. E. Grant. 1986. Fire ant-induced trap mortality of small mammals in east-central Texas. *Southwest. Natur.* 31: 540-542.

McDaniel, S. G., and W. L. Sterling. 1979. Predator determination and efficiency on *Heliothis virescens* eggs in cotton using ^{32}P. *Environ. Entomol.* 8: 1083-1087.

McDaniel, S. G., and W. L. Sterling. 1982. Predation of *Heliothis virescens* (F.) eggs on cotton in east Texas. *Environ. Entomol.* 11: 60-66.

Mills, H. B. 1967. (Chairman) National Acad. Sci. Committee on imported fire ant. Report to Administrator, U. S. Dept. Agric., Agric. Res. Serv.

Mirenda, J. T., and S. B. Vinson. 1982. Single and multiple queen colonies of imported fire ants in Texas. *Southwest. Entomol.* 7: 135-141.

Moody, J. V., O. F. Franke, and F. W. Merickel. 1981. The distribution of fire ants, *Solenopsis* (Hymenoptera: Formicidae), in western Texas. *J. Kansas Entomol. Soc.* 54: 469-480.

Mount, R. H. 1981. The red imported fire ant, *Solenopsis invicta* (Hymenoptera: Formicidae), as possible serious predator on some native southeastern vertebrates: Direct observations and subjective impressions. *J. Alabama Acad. Sci.* 52: 71-78.

Mount, R. H., S. E. Trauth, and W. H. Mason. 1981. Predation by red imported fire ant, *Solenopsis invicta* (Hymenoptera: Formicidae), on eggs of the lizard *Cnemidophorus sexlineatus* (Squamata: Teiidae). *J. Alabama Acad. Sci.* 52: 66-70.

Nordlund, D. A. 1988. The imported fire ant and naturally occurring and released beneficial insects. Pp. 51-54. In: S. B. Vinson and J. G. Teer [eds.].The Imported Fire Ant: Assessment and Recommendations. Proc. Governor's Conference. Sportsmen Conservationists of Texas, Austin, TX.

Parker, J. W. 1977. Mortality of nestling Mississippi kites by ants. *Wilson Bull.* 89: 1-76.

Parsons, P. A. 1983. *The evolutionary biology of colonizing species.* Cambridge Univ. Press, England.

Phillips, S. A., S. R. Jones, and D. M. Claborn. 1986. Temporal foraging patterns of *Solenopsis invicta* and native ants of central Texas. Pp. 114-122. In: C. S.

Lofgren and R. K. Vander Meer [eds.]. *Fire ants & leaf-cutting ants: Biology & Management.* Westview Press, Boulder, CO.

Porter, S. D., A. Bhatkar, R. Mulder, S. B. Vinson, and D. J. Clair. 1991. Distribution and density of polygyne fire ants (Hymenoptera: Formicidae). Texas. *J. Econ. Entomol.* 84: 866-874.

Porter, S. D., H. G. Fowler, and W. P. McKay. 1992. Fire ant mound densities in the United States and Brazil (Hymenoptera: Formicidae). *J. Econ. Entomol.* 85: 1154-1161.

Porter, S. D., and D. A. Savignano. 1990. Invasion of polygyne fire ants decimates native ants and disrupts arthropod community. *Ecology* 71: 2095-2106.

Porter, S. D., B. Van Eimeren, and G. Lawrence. 1988. Invasion of red imported fire ants (Hymenoptera: Formicidae): Microgeography of competitive replacement. *Ann. Entomol. Soc. Am.* 81: 913-918.

Reagan, T. E., G. Coburn, and S. D. Hensley. 1972. Effects of mirex on the arthropod fauna of a Louisiana sugarcane field. *Environ. Entomol.* 1: 588-591

Rettenmeyer, C. W., R. Chadab-Crepet, M. G. Naumann, and L. Morales. 1983. Comparative foraging by neotropical army ants. Pp. 59-73. In: P. Jaisson [ed.]. *Social Insects in the Tropics Vol. 2.* Univ. Paris-Nord, Paris.

Ricks, B. L., and S. B. Vinson. 1970. Feeding acceptability of certain insects and various water-soluble compounds to two varieties of the imported fire ant. *J. Econ. Entomol.* 63: 145-148.

Ridlehuber, K. T. 1982. Fire ant predation on wood duck ducklings and pipped eggs. *Southwest. Natur.* 27: 222.

Risch, S. J., and C. R. Carroll. 1983. Effect of a keystone predaceous ant, *Solenopsis geminata*, on arthropods in a tropical agroecosystem. *Ecology* 63: 1979-1983.

Roe, R. L., II. 1973. A biological study of *Solenopsis invicta* Buren, the red imported fire ant, in Arkansas, with notes on related species. M. S. Thesis, Univ. Arkansas, Fayetteville, AR.

Ross, K. G., D. J. C. Fletcher, and B. May. 1985. Enzyme polymorphisms in the fire ant, *Solenopsis invicta* (Hymenoptera: Formicidae). *Biochem. System. Ecol.* 13: 29-33.

Savolainen, R., and K. Vepsäläinen. 1988. A competition hierarchy among boreal ants: impact on resource partitioning and community structure. *Oikos* 51: 135-155.

Sikes, P. J., and K. A. Arnold. 1986. Red imported fire ant *(Solenopsis invicta)* predation on cliff swallow *(Hirundo pyrrhonota)* nestlings in east-central Texas. *Southwest. Natur.* 31: 105-106.

Smith, T. S. 1988. The influence of red imported fire ants *(Solenopsis invicta)* on rodent habitation of an old-field community. M. S. Thesis, Texas A&M Univ., College Station, TX.

Sterling, W. L. 1978. Fortuitous biological suppression of the boll weevil by the red imported fire ant. *Environ. Entomol.* 7: 564-568.

Sterling, W. L., D. Jones, and D. A. Dean. 1979. Failure of the red imported fire ant to reduce entomophagous insect and spider abundance in a cotton agroecosystem. *Environ. Entomol.* 8: 976-981.

Stewart, J. W., and S. B. Vinson. 1991. Red imported fire ant damage to commercial cucumber and sunflower plants. *Southwest. Entomol.* 16: 168-170.

Summerlin, J. W., and L. R. Green. 1977. Red imported fire ant, a review on invasion, distribution and control in Texas. *Southwest. Entomol.* 2: 94-101.

Summerlin, J. W., A. C. F. Hung, and S. B. Vinson. 1977a. Residues in non-target ants, species simplification and recovery of populations following aerial applications of mirex. *Environ. Entomol.* 6: 193-197.

Summerlin, J. W., and S. E. Kunz. 1978. Predation of the red imported fire ant on stable flies. *Southwest. Entomol.* 3: 260-262.

Summerlin, J. W., J. K. Olson, R. R. Blume, A. Aga, and D. E. Bay. 1977b. Red imported fire ants: Effects on *Onthophagus gazella* and the horn fly. *Environ. Entomol.* 6: 440-442.

Summerlin, J. W., H. D. Petersen and R. L. Harris. 1984. Red imported fire ant (Hymenoptera: Formicidae): Effects on the horn fly (Diptera: Muscidae) and coprophagous scarabs. *Environ. Entomol.* 13: 1405-1410.

Tedders, W. L., C. C. Reilly, B. W. Wood, R. K. Morrison and C. S. Lofgren. 1990. Behavior of *Solenopsis invicta* (Hymenoptera: Formicidae) in pecan orchards. *Environ. Entomol.* 19: 44-53.

Tevis, L., Jr. 1958. Interrelations between the harvester ant *Veromessor pergandei* (Mayr) and some desert ephemerds. *Ecology* 39: 695-704.

Travis, B. V. 1938. The fire ant (*Solenopsis* spp.) as a pest of quail. *J. Econ. Entomol.* 31: 649-652.

Tschinkel, W. R. 1986. The ecological nature of the fire ant: some aspects of colony function and some unanswered questions. Pp. 72-87. In: C.S. Lofgren and R.K. Vander Meer [eds.]. *Fire ants & leaf-cutting ants: Biology & Management*. Westview Press, Boulder, CO.

Tschinkel, W. R. 1988. Distribution of the fire ant *Solenopsis invicta* and *S. geminata* (Hymenoptera: Formicidae) in northern Florida in relation to habitat disturbance. *Ann. Entomol. Soc. Am.* 81: 71-81.

Ulloa-Chacon, P., and D. Cherix. 1990. The little fire ant, *Wasmannia auropunctata (R.)* (Hymenoptera: Formicidae). Pp. 281-289. In: R.K. Vander Meer, K. Jaffe and A. Cedeno [eds.]. *Applied Myrmecology: a World Perspective*. Westview Press, Boulder, CO.

U.S.D.A. 1958. Observations on the biology of the imported fire ant. *USDA, ARS* 33-49.

Vinson, S. B. 1972. Imported fire ant feeding on Paspalum seed. *Am. Entomol. Soc. Am.* 65: 988.

_____. 1991a. Population growth of the Imported fire ant. Pp. 70-73. In: *Texas Environmental Guide*. Holt, Rinehart and Winston, Inc., and Harcourt, Brace Jovanovich, Inc., Austin, TX.

_____. 1991b. Effect of the red imported fire ant (Hymenoptera: Formicidae) on a small plant-decomposing arthropod community. *Environ. Entomol.* 20: 98-103.

Vinson, S. B., and T. A. Scarborough. 1989. Impact of the imported fire ant on laboratory populations of cotton aphid (Aphis gossyppii) predators. *Florida Entomol.* 72: 107-111.

_____. 1991. Interactions between *Solenopsis invicta* (Hymenoptera: Formicidae), *Rhopalosiphum maidis* (Hymenoptera: Aphididae), and the parasitoid *Lysiphlebus testaceipes* Cresson (Hymenoptera: Aphididae). *Ann. Entomol. Soc. Am.* 84: 158-164.

Walker, T. D., and J. W. Valentine. 1984. Equilibrium models of evolutionary species diversity and the number of empty niches. *Am. Natur.* 124: 887-897.

Ward, P. S. 1987. Distribution of the introduced Argentine ant *(Iridomyrmex humilis)* in natural habitats of the lower Sacramento Valley and its effects on the indigenous ant fauna. *Hilgardia* 55: 1-16.

Went, F. W., J. Wheeler, and G. C. Wheeler. 1972. Feeding and digestion in some ants *(Veromessor* and *Manica). BioScience* 22: 82-88.

Whitcomb, W. H. 1980. Expedition into the Pantanal. *Proc. Tall Timbers Conf. Ecol. Anim. Control Habitat Manage.* 7: 113-122.

Whitcomb, W. H., H. A. Denmark, A. P. Bhatkar, and G. L. Greene. 1972. Preliminary studies on the ants of Florida soybean fields. *Florida Entomol.* 55: 129-142.

Williams, R. N. 1980. Insect natural enemies of fire ants in South America with several new records. *Proc. Tall Timbers Conf. Ecol. Anim. Control Habitat Manag.* 7: 123-134.

Williams, R. N., M. de Menezes, G. E. Allen, W. F. Buren, and W. H. Whitcomb. 1975. Observacões ecológicas sobre a formiga lava-pé, *Solenopsis invicta* Buren. 1972 (Hymenoptera: Formicidae). *Rev. Agr. Piracicaba* 50: 9-22.

Wilson, E. O. 1951. Variation and adaptation in the imported fire ant. *Evolution* 5: 68-79.

Wilson, E. O., and W. L. Brown. 1958. Recent changes in the introduced population of the fire ant *Solenopsis saevissima* (Fr. Smith). *Evolution* 12: 211-218.

Wilson, E. O., and J. H. Eads. 1949. A report on the imported fire ant *Solenopsis saevissima* var. *richteri* Forel in Alabama. *Alabama Dep. Conserv. Spec. Rep.* 1-53, 13 plates [Mimeographed].

Wilson, N. L., and A. D. Oliver. 1969. Food habits of the imported fire ant in pasture and pine forest areas in southeastern Louisiana. *J. Econ. Entomol.* 62: 1268-1271.

Wojcik, D. P. 1983. Comparison of the ecology of red imported fire ants in North and South America. *Florida Entomol.* 66: 101-111.

Zenner-Polania, I. 1990. Biological aspects of the "Hormiga loca" *Paratrichina* (Nylanderia) *fulva* (Mayr), in Colombia. Pp. 290-297. In: R. K. Vander Meer, K. Jaffe and A. Cedeno [eds.]. *Applied Myrmecology: a World Perspective.* Westview Press, Boulder, CO.

22

Impact of Red Imported Fire Ants on the Ant Fauna of Central Texas

Rafael Jusino-Atresino and Sherman A. Phillips, Jr.

Introduction

Ants are present in almost all terrestrial habitats and occupy a great diversity of niches within communities. Interspecific interactions within these communities are the main factors that determine species distribution, abundance, and behavior. In undisturbed communities, different foraging strategies may evolve, enabling species to avoid interspecific competition. Marsh (1988) elaborated on this fact in his discussion of the activity patterns of eight species of ants in the Namib Desert of southwestern Africa. Seven of the eight species showed a change in their activity pattern among seasons. Furthermore, these differences became more apparent during the summer months when the activity of all the species was greater. He pointed out that these differences were mostly caused by abiotic factors such as navigational constraints and thermal preferenda, but furthermore suggested that other factors may be more important for the evolution of interspecific differences, such as aggression.

When an ant community is disturbed by an intruder ant species, activity patterns already established within the community may be disrupted, and new interspecific interactions within the community may develop. The amount of change in activity patterns may be directly related to the competitive ability of the new intruder and may become evident in several possible ways. The first possibility is the extinction of weaker competitors. An example can be found in the social organization and behavior of the boreal ants in Finland (Savolainen and Vepsalainen

1988). They categorized the ant fauna into three different groups: "territorials", "encounterers", and "submissives". Their study showed that encounterers and territorials were very competitive and could not coexist and, following Gause's principle, the encounterers became extinct when in the presence of territorials. The second possibility is for a weaker competitor to be displaced from the habitat rather than become extinct. Ward (1987) showed displacement of native ant species from the Sacramento Valley in California by the Argentine ant, *Linepithema humile* [=*Iridomyrmex humilis* (Mayr)], and Tschinkel (1988), as another example, showed the displacement of the fire ant, *Solenopsis geminata* (Fabricius), from highly disturbed habitats in Florida by the red imported fire ant, *Solenopsis invicta* Buren. The third possibility is avoidance of interspecific competition by the species remaining within the habitat. Fellers (1987) evaluated the effect of exploitation and interference on the foraging behavior of nine species of ants in a Maryland woodlot by first categorizing them into two groups: "dominants" and "subordinates". Her study demonstrated that species, both between and within these groups, overlapped substantially in size and type of prey taken, and that subordinate species significantly reduced their feeding time in the presence of other ant species. Also, Savolainen and Vepsalainen (1988) showed that submissives can coexist with territorials and encounterers, but only with a reduction in nest density and forager numbers. In addition, submissives shifted their diet from a protein to a carbohydrate source.

The red imported fire ant is a very good competitor, and when introduced into a new area, it becomes the dominant species in a short period of time. In 1990, Camilo and Phillips studied the influence of *S. invicta* on the ant community structure of central Texas. They showed that three species, the pyramid ant, *Conomyrma insana* (Buckley), the little black ant, *Monomorium minimum* (Buckley), and *Forelius pruinosus* (Roger) decreased in number of individuals at a slower rate than the other ants present within the community. However, whether these species maintained or modified their innate foraging activity was unknown. Based on their study, our objective was to determine if significant differences exist among species diversity and foraging activities of native Texas ants as a result of red imported fire ant invasion.

Materials and Methods

The study site was located within the Rolling Plains vegetational region. This region is part of the Southern Great Plains of the central United States that covers approximately 10 million ha. of mostly rangeland with gently rolling to moderately rough topography. Precipitation is highly variable with an average rainfall of 66 cm. The original prairie vegetation consisted of tall- and midsize-grasses, but now consists of short-, mid-, and tall-grasses and weedy plants such as mesquite, *Prosopis glandulosa* Torr. (Correll and Johnston 1970). All collections were made in a rangeland area of approximately 5 acres located at the southeastern edge of Lake Kirby in Abilene, Texas. Vegetation consisted of short-, mid-, and tall-grasses with mesquite trees to the east and west of the area.

The first part of this study was designed to determine the ant species present and the distribution of the red imported fire ant within the area. Pitfall traps were used to sample the ant fauna. They consisted of 473 ml plastic cups containing ethylene glycol (one third volume) as a preservative. Traps were placed in the ground for seven days every other week during the months of June through September of 1989 and were arranged in 23 parallel rows of four pitfall traps, each arranged along an east-west transect. The distance between rows was 20 m, and the distance between traps within rows was 10 m. Based on data obtained, the boundary between the infested (presence of *S. invicta*) and uninfested areas (absence of *S. invicta*) was established, and species were determined. The Shannon-Wiener species diversity index (Southwood 1978) was used to measure community complexity and organization based on species richness and evenness.

In the second half of the study, bait traps were utilized to monitor foraging activity of ant species within the *S. invicta* infested and the uninfested areas. Each bait trap consisted of a gel mixture (approx. 1.0 cm^3) of albacore and grape jelly (1:1 by wt.) within 30 ml clear, plastic cups. Data were obtained every second week during the months of June through September of 1989. Each collection day consisted of taking 600 samples by placing 25 bait cups randomly in each site (infested and uninfested) every two hours for a 24-hour period. Bait cups were left on the ground for 15 minutes before collection. Each sample date began at 1100 hrs and ended at 0900 hrs the next day. Ants were identified and counted to determine the frequency of individuals per species at each collection interval. A goodness of fit test or G-test (Sokal and Rohlf 1981) was used to compare the intraspecific frequency of foragers for each time period between infested and uninfested areas.

Results and Discussion

Pitfall trapping was the only method used to determine species diversity because our focal point was to identify ground foraging

TABLE 22.1. Ant species collected using pitfall traps within red imported fire ant infested and uninfested areas; June, July, August, and September 1989; Taylor Co., Texas.

Taxa	*Number Collected*	
	Uninfested	*Infested*
Subfamily: *Ecitoninae*		
Neivamyrmex nigrescens	60	36
Subfamily: Dolichoderinae		
Conomyrma flava	18	71
Conomyrma insana	27	42
Forelius pruinosus	212	151
Forelius foetidus	107	26
Subfamily: *Ponerinae*		
Leptogenys elongata elongata	2	7
Subfamily: *Myrmicinae*		
Crematogaster laeviuscula	263	204
Crematogaster punctulata	1	31
C. minutissima missuriensis	7	0
Monomorium minimum	540	128
Solenopsis invicta	28	3664
Solenopsis molesta	2	0
Pheidole dentata	267	95
Pheidole lamia	9	0
P. metallescens metallescens	43	3
Pogonomyrmex barbatus	33	14
Subfamily: *Formicinae*		
Camponotus vicinus	3	1
Camponotus festinatus	2	6
Formica schaufussi	11	72
Formica gnava	0	5
Paratrechina terricola	84	54

species that can be influenced by *S. invicta* and not the total ant fauna for the area. A total of 21 species was collected from pitfall traps (Table 22.1). Three species were collected only from the uninfested area: *Pheidole lamia* Wheeler, *Crematogaster minutissima missouriensis* Emery, and the thief ant, *Solenopsis molesta* (Say). A low number of individuals were collected for these three species, probably an indication of a small population within the original community that became extinct at the time of infestation. One species was exclusively collected from the infested area: *Formica gnava* Buckley. This species, a tree forager, was collected in pitfall traps only near mesquite trees and sunflower plants. Red imported fire ants do not forage in trees unless the amount of food available is almost depleted. This fact might explain why *F. gnava* can survive within the infested area. Seventeen species were collected from both areas. Most of the species common to both areas were adversely affected by *S. invicta,* with an intraspecific reduction in excess of 20% in number of individuals between infested and uninfested areas. The little black ant, *Monomorium minimum* (Buckley), was most affected with a reduction of 76%, and the least affected was *Crematogaster laeviuscula* Emery, with a reduction of 22% in number of individuals collected from the infested area. However, four species, *Conomyrma insana* (Buckley), *Conomyrma flava* (McCook), *Leptogenis elongata elongata* (Buckley), and *Formica schaufussi* Mayr were collected in larger numbers within the infested area. Of this group *C. insana,* as demonstrated by Camilo and Phillips 1990, decreased in number at a slower rate within the infested area. The latter two species forage on plants, and thus do not interfere with foraging by *S. invicta.* Species richness for the uninfested and infested areas was 20 and 18, respectively. The distribution of number of individuals within species per area was more uniform within the uninfested area. In the infested area 14 of the 18 species were collected in individual numbers under 100, whereas *S. invicta* was collected in excess of 3,000 individuals. Results of the Shannon-Wiener species diversity index significantly demonstrate that communities within the infested and uninfested areas were not equally diverse (cal. $C = 756$; $v = 1853$; $P < 0.001$).

A total of 18 ant species was collected from bait traps (Table 22.2). These species were divided into four groups depending on collection locality and consistency of occurrence. Three species occurred exclusively in the infested area: *Formica gnava* Buckley, *Camponotus festinatus* (Buckley), and *Pheidole lamia* Wheeler. Of this group, *F. gnava* and *C. festinatus* forage on plant material, and the grape jelly used in the baits might have attracted foragers of these species. *P. lamia* was represented by 14 individuals collected in one sample. It was never collected in pitfall traps within this area. Three other species were collected exclusively from the uninfested area: the red harvester ant, *Pogonomyrmex barbatus* (F. Smith), *Pheidole metallescens metallescens* Emery, and *Prenolepis imparis* (Say). These three species are ground foragers that indicates they were in competition with *S. invicta.* This probably indicates these species were displaced from the infested area. Nine species were common to both localities, and four of these species were collected in larger numbers within the infested area: *C. laeviuscula, F. schaufussi, C. flava,* and *C. insana.* The first two species are plant foragers and do not interfere with *S. invicta* unless food is almost depleted; whereas, the last two may withstand the presence of *S. invicta* for a longer period of time (Camilo and Phillips 1990). These nine

TABLE 22.2. Ant species collected from bait traps within red imported fire ant infested and uninfested areas; June, July, August and September 1989; Taylor Co., Texas.

	Number Collected	
Taxa	*Uninfested*	*Infested*
Subfamily: *Dolichoderinae*		
Conomyrma flava	195	918
Conomyrma insana	13	451
Forelius foetidus	1408	1244
Forelius pruinosus	14840	5436
Subfamily: *Myrmicinae*		
Crematogaster laeviuscula	4731	7730
Crematogaster punctulata	35	25
Monomorium minimum	19727	2562
Solenopsis invicta	91	86268
Leptothorax pergandei	2	1
Pheidole dentata	1306	77
P. metallescens metallescens	41	0
Pheidole lamia	0	14
Formica schaufussi	29	7275
Pogonomyrmex barbatus	547	0
Subfamily: *Formicinae*		
Camponotus festinatus	0	3
Formica gnava	0	13
Paratrechina terricola	3455	1594
Prenolepsis imparis	1	0

species were not collected on one or more of the collection days, or they were not found in one of the areas on one or more of the collection days. Because of this inconsistency they were not used for the statistical analysis. Two ant species were collected consistently throughout the study and were used for statistical analysis: *Monomorium minimum* and *Forelius pruinosus*.

The foraging activity of *M. minimum* (Figure 22.1A) began at 0500 hrs for both localities. The number of foragers within the uninfested area was relatively constant during the entire foraging period, with two peak activity periods at 1100 and 1700 hrs. In contrast, within the infested area, foraging was erratic throughout the activity period, with three distinct peaks at 0900, 1300, and 1700 hrs. The number of foragers within the infested area increased and decreased very rapidly before and after each activity peak. Foraging activity ended at 2300 hrs within the uninfested area; whereas, foraging activity in the infested area ended at 0300 hrs.

FIGURE 22.1A and B. Temporal foraging frequences for *M. miniumum* and *F. pruinosus* based on total number of ants collected using bait cups within red imported fire ant infested area (IN) and uninfested area (UN); June, July, August, and September 1989; Taylor Co., Texas.

A. Monomorium minumum

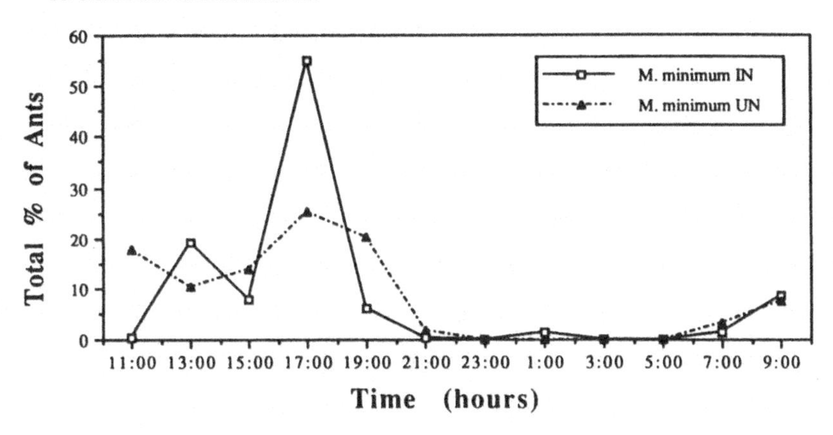

B. Forelius pruinosus

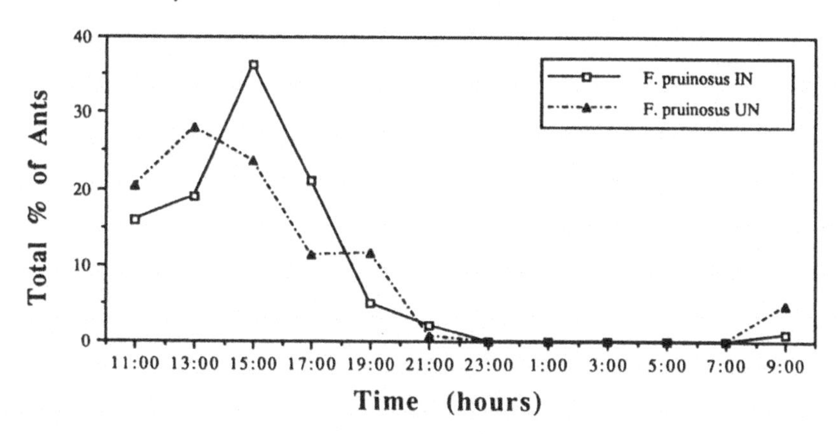

Foraging activity of *F. pruinosus* (Figure 22.1B) began at 0700 hrs in both areas. Within the uninfested area, this species showed greater foraging activity after the major activity peak at 1300 hrs; whereas, in the infested area it showed greater foraging activity before the major activity peak at 1500 hrs. For both localities foraging activity ended at 2300 hrs.

The goodness of fit test or G-test comparisons for the intraspecific frequency of foragers for each time between infested and uninfested areas indicates that foraging activities were highly dependent on locality

FIGURE 22.2A and B. Temporal foraging frequencies based on total number of ants collected using bait cups within red imported fire ant infested and uninfested areas; June, July, August, and September 1989; Taylor Co., Texas.

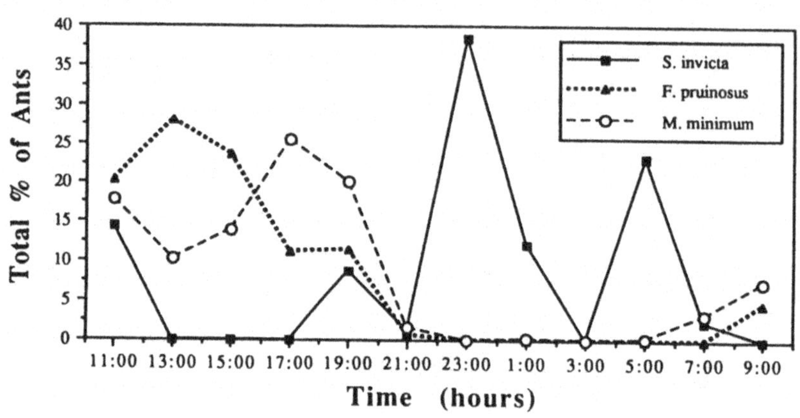

B. Infested Area

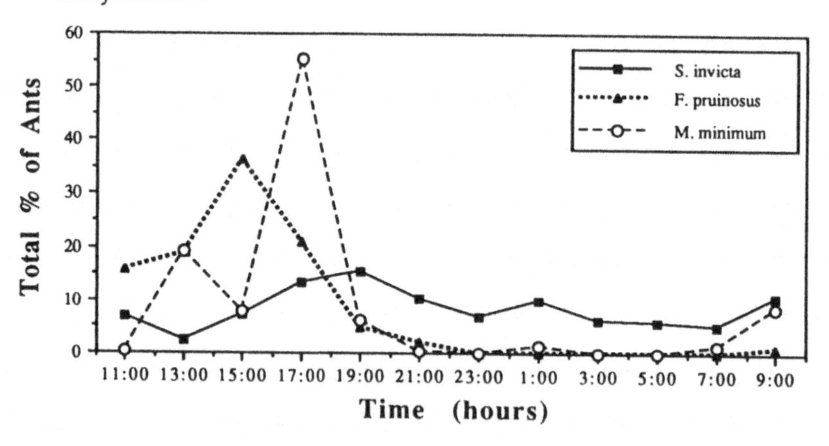

(d.f. = 11; $P < 0.001$). Within the uninfested area (Figure 22.2A), greater numbers of *M. minimum* foragers were collected during the cool temperatures of morning and evening. *F. pruinosus* was collected in greater numbers during the warmer part of the day. *S. invicta* was captured only when other species were not extensively foraging and only within the edge of the uninfested area nearest the infested area. Also, the total number of *S. invicta* specimens collected within the uninfested area was 91 individuals compared to 86,268 specimens collected within the infested area.

For the infested area (Figure 22.2B), the number of *S. invicta* foragers fluctuated minimally, with a noticeable drop only during the warmer part of the day (mid-afternoon). Possibly, because of the decrease in foraging of *S. invicta* during this activity period, *M. minimum* and *F. pruinosus* could increase their foraging activity to a maximum at 1700 and 1500 hrs, respectively. Compared to the uninfested area, the foraging time of these species was reduced, but an increased number of foragers may have compensated for the shorter foraging period. Later in the afternoon (1500 hrs), when *S. invicta* resumed its foraging activity, fewer *M. minimum* and *F. pruinosus* foragers were collected, and forager numbers decreased to zero once *S. invicta* again became the dominant forager (1900 hrs).

In conclusion, significant differences in foraging activity were found between the infested and uninfested areas, suggesting an adaptation that enables *M. minimum* and *F. pruinosus* to survive within the red imported fire ant territory. This finding, with that of Camilo and Phillips (1990), demonstrates that some species remaining within the *S. invicta* territory are adapting to the presence of the red imported fire ant and thus may be avoiding competition.

Resumen

Se ha demostrado que la hormiga brava importada, *Solenopsis invicta* Buren altera la estructura basica de comunidades ya establecidas por largo tiempo, pero casi no se conocen las caracterasticas de nicho de las especies que pueden permanecer dentro de las comunidades alteradas. Por esta razon se empezo un estudio para determinar si la actividad de forajeo a travez del tiempo de las especies de hormigas que permanecen dentro de las zonas alteradas ha sido modificada por causa de la hormiga brava importada. Trampas "pitfall" fueron empleadas para determinar la estructura de especies en la comunidad y el area ya invadida por la hormiga brava importada. La actividad de forajeo para las diferentes especies fue determinada por medio de trampas de comida. Usando el indice de diversidad Shannon-Wienner, se encontro una diferencia significativa para la estructura de especies entre las zonas infestada y no infestada. Tambien la actividad de forajeo para las especies comunes entre las dos zonas fueron significativas de acuerdo con la prueba G o "goodness of fit test". Estos resultados proponen que estas especies que permanecen dentro de las zonas infestadas se estan adaptando y estan alterando su actividad de forajeo en respuesta a la precencia de la hormiga brava importada.

References

Camilo, G. R. and S. A. Phillips, Jr. 1990. Evolution of ant communities in response to invasion by the fire ant *Solenopsis invicta*. Pp. 190-198. In: R.K. Vander Meer, K. Jaffe and A. Cedeño [eds.]. *Applied Myrmecology: a World Perspective*. Westview Press, Boulder, CO.

Correll, D. S. and M. C. Johnston. 1970. *Manual of the vascular plants of Texas*. Texas Res. Found., Renner, TX.

Fellers, J. H. 1987. Interference and exploitation in a guild of woodland ants. *Ecology* 68: 1466-1478.

Marsh, A. C. 1988. Activity patterns of some Namib desert ants. *J. Arid Environ.* 14: 61-73.

Savolainen, R. and K. Vepsalainen. 1988. A competition hierarchy among boreal ants: impact on resource partitioning and community structure. *Oikos* 51: 135-155.

Sokal, R. R. and F. J. Rohlf. 1981. *Biometry, the principles and practice of statistics in biological research*. Pp. 691-778. W. H. Freeman and Co., NY.

Southwood, T. R. E. 1978. *Ecological methods, with particular reference to the study of insect populations*. Pp. 420-437. Chapman and Hall, NY.

Tschinkel, W. R. 1988. Distribution of the fire ants *Solenopsis invicta* and *S. geminata* (Hymenoptera: Formicidae) in northern Florida in relation to habitat and disturbance. *Ann. Entomol. Soc. Am.* 81: 76-81.

Ward, P. S. 1987. Distribution of the introduced Argentine ant *Iridomyrmex humilis* in natural habitats of the lower Sacramento Valley California, USA and its effects on the indigenous ant fauna. *Hilgardia* 55(2): 1-16.

23

Impact of the Red Imported Fire Ant on Native Ant Species in Florida

Daniel P. Wojcik

Introduction

Invasion by ants and their subsequent displacement of native fauna is a well known phenomenon (Lubin 1984, Ward 1987, Haskins and Haskins 1988). These invasions are most devastating on islands (Lubin 1984, Haskins and Haskins 1988). The published studies have not been conducted continuously over a long period of time, although the displacement of *Pheidole megacephala* by *Linepithema humile* in the Bermuda islands has been documented periodically through the years (Haskins and Haskins 1988).

The red imported fire ant, *Solenopsis invicta* Buren, is a pugnacious, immensely successful invader of disturbed habitats throughout the Southeastern United States (Lofgren 1986, Tschinkel 1987). Changes in native ant fauna resulting from insecticide treatment for *S. invicta* control have been documented (Apperson et al. 1984, Edmonson 1981, Markin et al. 1974), along with subsequent reinfestation by *S. invicta* (Summerlin et al. 1977). The effects of *S. invicta* invasion on native ant fauna (in the absence of insecticide treatments) have usually been studied by comparing habitats with and without *S. invicta* (Camilo and Phillips 1990, Fowler et al. 1990, Phillips et al. 1987, Porter and Savignano 1990, Porter et al. 1988, Whitcomb et al. 1972). However, after invasion by *S. invicta*, some native and other introduced species of ants are known to persist (Baroni-Urbani and Kannowski 1974, Glancey et al. 1976, Whitcomb et al. 1972).

The successful movement of *S. invicta* into uninfested areas by taking advantage of man-made ecological disturbances, other than large-scale insecticidal treatments, has not been properly documented. Glancey et al. (1976) examined the ant fauna of Mobile Co., Alabama, but their pre-*S. invicta* invasion comparisons were based on older publications. *S. invicta* had apparently displaced the introduced Argentine ant, *L. humile* (Mayr) (formerly *Iridomyrmex humilis*), and 2 native fire ants, *Solenopsis geminata* (F.) and *S. xyloni* McCook. An interesting observation on their data, although not realized at the time, was the continued presence of non-target native and introduced species.

Gainesville, Florida, an area that has never received any large-scale insecticide treatment, first became lightly infested with *S. invicta* in late 1971. In March 1972, a long-term study was initiated to assess the long-term effects of the ensuing population increases of *S. invicta* on non-target ants in the absence of insecticide treatments.

Materials and Methods

One hundred permanent bait stations were designated and marked about every 0.8 km (0.5 mile) along 4 paved roads in Gainesville, Alachua Co., Florida (Wojcik 1983). The baits [a 1.9 cm, (0.7 inch) diameter ground-beef ball and a 1.3 cm (0.5 inch) honey-agar cube] were individually placed on 2.5 cm (1 inch) aluminum squares. In the field, the 2 baits were placed at bait stations in the same relative positions, 0.3 to 1 m (1-3 ft) apart, at 9 AM each sampling period (Wojcik et al. 1975). The baits were collected after one hour, the ants were preserved in alcohol, sorted to species and counted. Sampling of the ant populations has been conducted periodically for 21 years (March 1972 to September 1992). Initially, sampling was done every other month, but it varied with the demands of other projects and the rate of increase of *S. invicta* populations. The areas of Gainesville sampled by the transect are occupied only by monogyne *S. invicta* colonies.

Number of sites is defined as the number of bait stations out of 100 bait stations (each had 2 baits) for each sampling period for a species. For each species, percent occurrences is defined as the number of baits with a given ant species out of 200 baits per sampling period. For each species, percent specimens is defined as the percent of all ant specimens collected per sampling period.

Data were transformed to rank values for each species with means for ties (Conover and Iman 1976). Correlation was determined between the ranked values of monthly percent occurrence of the various species to ranked values of monthly percent *S. invicta* occurrence. Correlation coefficients were calculated using linear regression analysis with Lotus 123 ver. 3.3 and significance determined by t-test (Steel and Torrie 1960).

Results and Discussion

During 68 sampling periods over 21 years, 13,600 bait samples and 990,079 specimens were collected and identified. More than one ant species was collected on approximately 33% of the baits; 16% of the baits were blank. *S. invicta, S. geminata* and *Pheidole dentata* were the 3 major species collected on the transect over the years in total occurrences (58.33%) and specimens (82.5%). Through September 1992, the following 55 species of ants from 5 subfamilies and 22 genera have been collected on the transect: *Aphaenogaster ashmeadi* Emery, *A. flemingi* Smith, *A. floridana* Smith, *A. fulva* Roger, *A.* near *rudis, Brachymyrmex depilis* Emery, *Camponotus decipiens* Emery, *C. floridanus* (Buckley), *C. socius* Roger, *Cardiocondyla ectopia* Snelling, *C. emeryi* Forel, *C. nuda* (Mayr), *C. venustula* Wheeler, *C. wroughtoni* (Forel), *Conomyrma bureni* Trager, *C. medeis* Trager, *Crematogaster ashmeadi* Mayr, *C. atkinsoni* Wheeler, *C. lineolata* (Say), *C. pilosa* Emery, *Cyphomyrmex rimosus* (Spinola), *Forelius pruinosus* (Roger), *Formica archboldi* Smith, *F. pallidefulva* Latreille, *Hypoponera opaciceps* (Mayr), *Lasius alienus* (Foerster), *Leptothorax pergandei* Emery, *Monomorium viridum* Brown, *Odontomachus brunneus* (Patton), *Paratrechina bourbonica* (Forel), *P. concinna* Trager, *P. faisonensis* (Forel), *P. longicornis* (Latreille), *P. parvula* (Mayr), *P. vividula* (Nylander), *Pheidole carrolli* Naves, *P. dentata* Mayr, *P. floridana* Emery, *P. metallescens* Emery, *P. moerens* Wheeler, *P. morrisi* Forel, *P. vinlandica* Forel, *Prenolepis imparis* (Say), *Pseudomyrmex ejectus* F. Smith, *Solenopsis geminata* (F.), *S. invicta* Buren, *S. globularia littoralis* Creighton, *S. picta* Emery, *S.* (*Diplorhoptrum*) spp., *S.* near *truncorum, Tetramorium bicarinatum* (F.), *T. lanuginosum* Mayr, *T. simillimum* (F. Smith), *Trachymyrmex septentrionalis* (McCook), *Trichoscapa membranifera* (Emery).

The *S. invicta* population has gradually increased until in September 1992 it now dominates the ant fauna: 43.3% of the sample occurrences (Figure 23.1A, maximum 55.8% in March 1990 and April 1992), 63.1% of the sample specimens (Figure 23.1B, maximum 74.3% in April 1992), and 50 sites (Figure 23.1C, maximum 59 sites in April 1991). This increase has undoubtedly been aided by habitat disturbances and habitat simplification (the process of urbanization) which have gradually occurred in Gainesville. The increase occurred in spite of high

FIGURE 23.1 Percent occurrences (A), percent specimens (B), and number of sites (C) for major species of ants collects on the Gainesville transect from March 1972 to September 1992.

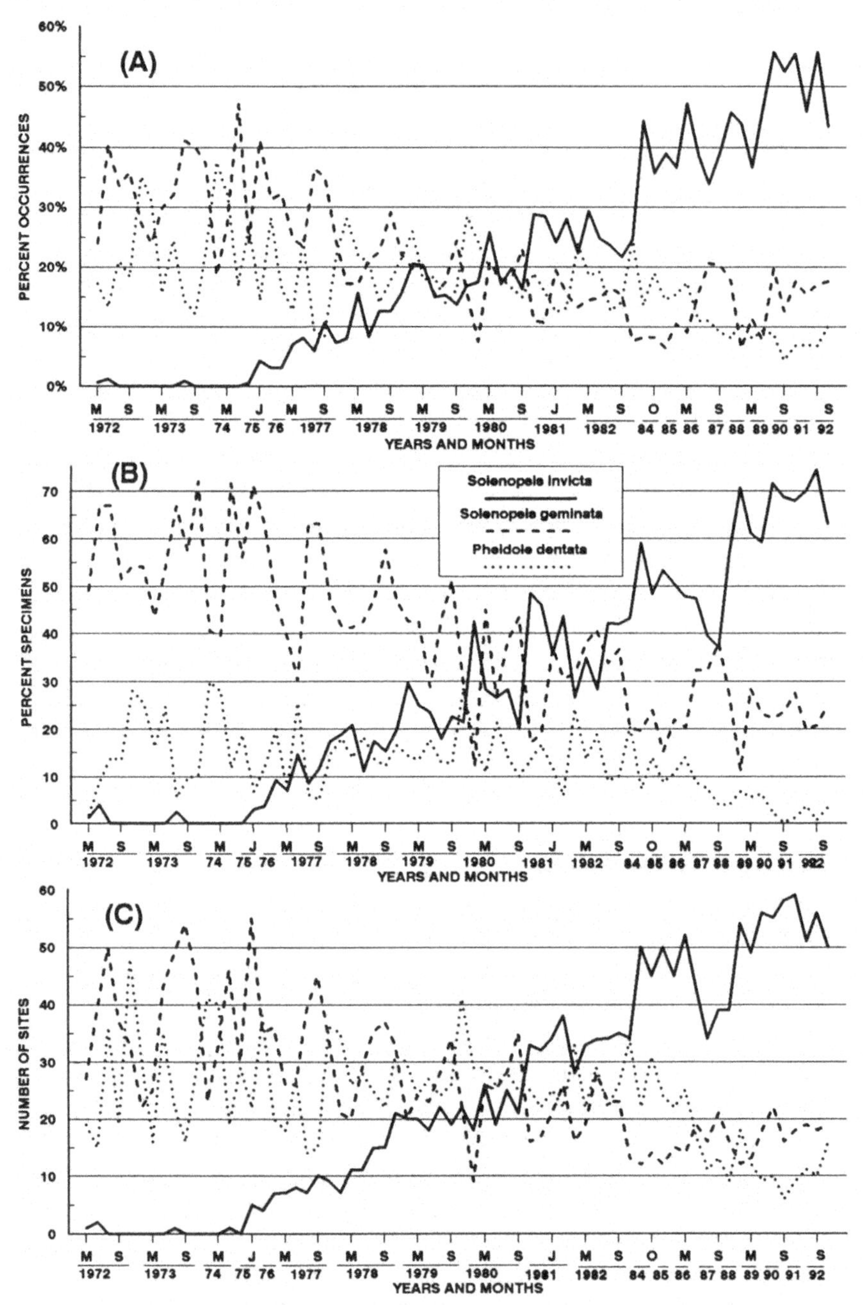

populations of *S. geminata* and *P. dentata,* two species which are predaceous on newly mated *S. invicta* queens (Whitcomb et al. 1972). *S. invicta* is an r-strategist or weed species with great reinfestation abilities, enabling it to invade, establish, and rebuild populations quickly (Buren and Whitcomb 1977, Tschinkel 1987). Once established, it persists and dominates its habitat, becoming a keystone species and influencing community structure. Like its congener, *S. geminata* (F.) (Risch and Carroll 1982, 1983), *S. invicta* dominates the ant fauna numerically and may affect other arthropod populations as well (Porter and Savignano 1990).

Both *S. geminata* and *P. dentata* showed significant negative correlations ($P<0.01$) when compared to *S. invicta* in percent occurrences (Figure 23.1A), in specimens (Figure 23.1B), and in number of sites occupied (Figure 23.1C). Tschinkel (1988) found that the outcome of the competition between *S. invicta* and *S. geminata* is usually mediated by the degree of disturbance in the environment. *S. geminata* was able to overcome moderate disturbance and persist and flourish in the presence of the *S. invicta* invasion. *S. geminata* populations can return to pre-disturbance levels if they are not displaced by competitors (Risch 1981). Whitcomb et al. (1972) reported that *S. geminata* was one of the first species to decrease or disappear after *S. invicta* invasion of soybean fields, but pesticide usage was not considered and no time parameters were given. Despite having an alarm-recruitment defense system specific to *Solenopsis* species ants (Wilson 1976), *P. dentata* has decreased wherever it has been studied following *S. invicta* invasion (Glancey et al. 1976, Long et al. 1987, Cherry and Nuessly 1992). This decrease is at least partially attributable to the superior recruitment and displacement abilities of *S. invicta* over *P. dentata* (Fraelich 1991).

An additional 13 species occurred often enough to allow calculation of significant correlations against *S. invicta* occurrences (Table 23.1). These species can be divided into native ants and introduced ants. Each group had species which were either negatively or positively correlated with the ranked percent occurrences of *S. invicta* (Table 23.1).

TABLE 23.1. Correlation of species ranked percent occurrences to ranked percent *S. invicta* occurrences (df = 66).

Species	*R*	*Number of Occurrences*	*Percent of Occurrences*
Solenopsis invicta	——	2659	20.27
Solenopsis geminata	-0.782**	2745	20.93
Pheidole dentata	-.543**	2247	17.13
Pheidole metallescens	-.589**	777	5.92
Pheidole floridana	-.551**	191	1.46
Paratrechina longicornis	+.745**	131	1.00
Tetramorium simillimum	+.554**	219	1.67
Pheidole morrisi	-.489**	537	4.09
Pheidole moerens	+.421**	144	1.10
Odontomachus brunneus	+.410**	251	1.91
Paratrechina vividula	-.400**	144	1.10
Forelius pruinosus	-.389**	83	.63
Cardiocondyla emeryi	-.376**	194	1.48
Crematogaster ashmeadi	-.353*	27	.21
Monomorium viridum	-.349*	80	.61
Conomyrma bureni	-.298*	1216	9.27

*significant at the 0.05% level   **significant at the 0.01% level

Native species which occur in habitats and niches similar to *S. invicta* have generally been negatively affected by the habitat disturbances in Gainesville and the corresponding increases in *S. invicta* populations. The population changes, as reflected in the percent changes in the ranked occurrences, of *Pheidole metallescens, Pheidole floridana, Pheidole morrisi* (Figure 23.2), *Paratrechina vividula, Forelius pruinosus,* and *Monomorium viridum* (Figure 23.3) are all significant (P<0.01) and represent real population decreases of these ants. Although many of these species are predators and will attack newly mated *S. invicta* queens, Whitcomb et al. (1972) report decreases in their populations as a result of *S. invicta* invasion. The smaller nest sizes, seasonally restricted mating flights, and limited ability to withstand habitat disturbance (Naves 1985, Smith 1965, Trager 1984, Harada 1990, DuBois 1986) puts these species at a disadvantage in relation to *S. invicta*.

FIGURE 23.2. Percent occurrences of 6 species of ants collected on the Gainesville transect from March 1972 to September 1992.

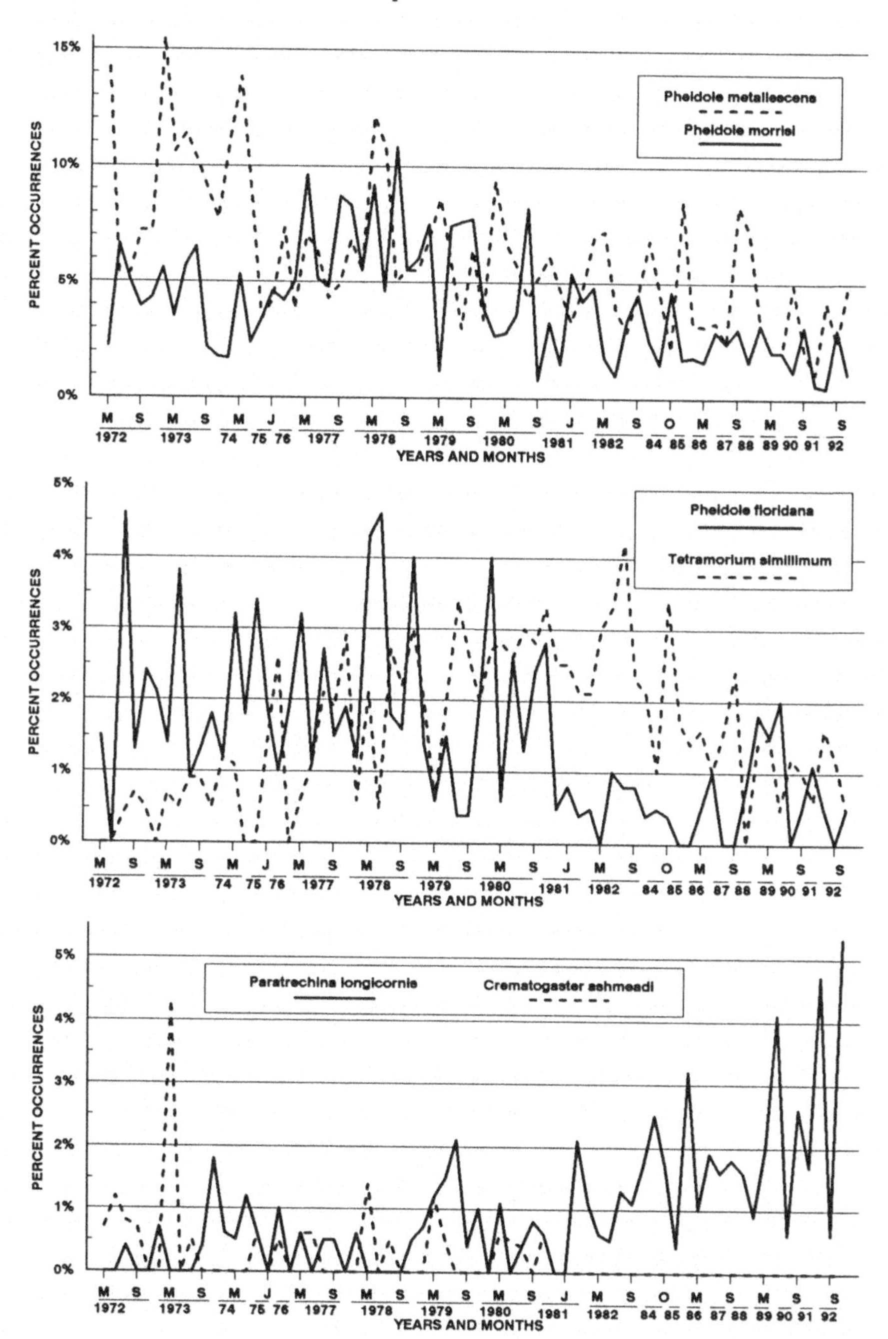

FIGURE 23.3. Percent occurrences of 6 species of ants collected on the Gainesville transect from March 1972 to September 1992.

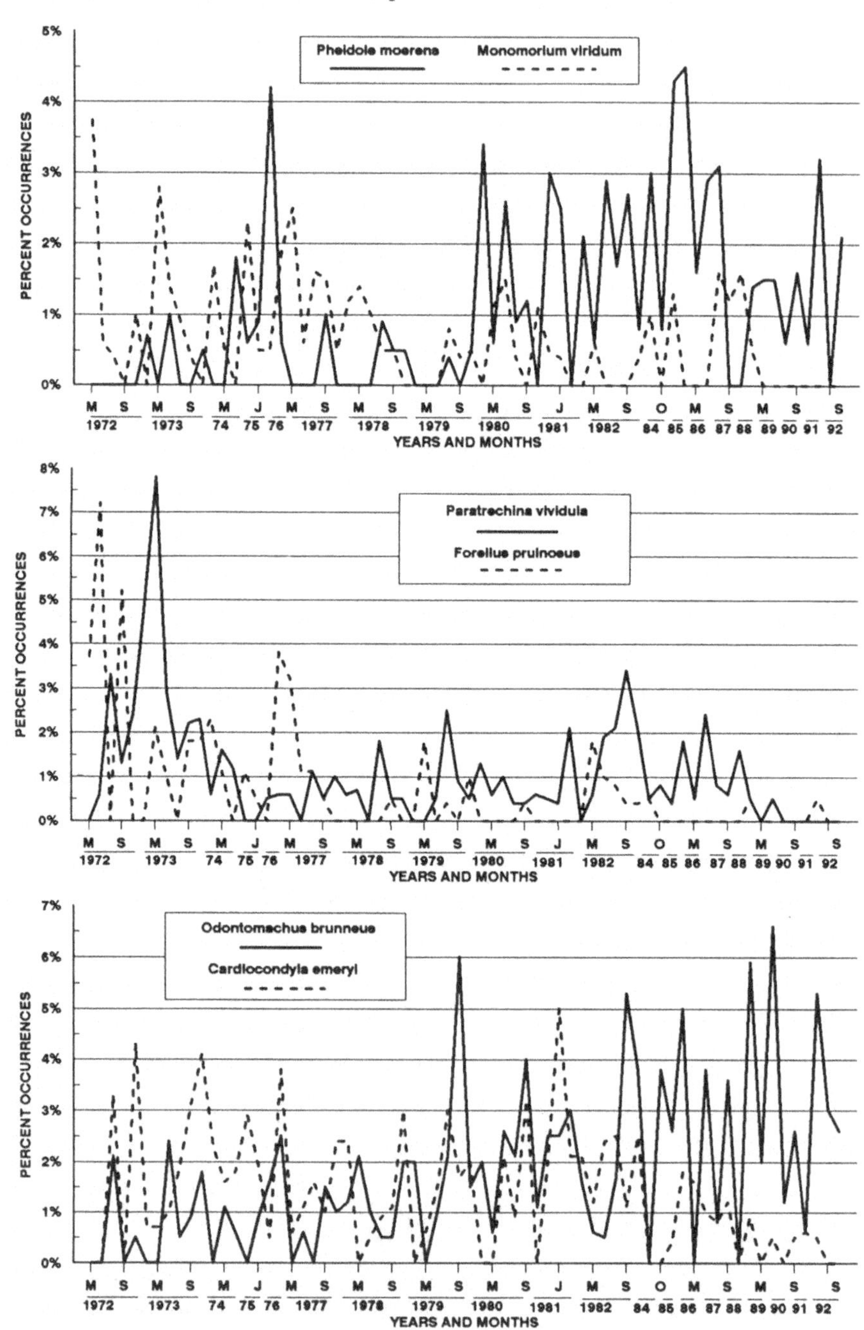

The decrease in *Crematogaster ashmeadi* occurrences (Figure 23.2) is statistically significant (P<0.05). The change in *C. ashmeadi* occurrences probably represents habitat changes as this ant nests in trees, shrubs, and vines (Johnson 1988). This ant has not been collected on the transect since 1981, but it still can be found in wooded areas in and around Gainesville. *Conomyrma bureni* occurrences (not figured) are weakly negatively correlated (P<0.05) with *S. invicta* occurrences. This native species is an ant of disturbed areas (Trager 1988), preferring small areas of bare soil for its nests. The decreasing number of occurrences probably do not represent a real population decrease as this species and its congeners can exist quite well in areas infested with *S. invicta* (Hung 1974, Claborn et al. 1988, Trager 1988).

Odontomachus brunneus occurrences (Figure 23.3) showed an unexpected positive correlation (P<0.05) with increasing *S. invicta* occurrences. This native predator nests in small colonies (Creighton 1950, Brown 1976), which it successfully defends against *S. invicta* and other ants with a unique defensive mechanism (Carlin and Gladstein 1989). This species has an excellent ability to defend itself against *S. invicta* in direct confrontation as measured by Bhatkar (1988). The dynamics of habitat disturbance and urbanization have somehow improved the habitat for this ant in spite of the increases in *S. invicta* population. This species has been collected on both meat and honey-agar baits belying its strictly predatory reputation (Creighton 1950, Whitcomb et al. 1972).

The second group of ants consists of introduced species most of which show a positive correlation (P<0.01) in the percent changes in the ranked occurrences (Table 23.1). This indicates their populations increases with corresponding increases in *S. invicta* populations. The habitat disturbance and urbanization of Gainesville should favor population increases in these species. *Paratrechina longicornis* (Figure 23.2) has the largest positive correlation with *S. invicta*. This introduced species is a structural pest which can nest outdoors in northern Florida (Trager 1984). It coexists with *S. invicta,* even in its Brazilian homeland (Banks and Williams 1989). *Tetramorium simillimum* (Figure 23.2) and *Pheidole moerens* (Figure 23.3) are introduced pests which show weaker positive correlations (P<0.01) with *S. invicta* occurrences. These introduced species are structural pests which can nest outdoors in northern Florida (Smith 1965, Naves 1985).

Cardiocondyla emeryi was the only other introduced species collected in sufficient numbers for analysis (Table 23.2). This tiny ant is generally ignored by other ants and occupies slightly to heavily disturbed habitats (Creighton 1950). It showed a negative correlation (P<0.05) (Figure 23.3).

As this species nests in plants as well as under items on the ground, the reasons for its decrease probably are similar to those responsible for the decrease shown by *Crematogaster ashmeadi*.

The habitat disturbance and urbanization of Gainesville has included the widening of streets and right-of-ways, which has resulted in the removal of native vegetation on the transect since the study began. These changes will undoubtedly continue in the future. The lack of a large area-wide insecticide treatment for *S. invicta* has allowed us to study the gradual changes in other ant populations which have taken place over the last 21 years. The processes operating on the ant populations in Gainesville are also at work throughout the southeastern United States. In very few areas have the *S. invicta* populations reached stable levels. In the absence of large area-wide insecticide treatments, *S. invicta* populations may take years to reach their peak and stabilize. The Gainesville populations of *S. invicta* have not stabilized (reached the carrying capacity) and will continue to be monitored in the future.

Acknowledgments

The following present or past USDA or University of Florida staff deserve thanks for their help in conducting the transect for the past 21 years: W. Banks, C. Blanton, R. Burges, L. Davis, N. Hicks, D. Jouvenaz, T. Krueger, D. Labella, A. Lemire, J. Plumley, J. Sullenger, D. Weigle, and a host of summer and temporary employees. Dr. C. S. Lofgren deserves special acknowledgement for enabling me to conduct this long-term study.

Resumen

Se discute en este capítulo los resultados de muestreos de hormigas realizados en sitios localizados a la orilla del camino en el area de Gainesville, Florida mediante el uso de cebos con carne y miel-agar, realizados desde Marzo, 1972 a Septiembre, 1992. Al usar 13,600 colecciones de cebos, se encontraron 990,079 especimenes. Esta colección representa 55 especies en 22 generos y 5 subfamilias. *Solenopsis invicta, S. geminata,* y *Pheidole dentata* fueron las 3 especies mayores colectadas en estos 21 años con una ocurrencia total (58.33%) y un número total de especimenes del 82.5%. *S. invicta* incrementó hasta Septiembre 1992 fué del 43.3% en ocurrencia y representó el 63.1% de los especimenes. Durante 1992, *S. invicta* fué colectada en el 65% de los sitios. Este incremento fué ayudado por la urbanización del habitat así como también su pertubabilidad. Tanto *S. geminata* como *P. dentata* mostraron corelaciones negativas cuando se comparó con el porcentaje de ocurrencia, porcentaje de especimenes y número de sitios ocupados por *S. invicta*. Ocho especies de hormigas mostraron decremento significativo, tales como *Pheidole metallescens* Emery, *Pheidole floridana* Emery, *Pheidole morrisi* Forel, *Paratrechina vividula* (Nylander), *Forelius pruinosus* (Roger), *Monomorium viridum* (Brown), *Crematogaster ashmeadi* Mayr, *Conomyrma bureni* Trager. La hormiga nativa, *Odontomachus brunneus* (Patton), mostró una corelación positiva con el incremento de las ocurrencias de *S. invicta*. Tres especies introducidas, *Paratrechina longicornis* (Latreille), *Tetramorium simillimum* (F. Smith), *Pheidole moerens* Wheeler mostraron incrementos significativos en ocurrencia y *Cardiocondyla emeryi* Forel mostró una reducción significativa.

References

Apperson, C. S., R. B. Leidy, and E. E. Powell. 1984. Effects of Amdro on the red imported fire ant (Hymenoptera: Formicidae) and some nontarget ant species and persistence of Amdro on a pasture habitat in North Carolina. *J. Econ. Entomol.* 77: 1012-1018.

Banks, W.A., and D. F. Williams. 1989. Competitive displacement of *Paratrechina longicornis* (Latreille) (Hymenoptera: Formicidae) from baits by fire ants in Mato Grosso, Brazil. *J. Entomol. Sci.* 24: 381-391.

Baroni-Urbani, C., and P. B. Kannowski. 1974. Patterns in the red imported fire ant settlement of a Louisiana pasture: Some demographic parameters, interspecific competition and food sharing. *Environ. Entomol.* 3: 755-760.

Bhatkar, A. P. 1988. Confrontation behavior between *Solenopsis invicta* and *S. geminata*, and competitiveness of certain Florida ant species against *S. invicta*, Pp. 445-464. In: J.C. Trager [ed.]. *Advances in Myrmecology*. E.J. Brill, New York.

Brown, W. L., Jr. 1976. Contribution toward a reclassification of the Formicidae. Part VI. Ponerinae, Tribe Ponerini, subtribe Odontomachiti. Section A. Introduction, subtribal characters, Genus *Odontomachus*. *Stud. Entomol.* 19: 67-171.

Buren, W. F., and W. H. Whitcomb. 1977. Ants of citrus: some considerations. *Proc. Int. Soc. Citriculture* 2: 496-498.

Camilo, G. R., and S. A. Phillips, Jr. 1990. Evolution of ant communities in response to invasion by the fire ant *Solenopsis invicta*, Pp. 190-198. In: R. K. Vander Meer, K. Jaffe, and A. Cedeno [eds.]. *Applied Myrmecology, A World Perspective*. Westview Press, Boulder, CO.

Carlin, N. F., and D. S. Gladstein. 1989. The "bouncer" defense of *Odontomachus ruginodis* and other Odontomachine ants (Hymenoptera: Formicidae). *Psyche* 96: 1-19.

Cherry, R. A., and G. S. Nuessly. 1992. Distribution and abundance of imported fire ants (Hymenoptera: Formicidae) in Florida sugarcane fields. *Environ. Entomol.* 21: 767-770.

Claborn, D. M., S. A. Phillips, Jr., and H. G. Thorvilson. 1988. Diel foraging activity of *Solenopsis invicta* and two native species of ants (Hymenoptera: Formicidae) in Texas. *Texas J. Sci.* 40: 93-99.

Conover, W. J., and R. L. Iman. 1976. On some alternative procedures using ranks for the analysis of experimental designs. *Commun. Statist.-Theor. Meth.* A5: 1349-1368.

Creighton, W. S. 1950. The ants of North America. *Bull. Mus. Comp. Zool. Harvard Coll.* 104: 1-585, plates 1-57.

DuBois, M. B. 1986. A revision of the native New World species of the ant genus *Monomorium* (*minimum* Group) (Hymenoptera: Formicidae). *Univ. Kansas Sci. Bull.* 53: 65-119.

Edmonson, M. B. 1981. The effect of Amdro on nontarget ant species associated with *Solenopsis invicta* Buren in Florida. M.S. thesis, Univ. of Florida, Gainesville, FL.

Fraelich, B. A. 1991. Interspecific competition for food between the red imported fire ant, *Solenopsis invicta,* and native Floridian ant species in pastures. M.S. Thesis, Univ. of Florida.

Fowler, H. G., J. V. E. Bernardi, and L. F. T. di Romagnano. 1990. Community structure and *Solenopsis invicta* in Sao Paulo, Pp. 199-207. In: R. K. Vander Meer, K. Jaffe, and A. Cedeno [eds.]. *Applied Myrmecology, A World Perspective.* Westview Press, Boulder, CO.

Glancey, B. M., D. P. Wojcik, C. H. Craig, and J. A. Mitchell. 1976. Ants of Mobile County, AL, as monitored by bait transects. *J. Georgia Entomol. Soc.* 11: 191-197.

Harada, A. Y. 1990. Ant pests of the Tapinomini tribe. Pp. 298-315. In: R. K. Vander Meer, K. Jaffe, and A. Cedeno [eds.]. *Applied Myrmecology, A World Perspective.* Westview Press, Boulder, CO.

Haskins, C. P., and E. F. Haskins. 1988. Final observations on *Pheidole megacephala* and *Iridomyrmex humilis* in Bermuda. *Psyche* 95: 177-184.

Hung, A. C. F. 1974. Ants recovered from refuse pile of the pyramid ant *Conomyrma insana* (Buckley) (Hymenoptera: Formicidae). *Ann. Entomol. Soc. Am.* 67: 522-523

Johnson, C. 1988. Species identification in the eastern *Crematogaster* (Hymenoptera: Formicidae). *J. Entomol. Sci.* 23: 314-332.

Lofgren, C. S. 1986. History of imported fire ants in the United States, Pp. 36-47. In: C. S. Lofgren and R. K. Van der Meer [eds.]. *Fire Ants & Leaf-Cutting Ants; Biology & Management.* Westview Press, Boulder, CO.

Long, W. H., L. D. Nelson, P. J. Templet, and C. P. Viator. 1987. Abundance of foraging ant predators of the sugarcane borer in relation to soil and other factors. *J. Am. Soc. Sugar Cane Techn.* 7: 5-14.

Lubin, Y.D. 1984. Changes in the native fauna of the Galapagos Islands following invasion by the little red fire ant, *Wasmannia auropunctata. Biol. J. Linn. Soc.* 21: 229-242.

Markin, G. P., J. O'Neal, and H. L. Collins. 1974. Effects of mirex on the general ant fauna of a treated area in Louisiana. *Environ. Entomol.* 3: 895-898.

Naves, M. A. 1985. A monograph of the genus *Pheidole* in Florida (Hymenoptera: Formicidae). *Insecta Mundi* 1: 53-90.

Phillips, S. A. Jr., W. M. Rogers, D. B. Wester, and L. Chandler. 1987. Ordination analysis of ant faunae along the range expansion front of the red imported fire ant in South-central Texas. *Texas J. Agric. Natur. Res.* 1: 11-15.

Porter, S.D., and D. A. Savignano. 1990. Invasion of polygyne fire ants decimates native ants and disrupts arthropod community. *Ecology* 71: 2095-2106.

Porter, S. D., B. Van Eimeren, and L .E. Gilbert. 1988. Invasion of red imported fire ants (Hymenoptera: Formicidae): microgeography of competitive replacement. *Ann. Entomol. Soc. Am.* 81: 913-918.

Risch, S. J. 1981. Ants as important predators of rootworm eggs in the Neotropics. *J. Econ. Entomol.* 74: 88-90.

Risch, S. J., and C. R. Carroll. 1982. The ecological role of ants in 2 Mexican agroecosystems. *Oecologia* 55: 114-119.

_____ . 1983. Effect of a keystone predaceous ant, *Solenopsis geminata*, on arthropods in a tropical agroecosystem. *Ecology* 63: 1979-1983.

Smith, M. R. 1965. House-infesting ants of the eastern United States. *USDA Tech. Bull.* 1326.

Steel, R. G. D., and J. H. Torrie. 1960. *Principles and procedures of statistics.* McGraw-Hill Book Co., Inc. New York.

Summerlin, J. W., A. C. F. Hung, and S. B. Vinson. 1977. Residues in nontarget ants, species simplification and recovery of populations following aerial applications of mirex. *Environ. Entomol.* 6: 193-197.

Trager, J. C. 1984. A revision of the genus *Paratrechina* (Hymenoptera: Formicidae) of the continental United States. *Sociobiology* 9: 49-162.

_____ . 1988. A revision of *Conomyrma* (Hymenoptera: Formicidae) from the southeastern United States, especially Florida, with keys to the species. *Florida Entomol.* 71: 11-29. Errata *Florida Entomol.* 71: 219.

Tschinkel, W. R. 1987. The fire ant, *Solenopsis invicta*, as a successful weed. Pp. 585-588. In: J. Eder, and H. Rembold [eds.]. *Chemistry and Biology of Social Insects.* Verlag J. Peperny, München.

_____ . 1988. Distribution of the fire ants *Solenopsis invicta* and *S. geminata* (Hymenoptera: Formicidae) in northern Florida in relation to habitat and disturbance. *Ann. Entomol. Soc. Am.* 81: 76-81.

Ward, P. S. 1987. Distribution of the introduced Argentine ant (*Iridomyrmex humilis*) in natural habitats of the lower Sacramento Valley and its effects on the indigenous ant fauna. *Hilgardia* 55(2): 1-16.

Whitcomb, W. H., H. A. Denmark, A. P. Bhatkar, and G. L. Greene. 1972. Preliminary studies on the ants of Florida soybean fields. Florida Entomol. 55: 129-142.

Wilson, E. O. 1976. The organization of colony defense in the ant *Pheidole dentata* Mayr (Hymenoptera: Formicidae). *Behav. Ecol. Sociobiol.* 1: 63-81.

Wojcik, D. P. 1983. Comparison of the ecology of red imported fire ants in North and South America. *Florida Entomol.* 66: 101-111.

Wojcik, D. P., W. A. Banks, and W. F. Buren. 1975. First report of *Pheidole moerens* in Florida (Hymenoptera: Formicidae). *USDA Coop. Econ. Insect Rpt.* 25: 906.

24

Control of the Introduced Pest *Solenopsis Invicta* in the United States

David F. Williams

Introduction

Two species of imported fire ants were introduced into the U.S. at Mobile, Alabama (Lofgren et al. 1975). The black imported fire ant, *Solenopsis richteri* (Forel), was introduced around the early 1900s while the red imported fire ant, *Solenopsis invicta* (Buren) probably entered in the late 1930s or early 1940s (Buren et al. 1974). The red imported fire ant (RIFA) is the most widespread of the two and presents the greatest problem. From Mobile, the RIFA have spread naturally by such means as mating flights and floating of colonies on rivers and streams after floods. Most importantly, the ant has spread artificially with the aid of man during shipment of nursery stock containing queens and small colonies. Currently, the RIFA infest more than 106 million hectares in 11 southern states and Puerto Rico.

Since their introduction, populations of *S. invicta* have not only expanded their range throughout the southeastern and southwestern United States, but the total number of colonies (mounds) has greatly increased, especially in states such as Texas which has a greater proportion of multiple queen (polygynous) colonies than colonies with a single queen (monogynous) (Porter et al. 1991).

S. invicta has become a serious pest of man throughout its range (Lofgren et al. 1975, Lofgren 1986, Adams 1986). They continue to spread with small infestations recently appearing in New Mexico, Arizona, California, and Virginia. The increasing incidence of the polygyne (multiple queen) form poses additional problems not only to humans and agricultural crops, but also to wildlife, especially surface-active animals (Porter and Savignano 1990; see also Vinson, Chapter 21; Jusino-Atresino and Phillips, Chapter 22). The most serious problems occur when people who have been stung become hypersensitive to the proteinaceous component of the venom resulting in anaphylactic shock and death. For additional reviews on the impact of fire ants on man and his environment see Adams and Lofgren (1981), Adams (1986), Lofgren (1986), deShazo et al. (1990), and MacKay et al. (1992).

Wherever these ants have become established, they have caused serious problems by their stinging, mound-building, and feeding habits. Because of this, they have been the object of research and control efforts for more than 4 decades. When one considers the many problems that *S. invicta* have caused people, it is little wonder that methods for their control are very important.

Control

It seems that there are as many control recommendations for fire ants as there are fire ant colonies. These range from "home remedies" to "high tech". Most home remedies involve treating individual mounds with an assortment of products including gasoline and other petroleum derivatives, solutions of soaps and detergents, bleaches, wood ashes, vinegar, grits, yeasts, citrus peels and watermelon rinds. The "high tech" solutions involve the use of microwaves, electrical probes, and explosives, all of which are of dubious value in controlling fire ants. Also, another remedy is to dig up a mound and place it on top of another mound with the expectation that the ants will fight, thus eliminating both colonies. As with most home remedies, the large majority simply do not work. Some are not only ineffective, but can be dangerous to use and can cause damage to the environment (e.g. gasoline, other petroleum products, lye, bleaches, microwaves, and explosives).

Chemical. Chemicals are the most widely used and, for the present time, most effective control method available against fire ants. There are many chemicals and they can be applied in several ways but generally 2 approaches are used: (1) contact insecticide treatments with drenches, sprays, dusts, granules, aerosols, and fumigants, and (2) toxic baits. Both contact insecticide treatments and baits have advantages and disadvantages with the specific situation determining which to use.

Contact insecticide treatments. The first use of chemicals as contact insecticides for the control of fire ants began as early as 1937 with the use of calcium cyanide dust applied to individual mounds (Eden and

Arant 1949). Since that time numerous chemicals have been tried with mixed results. For a historical perspective of the early control programs with contact insecticides, see Collins 1992. Presently, there are several chemicals used against *S. invicta* as contact insecticides (Drees and Vinson 1991). The majority of contact insecticides presently registered for fire ant control are used in emulsifiable concentrate form as drenches. Other formulations of chemicals are used as pressurized sprays, dusts, granules, and fumigants. Boiling water can also be used to treat ant mounds (Tschinkel and Howard 1980), but gave poor control in tests against large mounds (D. F. Williams, unpublished data) and would be ineffective against multiple queen colonies. As always, caution should be used in handling all chemicals including hot water, which can cause serious burns. The following is a list of the advantages and disadvantages of using contact insecticides in treatments against fire ants. The advantages include: (1) fast kill; (2) only target ants are affected; (3) moisture and rainfall usually have little affect on treatments; (4) special equipment is not needed; and (5) excellent shelf-life. The disadvantages include: (1) too labor intensive for treating large areas; (2) soil type and moisture may affect treatment; (3) seasonal weather effects can influence control; (4) it is not easy to kill the queen(s) in the colony; (5) only ants in the mound are contacted, foraging workers usually are not killed; (6) it causes frequent colony movement requiring retreatment, and (7) a large amount of pesticide (active ingredient) is applied.

Broadcast treatments using toxic baits. Broadcast application of toxic baits is generally considered the most effective and efficient method to control numerous colonies over a large area and to maintain control for a long period (Williams 1983, Banks 1990, Collins 1992). The method of using toxic baits usually results in killing or sterilizing the colony queen(s) and this eliminates the entire colony. Because large areas are treated, this method can slow down reinfestation by the migration of colonies from untreated areas. Although mating flights will occur and newly-mated queens from these flights will inundate a recently treated area, several months are required before these new queens will produce incipient colonies of sufficient size to present a problem. For an extensive review of the development of toxic baits, the reader should refer to Lofgren et al. 1975, Williams 1983, Banks et al. 1985, Lofgren 1986, Banks 1990, and Collins 1992.

In 1957, the U.S. Department of Agriculture began a comprehensive search for an effective toxic bait against fire ants which resulted in the development of a bait containing the chemical, mirex (Lofgren et al. 1962, 1963, 1964).

In 1978, serious concerns regarding mirex residues in the environment led the U.S. Environmental Protection Agency to ban its use (Johnson 1976). The loss of mirex for the control of fire ants initiated a concerted effort to discover new chemicals, especially ones that were more environmentally acceptable. The difficulty of discovering chemicals for use as baits for fire ants can be explained by the fact that although more than 7,100 chemicals have been evaluated in the USDA's laboratory since 1958 (Williams 1983, Banks et al. 1992), only 5 have been commercially developed (Collins 1992).

New baits developed in the early to mid-1980s were based on a formulation similar to that of mirex bait, i.e., a chemical (the toxicant) dissolved in once-refined soybean oil and applied to a corn grit carrier (Williams et al. 1980; Lofgren and Williams 1982; Williams 1983; Banks et al. 1983, 1988). Because they degrade rapidly and leave no residues, the new toxicants are less hazardous to the environment. Some give excellent control when used as broadcast treatments against *S. invicta* (Collins et al. 1992). However, like all baits, they still present problems such as (1) the formulations are attractive to nontarget ants that may also feed on the bait and become similarly affected (Williams 1986), (2) the new baits are formulated with increased amounts of soybean oil (20-30% soybean oil versus 15% with mirex) which can cause dispersal and flowability problems (D. F. Williams, unpublished data), (3) the extremely small amount of total baits needed per hectare requires special equipment for application (Williams et al. 1983), (4) rancidity of the soybean oil can result in poor shelf-life of the formulations, and (5) the newer baits are higher in cost.

The toxic baits registered and currently available for control of imported fire ants are as follows: (1) Amdro (a.i., hydramethylnon--American Cyanamid,Wayne, NJ, USA); (2) Affirm or Ascend (a.i., abamectin--Merck & Co, Rahway, NJ, USA); (3) Bushwhacker (a.i., boric acid--Bushwhacker Associates, Inc., Galveston, TX, USA); and (4) Logic or Award (a.i., fenoxycarb--Ciba-Geigy, Greensboro, NC, USA). Three other chemicals (pyriproxyfen--Sumitomo Chemical Co., Osaka, Japan); sulfluramid (Griffin Corporation, Valdosta, GA); and teflubenzuron (Shell International Chemical Co., London, UK) have given excellent results in tests against *S. invicta*. However, none are currently registered for use against fire ants. Efficacy tests by researchers with the U.S. Department of Agriculture of several of these chemicals as baits in broadcast application is shown in Table 24.1. Most of the baits above are formulated by dissolving the active ingredient in once-refined soybean oil which is then impregnated on a corn pellet carrier for dispersal. These baits can act in several ways such as a stomach poison (Amdro,

TABLE 24.1. Efficacy of some chemicals tested as baits in broadcast applications against natural infestations of the red imported fire ant, *S. invicta*.

Chemical Name[a]	*Tradename(s)*	N[b]	*AI (g/ha)*	*% reduction in population index after weeks indicated.*[c] 6-10	12-14	16-18	>20	*References*
abamectin	Affirm, Ascend	4	0.015-0.49	85	87	91	--[d]	Lofgren and Williams 1982
fenoxycarb	Logic, Award	11	6.2-25.1	94	96	84	92	Banks et al. 1983 Callcott and Collins 1992
hydramethylnon	Amdro	9	4.2-10.4	86	91	64	79	Williams (unpublished data) Banks et al. 1992
pyriproxyfen	Nylar	12	5.3-24.5	83	89	--[d]	87	Banks and Lofgren 1991
sulfluramid	----	8	6.7-10.1	93	92	92	79	Banks et al. 1992
teflubenzuron	----	3	0.051-0.2	77	85	83	--[d]	Williams and Banks (unpublished data)

[a] All baits were formulated in a soybean oil-pregel defatted corn grit mixture containing the active ingredients.
[b] N=Number of tests.
[c] See Lofgren and Williams 1982 for explanation of method of determining the population index. Percentages are means of the population index.
[d] Data not recorded for this period.

boric acid, and sulfluramid), an insect growth regulator (Logic or Award) or a reproductive inhibitor (Affirm or Ascend).

The advantages of using baits include: (1) they are easy to use; (2) soil types do not effect efficacy; (3) one or two treatments is generally sufficient for long term (several months to a year) control; (4) the toxicant is spread to all members of the colony therefore, colony movement is not a problem; and (5) treatment requires a very small amount of toxicant, thus, less contamination of the environment. The disadvantages include: (1) most baits currently on the market give very slow kill or control; (2) nontarget ants may feed on bait; (3) the time of application in relation to temperature may be critical; (4) moisture and rainfall may affect ability of the ants to harvest the bait; (5) special equipment is needed to apply extremely small quantities of bait; and (6) poor shelf life.

Mechanical and cultural control. Several mechanical and electrical devices have been developed to control individual mounds of fire ants. These include microwave units, electrical probes, heating elements, explosive charges, steam probes, and mechanical borers. Although some of these devices will kill individual colonies of fire ants, most are expensive, labor intensive, and of questionable value (Hamman et al. 1986, Drees and Vinson 1991). Various cultural methods have been tried in which mounds were knocked down during cooler months by dragging steel beams across fields, using different tillage methods prior to planting crops, or burning fields in efforts to reduce or eliminate *S. invicta* (Blust et al. 1982, Morrill and Green 1975, Sauer et al. 1982, Collins 1992). In most cases, little reduction of the population occurred with any of these methods, and in those few circumstances where a small reduction did occur, the populations quickly returned to previous levels. Physical removal of colonies from an area by digging them up is effective if the queen is also removed with the colony. However, this is a very labor intensive method and control would be limited because the probability of removing the queen is small, and in the case of multiple queen colonies, there is almost no chance of removing all of the queens. Finally, this method is potentially hazardous because of the danger of incurring fire ant stings.

Biological control. Although research in the area of biological control of fire ants has been underway for several years, the results have been disappointing so far. For example, tests have shown that a parasitic mite, *Pyemotes tritici*, which has been marketed for fire ant control, is ineffective (Jouvenaz and Lofgren 1986). Several organisms kill newly-mated fire ant queens as they alight from mating flights (Whitcomb et al. 1973, Nickerson et al. 1975), however, organisms that will eliminate an entire colony have not been found. Patterson (see Chapter 25) lists several pathogens and parasites that have been investigated as potential agents for the biological control of fire ants. Consequently, this method of control will not be discussed here.

Summary and Conclusion

Since its introduction into the United States over 50 years ago, *S. invicta* presently infests more than 106 million hectares in eleven states and Puerto Rico. More recently, colonies have been found in New Mexico, Arizona, California, and Virginia but reportedly have been eliminated in New Mexico and California.

This ant has had a substantial impact in the U.S. on humans, agriculture, wildlife and other organisms in the environment, and has caused damage to roads, electrical equipment, roofing materials, and telephone junction boxes. The most serious problem caused by this ant is its stinging of humans which in some cases, has caused serious injuries and even death of hypersensitive individuals.

Control of the fire ant usually consists of the use of chemicals using two approaches: (1) application of contact insecticide treatments and (2) broadcast treatments with toxic baits. Contact insecticide treatments are advantageous in that they act quickly (a few hours or days), and are applied directly on the mound, thus mainly affecting fire ants while minimizing exposure to non-target ants. The disadvantages are that the queens often escape treatment so complete elimination of the colony does not occur, small mounds are not seen and therefore not treated, and applying treatments is labor intensive.

The advantages of broadcast bait treatments are that they are more economical because they are less labor intensive, larger areas can be treated quickly, and small unseen colonies are also eliminated. The disadvantages are that the baits are relatively slow-acting (requiring several weeks), treatments can be greatly effected by weather conditions, and baits are not specific to fire ants and can harm nontarget ant species.

Imported fire ants are an increasing urban and public health problem in the southern United States due to a concurrent rise in both human and fire ant populations. This fact assures an increasing chance of contact between the two. These confrontations will result in the demand for additional measures to manage this pest. This will require control in a variety of situations and habitats, and their suppression or elimination will depend on better management techniques.

Research programs are focusing on studying the biology and behavior of this pest and developing newer methods of control that utilize biorational and other agents that will have less impact on the environment and are less dangerous to use. It is of utmost importance that safer methods be discovered, especially given that these ants are becoming a greater problem in our urban environment.

The development of newer, safer and more environmentally comptible chemicals, formulations, and methods of control, such as biological control, are needed and should be a high priority in fire ant research. Finally, continued research in basic biology, ecology, and population dynamics of this pest ant is mandatory if we hope to be able to implement a holistic approach for control.

Acknowledgments

The author wishes to thank C. S. Lofgren, W. A. Banks, D. P. Wojcik, H. L. Collins, D. H. Oi, R. J. Brenner, J. A. Hogsette, and K. M. Vail for critical review of the manuscript. Mention of proprietary names is for the purpose of identification only, and is not an endorsement by the United States Department of Agriculture.

Resumen

La hormiga roja de fuego, *Solenopsis invicta* Buren fue introducida accidentalmente en los Estados Unidos hace casi 50 años y actualmente infesta mas de 120 millones de hectareas en 11 estados y en Puerto Rico. Esta hormiga ha causado un gran impacto en los seres humanos, la agricultura, la fauna y otros organismos que componen el medio ambiente, y ha causado daño a las vías de comunicación y a muchas clases de equipo electronico. El daño mas grave es causado al picar los humanos lo cual a veces resulta en muerte.

El control de la hormiga de fuego consiste en (1) el uso de insecticidas de contacto y (2) tratamientos de cebo al voleo. Las ventajas de los insecticidas de contacto es que actuan rápidamente regularmente en pocas horas o días y usualmente solamente la hormiga es afectada. Las desventajas es que las reinas se escapan del tratamiento, y en consecuencia la colonia no es eliminada, los montículos pequeños al no ser observados fácilmente, no son tratados, y la aplicación es muy laboriosa, intensiva y puede ser usada únicamente en areas pequeñas.

Las ventajas del tratamiento de cebos aplicados al voleo es que son mas económicos, se pueden tratar rapidamente areas de gran extensión, requieren menor labor y las colonias poco visibles son eliminadas. Las desventajas son su lenta acción, lo cual requiere varias semanas para un control, los tratamientos son afectados por el estado del tiempo, y los cebos no son especificos unicamente para las hormigas de fuego y pueden en consecuencia afectar otras especies.

References

Adams, C. T. 1986. Agricultural and medical impact of the imported fire ants. Pp. 48-57. In: C. S. Lofgren and R. K. Vander Meer [eds.]. *Fire ants and leaf-cutting ants, Biology and management.* Westview Press, Boulder, CO.

Adams, C. T. and C.S. Lofgren. 1981. Red imported fire ants (Hymenoptera: Formicidae): Frequency of sting attacks on residents of Sumter County, *Georgia. J. Med. Entomol.* 18: 378-382.

Banks, W. A. 1990. Chemical control of the imported fire ants. Pp. 596-603. In: R. K. Vander Meer, K. Jaffe, and A. Cedeno [eds.], *Applied Myrmecology, a World Perspective.* Westview Press, Boulder, CO.

Banks, W. A., A. S. Las, C. T. Adams, and C. S. Lofgren. 1992. Comparison of several sulfluramid bait formulations for control of the red imported fire ant (Hymenoptera: Formicidae). *J. Entomol. Sci.* 27: 50-55.

Banks, W. A., and C. S. Lofgren. 1991. Effectiveness of the insect growth regulator pyriproxyfen against the red imported fire ant (Hymenoptera: Formicidae). *J. Entomol. Sci.* 26: 331-338.

Banks, W. A., C. S. Lofgren, and D. F. Williams. 1985. Development of toxic baits for control of imported fire ants. Pp. 133-143. In: T. M. Kaneko and L. D. Spicer [eds.]. Pesticide Formulations: 4th Symposium. *American Soc. Testing and Materials Publ.* 875. Philadelphia, PA.

Banks, W. A., L. R. Miles, and D. P. Harlan. 1983. The effects of insect growth regulators and their potential as control agents for imported fire ants (Hymenoptera: Formicidae). *Florida Entomol. Sci.* 66: 172-81.

Banks, W. A., D. F. Williams, and C. S. Lofgren. 1988. Effectiveness of fenoxycarb for control of red imported fire ants (Hymenoptera: Formicidae). *J. Econ. Entomol.* 81: 83-87.

Blust, W. E., B. H. Wilson, K. L. Koonce, B. D. Nelson, and J. E. Sedberry, Jr. 1982. The red imported fire ant: cultural control and effect on hay meadows. *Louisiana Agric. Exp. Sta. Bull.* 738.

Buren, W. F., G. E. Allen, W. H. Whitcomb, F. E. Lennartz, and R. N. Williams. 1974. Zoogeography of the imported fire ants. *J. New York Entomol. Soc.* 82: 113-124.

Callcott, A. A. and H. L. Collins. 1992. Temporal changes in a red imported fire ant (Hymenoptera: Formicidae) colony classification system following an insecticidal treatment. *J. Entomol. Sci.* 27: 345-353.

Collins, H. L. 1992. Control of imported fire ants: A review of current knowledge. *USDA, APHIS, Tech. Bull.* 1807.

Collins, H. L., A. A. Callcott, T. C. Lockley, and A. Ladner. 1992. Seasonal trends in effectiveness of hydramethylnon (Amdro) and fenoxycarb (Logic) for control of red imported fire ants (Hymenoptera: Formicidae). *J. Econ. Entomol.* 85: 2131-2137.

deShazo, R. D., B. T. Butcher, and W. A. Banks. 1990. Reactions to the stings of the imported fire ant. *New England J. Med.* 323: 462: 466.

Drees, B. M. and S. B. Vinson. 1991. Fire ants and their management. *Texas Agric. Extension Ser. B-1536, 20-M-3-91-revised.* Texas A&M Univ., College Station, TX.

Eden, W. G., and F. S. Arant. 1949. Control of the imported fire ant in Alabama. *J. Econ. Entomol.* 42: 976-979.

Hamman, P. J., B. M. Drees, and S. B. Vinson. 1986. Fire ants and their control. *Texas Agric. Extension Ser. B-1536, 40-M-3-86-revised.* Texas A&M Univ., College Station, TX.

Johnson, E. L. 1976. Administrator's decision to accept plan of Mississippi Authority and order suspending hearing for the pesticide chemical mirex. *Fed. Reg.* 41: 56694-56703.

Jouvenaz, D. P. and C .S. Lofgren. 1986. An evaluation of the straw itch mite, *Pyemotes tritici* (Acari: Pyemotidae) for control of the red imported fire ant, *Solenopsis invicta* (Hymenoptera: Formicidae). *Florida Entomol.* 69: 761-763.

Lofgren, C. S. 1986. The economic importance and control of imported fire ants in the United States. Pp. 227-256. In: S. B. Vinson [ed.]. *Economic impact and control of social insects.* Praeger, NY.

Lofgren, C. S., W. A. Banks, and B. M. Glancey. 1975. Biology and control of imported fire ants. *Ann. Rev. Entomol.* 20: 1-30.

Lofgren, C. S., F. J. Bartlett, and C. E. Stringer. 1963. Imported fire ant toxic bait studies: Evaluation of carriers for oil baits. *J. Econ. Entomol.* 56: 62-66.

Lofgren, C. S., F. J. Bartlett, C. E. Stringer, Jr., and W. A. Banks. 1964. Imported fire ant toxic bait studies: Further tests with granulated mirex-soybean oil bait. *J. Econ. Entomol.* 57: 695-698.

Lofgren, C. S., C. E. Stringer, Jr., and F. J. Bartlett. 1962. Imported fire ant toxic bait studies: GC-1283, a promising toxicant. *J. Econ. Entomol.* 55: 405-407.

Lofgren, C. S. and D. F. Williams. 1982. Avermectin B_1a: a highly potent inhibitor of reproduction by queens of the red imported fire ant (Hymenoptera: Formicidae). *J. Econ. Entomol.* 75: 798-803.

MacKay, W. P., S. Majdi, J. Irving, S. B. Vinson, and C. Messer. 1992. Attraction of ants (Hymenoptera: Formicidae) to electric fields. *J. Kansas Entomol. Soc.* 65: 39-43.

Morrill, W. L. and G. L. Greene. 1975. Reduction of red imported fire populations by tillage. *J. Georgia Entomol. Soc.* 10: 162-164.

Nickerson, J. C., W. H. Whitcomb, A. P. Bhatkar, and M. A. Naves. 1975. Predation on founding queens of *Solenopsis invicta* by workers of *Conomyrma insana. Florida Entomol. Sci.* 68: 75-82.

Porter, S. D., A. Bhatkar, R. Mulder, S. B. Vinson, and D. J. Clair. 1991. Distribution and density of polygyne fire ants (Hymenoptera: Formicidae) in Texas. *J. Econ. Entomol.* 84: 866-874.

Porter, S. D. and D. A. Savignano. 1990. Invasion of polygyne fire ants decimates native ants and disrupts arthropod community. *Ecology* 71: 2095-2106.

Sauer, R. J., H. L. Collins, G. Allen, D. Campt, T. D. Canerday, G. Larocca, C. Lofgren, G. Reagan, D. L. Shankland, M. Trostle, W. R. Tschinkel, and S. B.

Vinson. 1982. Imported fire ant management strategies. Pp. 91-110. In: S. L. Battenfield [ed.]. Proc. imported fire ant. USEPA, USDA-APHIS.

Tschinkel, W R., and D. F. Howard. 1980. A simple, non-toxic home remedy against fire ants. J. Georgia Entomol. 15: 102-105.

Whitcomb, W. H., A. Bhatkar, and J. C. Nickerson. 1973. Predators of *Solenopsis invicta* queens prior to successful colony establishment. *Environ. Entomol.* 2: 1101-1103.

Williams, D. F. 1983. The development of toxic baits for the control of the imported fire ant. *Florida Entomol. Sci.* 66: 162-172.

Williams, D. F. 1986. Chemical baits: Specificity and effects on other ant species. Pp. 378-386. In: C. S. Lofgren and R. K. Vander Meer [eds.]. *Fire ants and leaf-cutting ants, Biology and management.* Westview Press, Boulder, CO.

Williams, D. F., C. S. Lofgren, W. A. Banks, C. E. Stringer, and J. K. Plumley. 1980. Laboratory studies with nine amidinohydrazones, a promising new class of bait toxicants for control of red imported fire ants. *J. Econ. Entomol.* 73: 798-802.

Williams, D. F., C. S. Lofgren, J. K. Plumley, and D. M. Hicks. 1983. An auger-applicator for applying small amounts of granular pesticides. *J. Econ. Entomol.* 76: 395-397.

25

Biological Control of Introduced Ant Species

Richard S. Patterson

Introduction

It has been estimated that throughout the world there are over 20,000 species of ants in over 300 genera. Less than half of the ants have been described, because they are innocuous and, except for scientists, few people care (Lattke 1990). These ants are usually grouped according to genera that desperately need taxonomic revisions. The taxonomy of ants is even worse when it comes to describing potential biological control organisms. Often the biocontrol organism is identified to genus or species, but the affected ants are often identified only to genus or common names, such as harvester ant, black ant, etc. (Sweetman 1958). Almost no one is interested in what ails an ant in nature. Scientists who study ants, the true myrmecologists, are mainly interested in the social aspects of ant behavior: how ants function in the colony, how they interact with other ants and arthropods in the environment, and their taxonomy, morphology, and physiology. This is illustrated by the fact that the recent comprehensive book *The Ants* (Hölldobler and Wilson 1990) has less than one page of text on ant pathology in 732 pages on ant taxonomy, behavior, physiology, ecology, etc. Only a few economic entomologists, besides the general public, seem to be interested in developing safe and efficient means to suppress pest ant populations and then usually with pesticides (Lofgren 1990). There is a general perception that ants are not normally serious economic pests. They may become a troublesome nuisance if man significantly interferes with their normal habitat or accidentally transports them to new environments. For example, certain species of leaf-cutting ants are not a serious pest in their natural environment in the tropical forests, but can become a serious threat to large monocultures of cultivated crops when the natural restraints, mainly competitors and predators, are removed (Cherrett 1968 1986; Pollard 1982). The red imported fire ant, *Solenopsis invicta,* is an excellent example of a species that was accidentally introduced into an environment free of any biological restraints, and within a decade they had become a major economic pest (Wilson 1986). Actually, many of the major pest ants in North America were accidentally introduced (see Table 25.1) (Bennett et al. 1988, Smith and Whitman 1992).

The reason these few dozen species of exotic and native ants become pests while their fellow members in the same family and genus remain innocuous is simple. These species become pests when they are somehow released from their natural restraints and interfere with man's well

TABLE 25. 1. Common pest ants found in North America.

Common Name	*Scientific Name*	*Origin*
Acrobat Ant	*Crematogaster* spp.	Native
Allegheny Mound Ant	*Formica exsectoides*	Native
Argentine Ant	*Linepithema humile*	Introduced
Big-Headed Ants	*Pheidole* spp.	Native and Introduced
Black Imported Fire Ant	*Solenopsis richteri*	Introduced
Carpenter Ants	*Camponotus* spp.	Native and Introduced
Cornfield Ant	*Lasius alienus*	Native
Crazy Ant	*Paratrechina longicornis*	Introduced
Ghost Ant[1]	*Tapinoma melanocephalum*	Introduced
Harvester Ant	*Pogonomyrmex* spp.	Native
Large Yellow Ant	*Acanthomyops interjectus*	Native
Little Black Ant	*Monomorium minimum*	Native
Little Fire Ant	*Wasmannia auropunctata*	Introduced
Odorous House Ant	*Tapinoma sessile*	Native
Pavement Ant	*Tetramorium caespitum*	Introduced
Pharaoh Ant	*Monomorium pharaonis*	Introduced
Pyramid Ant	*Dorymyrmex* spp.	Native
Red Imported Fire Ant	*Solenopsis invicta*	Introduced
Small Honey Ant (winter ant)	*Prenolepis imparis*	Native
Thief Ant	*Solenopsis molesta*	Native
Velvety Tree Ant	*Liometopum* spp.	Native
Leaf-cutting Ants	*Atta* spp.	Native
Tropical Fire Ant	*Solenopsis geminata*	Introduced

[1]Not ESA approved common name

being. As stated earlier, leaf-cutting ants only become a problem when they destroy some valuable crop or trees of commercial value. In their native habitat as they go about collecting their leaves for their fungus gardens, their numbers are held in check by various predators, competitors and environmental conditions not found in artificial cropping systems, (Cherrett 1986). Several exotic ants that have been introduced on the Galapagos Islands affect the fragile ecosystem. Two ants which are creating a serious problem are the little fire ant, *Wasmannia auropunctata*, and the tropical fire ant, *Solenopsis geminata*. They have outcompeted the native ant fauna and reduced many indigenous arthropod populations on the various islands (Lubin 1984). These species also attack the young of some reptiles, such as the Galapagos tortoises and land iguanas, as well as stinging tourists and natives alike (Williams and Wilson 1987, Williams and Whelan 1991).

A similar situation has occurred with the red imported fire ant, *Solenopsis invicta*, and the black imported fire ant, *S. richteri*, in the United States. In undisturbed areas in their native homeland of South America, they are not a serious pest, but in the southeastern U.S., they are considered to be a major pest and millions of dollars are spent annually to control them (Lofgren 1986). In Florida, the red imported fire ant is the major ant pest that invades homes, hospitals and restaurants and is very difficult to control according to a survey of Pest Control Operators (Bieman and Wojcik 1990). Although the black imported fire ant, *S. richteri*, was first introduced into Mobile, Alabama in the 1920s, ironically its rapid spread may have been initially hampered by another South American ant accidentally introduced into the U.S.A., the Argentine ant, *Linepithema humile*. The black imported fire ant was, however, able to outcompete the Argentine ant and most of our native ants, and to establish itself in several thousand acres within two decades after introduction (Lofgren et al. 1975). Attempts to control this species with various pesticides met with limited success. Then in the late 1930s a more virulent form or species of fire ant, the red imported fire ant, later named *S. invicta*, was imported into Mobile, Alabama. Within twenty years it was the dominant ant in the region (Lofgren 1986). According to Porter et al. (1992), it now represents 97% of the ants collected along roadsides in the region, whereas in Brazil, fire ants represent only 23% of the ants collected and *S. invicta* is not necessarily the dominant fire ant species collected. There are numerous *Solenopsis* species in South America (Trager 1991). Wojcik, (see chapter 23) in a twenty year study in central Florida, showed that *S. invicta* had replaced the tropical fire ant, *S. geminata* and *Pheidole dentata* as the dominant ants found along roadsides in open areas. The impressive and rapid spread of *S. invicta* in the U.S.A. has been aided by such factors as favorable climate, the ant's biology, ecology, and the lack of competitors. Climatic conditions in the southeastern United States are ideal for the two species of fire ants introduced from Brazil and Argentina. *Solenopsis invicta* has a tremendous reproductive potential, more than any of our native ants. Individual or monogyne colonies may have up to several hundred thousand workers whereas polygyne colonies have as many as a million or more workers (Williams 1990). Just obtaining enough food to sustain these large colonies requires a tremendous amount of energy and aggressiveness. Fire ants, because of their need to support large colonies, are omnivorous feeders and very competitive with our native ants. Probably one of the biggest factors influencing the rapid spread of fire ants in the southeastern U.S.A. is the vast changes in the ecology of the region. Following World War II, there was a rapid population movement in the U.S. from the industrial areas of the North to the warm region of the South and West, "the Sunbelt". Forests and farms were eliminated to make room for new homes, recreational areas and industries. Since fire ants are a "weed species" (Tschinkel 1986) that thrives in recently opened disturbed areas, this newly disturbed environment was an ideal situation. In South America there are numerous closely related species of fire ants which may be equally or more competitive than the red imported fire ant (Trager 1991), plus in the tropics many other species of ants of different genera and families compete for the same space and food as the red imported fire ant. Thus, the general conception that the red imported fire ant is not a problem in its home land is true to some degree. However, in some areas of Brazil and Argentina, where climate, agricultural practices and land use are similar to the U.S.A., the red and/or black fire ant populations can be as high as in the U.S.A., often exceeding 200 colonies per acre (Banks et al. 1985). In one area of Argentina, covering several thousand hectares where *S. richteri* is the dominant species, the number of colonies per hectare exceeds 200; yet the farmers there do not consider fire ants a serious pest. Still, the overall population density of fire ants in South America is much less than in the U.S.A. (Porter et al. 1990). The reason for this is more complicated than just a single pathogen, parasite or predator limiting the ants distribution.

If one looks at ants in general, but especially the pest ants, they would seem superficially to be an ideal candidate for control and elimination by biological control agents (Jouvenaz 1986). They are very concentrated, feed one another, usually live in a colony which has a fairly high humidity, the immature stages are immobile in a concentrated mass, and only the queens are able to reproduce. However, over the years, ants seem to have developed techniques, probably through genetic selection, to prevent the invasion of any pathogen, parasite, or predator into their colony that would lead to its elimination. Ants in general are excellent housekeepers and personal groomers. They quickly recognize any invader through chemical cues and attack it in an attempt to kill it and/or remove it from the colony. Some arthropods and other organisms, such as myrmecophiles, have adapted and assumed the chemical cues of a particular fire ant colony and are thus able to survive quite well (Wojcik 1990). The effect of these myrmecophiles on colony survival is not well known, but it is assumed that they have some negative effect (Jouvenaz 1990). Glancey et al. (1981) reported that fire ants do filter out particles as small as 0.88 um, thus eliminating most bacterial and fungal spores before they can infect the ants. Ants are also capable of secreting from their venom and metapleural glands and other exocrine organs a number of chemicals which have antibiotic and antifungal properties. These secretions are spread from one ant to another and throughout the colony through grooming (Hölldobler and Wilson 1990, Obin and Vander Meer 1985, Jouvenaz et al. 1972). Since most soils harbor many organisms that are pathogenic to arthropods including ants, these secretions aid in the survival of the colony.

Few pathogens and parasites have been described in the literature for other species of ants other than fire ants and leaf-cutting ants (Kermarrec et al. 1986, Laumond et al. 1979). During the past several decades numerous surveys have been made in South America and throughout the southern United States for potential biological control organisms of fire ants (Jouvenaz et al. 1977, 1986). A limited number of organisms have been identified (Table 25.2) and many of these, such as the fungi of the genera *Beauveria* and *Metarhizium* (Stimac et. al 1989) are nonspecific for fire ants.

TABLE 25.2. Potential biological control organisms for red imported fire ant, *S. invicta*[1].

Organism	*Pathogen*	*Type*
Pseudomonas aeruginosa	Bacteria	non-specific
Serratia marcescens	Bacteria	non-specific
Bacillus thuringiensis	Bacteria	non-specific
Bacillus spheracus	Bacteria	non-specific
**Beauveria bassiana*	Fungus	non-specific
Metarhizium anisopliae	Fungus	non-specific
Aspergillus flavus	Fungus	non-specific
**Thelohania solenopsae*	Protozoa	specific
**Vairimorpha invicta*	Protozoa	specific
Mattesia geminata	Protozoa	specific
Burenella dimorpha	Protozoa	specific

(continues)

TABLE 25.2 *(continued)*

Organism	*Parasites*	*Type*
**Steinernema carpocapsae*	Nematode	non-specific
Heterorhabtidis heliothidis	Nematode	non-specific
**Tetradonema solenopsis*	Nematode	specific
Pyemotes tritici	Mite	non-specific
**Pseudacteon* spp.	Diptera (Phorid fly)	specific
**Apodicrania* spp.	Diptera (Phorid fly)	specific
Orasema spp.	Hymenoptera (wasp)	specific
**Solenopsis daguerrei*	Hymenoptera (ant)	specific

[1]Pathogens and parasites which have been investigated as potential fire ant biological control agents.

*Currently being investigated

The literature on pathogens and parasites of fire ants consists mainly of descriptions and how the organisms affects the individual ant. Limited research has been published on the impact of these organisms on the entire colony. Jouvenaz et al. 1981 did an excellent study of the pathogen, *Burenella dimorpha,* which affects only the tropical fire ant, *S. geminata.* Although usually less than 5% of colonies are infected, at times all the colonies at a single location will be infected (Jouvenaz 1986). Unfortunately, now that *S. geminata* populations have declined in central Florida, this disease is very scarce (Jouvenaz personal communications). The potential of this disease as a biological control agent was never tested in the field.

In an attempt to rectify this deficiency, a limited number of studies with biological organisms have been conducted on field populations of fire ants. Reported here are two studies which have been conducted on wild fire ant populations. The objective of the first study was to determine the impact of a nonspecific pathogen, *Beauveria bassiana,* on fire ant populations infesting potting soil used for nursery plants. The objective of the second study was to determine the impact of a naturally occurring pathogen, *Thelohania solenopsae,* on an indigenous population of fire ants in South America, and to determine its potential as a biological control organism for possible importation into the United States to suppress *S. invicta.*

Beauveria bassiana

Beauveria bassiana is a nonspecific pathogen that is capable of attacking many arthropods. Its potential as a biological control organism has been explored by numerous scientists for various insect pests over the years (Ferron 1978, 1981). Recently there has been renewed interest in this organism for fire ant control. Stimac et al. (1989, 1990) reported high colony mortality (80%) when a strain of this fungus from Brazil was applied to individual fire ant mounds. They also found that it was present in the soil 150 days after treatment. They treated the soils with 200g of a 5% fungal formulation of conidia on rice. Fire ant colony movement (forming of satellite colonies) increased following treatment with the fungal formulation. Because of these and other promising reports (Broome et al. 1976), a series of studies were designed to determine if *B. bassiana* could be used prior to shipment to prevent fire ant infestations in the potting soil used for nursery plants. The strain of *B. bassiana* used in these studies originated from leaf-cutting ants collected in Mexico. A number of concentrations of a dry mycelial formulation mixed with the soil and tested in the laboratory showed that a 0.5% formulation gave the best results. A large field study was set up using treated (0.5% dry mycelium/soil mixtures) and untreated soil in pots with and without plants. The test was run outside under ambient conditions. The pots were infested with polygyne colonies of fire ants that consisted of a set of 5 queens, 1 gram of brood and 3-4 thousand workers. To check if the fungus had any repellent effect, all pots were checked for colony movement. At set intervals for 30 days, the ant colonies in the soil from the pots were examined for any mortality. There was no significant mortality of the ant colonies from the fungus, but there was repellency. A second large study was run using various types of media (potting soil, sand, vermiculite, and various mixtures of these, both sterilized and unsterilized). The set up was basically the same as before, except no plants were used and the pots containing the soil were held at a constant temperature (80°F) and high humidity (80-90% R.H.). A check was made of the mortality of queens, brood and workers exposed to the fungi in the various media. High mortality of the queens, brood and workers occurred in the vermiculite medium (Table 25.3).

A mixture of vermiculite and sterile potting soil produced less mortality of the queens, brood and workers. Potting soil alone produced

TABLE 25.3. Effects of 0.5% *B. bassiana* on red imported fire ants in different substrates after 7 days.

Media		*% Mortality*	
	Queens	*Brood*	*Workers*
Control	9	0	0
Potting Soil	17	0	0
Sterile Potting Soil	25	95	0
Vermiculite	75	100	85
Sterile Potting Soil/ vermiculite 50/50	59	95	50

no effect on brood or workers and queen mortality was insignificant. Therefore, there must be something in the unsterilized soil that inhibits the *B. bassiana,* because almost all the brood was infected with fungus in the sterile potting soil. A series of tests to determine the cause of this lack of effectiveness of the fungus in nonsterilized soil to fire ants, were run. The results were inconclusive except to show that if the potting soil was sterilized, the fungus was effective. Also if a medium was used with a very low organic matter content, such as sand mixtures, the efficiency of the fungus also increased. A small field study in which individual mounds were treated with the fungus, showed that complete colony mortality did not occur, but treated colonies often moved.

To summarize, we found that *B. bassiana* will kill fire ants as reported previously (Stimac et al. 1990). The brood is very susceptible to the fungus, queens are less so and the workers are more resistant, probably because they are more adept at grooming themselves and one another of spores. The fungus does have a repellent effect, but it is not an extremely strong one. Although it can kill ants, unless the queen was infected, it is not efficient in eliminating an entire colony. At present, *B. bassiana* is not recommended for use in protecting potting soil which contains nursery stock from fire ant invasion prior to shipment.

Thelohania solenopsae

Although numerous surveys have been made to determine what pathogens and parasites attack fire ants, almost nothing is known of the effect such organisms have on indigenous ant populations (Jouvenaz et al. 1981). Therefore, a long term study was initiated in Argentina in 1988 to follow what happens to fire ant colonies infested with the microsporidian disease, *Thelohania solenopsae.* We worked with the black imported fire ant, *Solenopsis richteri,* however, some plots contained a few colonies of *S. quinquecuspis.*

The objectives of this study were to determine: (1) what effect, if any, this disease had on the total colony, not individual ants; (2) what percentage of the ants and castes had the disease; (3) does the disease cause the colony to move more frequently; (4) how the disease is spread within the colony from ant to ant and from colony to colony in the field; (5) do healthy ants perceive the infected ants in the colony and remove them, thus eliminating the disease; (6) does this disease require an intermediate host for transmission from ant to ant or colony to colony; (7) is the microsporidian disease, *T. solenopsae,* in *S. richteri* the same as the one described in *S. invicta* from Brazil and, finally, (8) does this disease have any potential as a biological control agent and should it be introduced into the U.S.A.?

Thelohania solenopsae was selected because it was the most common pathogen present in fire ant colonies examined in Argentina. The taxonomic description of *T. solenopsae* and the disease it caused in fire ants in Brazil was given by Knell et al. (1977). Its occurrence in host ant populations had been reported by numerous earlier scientists (Allen and Buren 1974, Allen and Silveira-Guido 1974, Jouvenaz et al. 1980). All attempts to transmit the disease to healthy colonies of fire ants have failed (Jouvenaz 1986). However, recent studies with another microsporidian disease, *Amblyosporidae*, in mosquitoes, have shown that an intermediate host is often required before transmission back into the mosquito population is possible, (Andreadis 1985, 1988; Sweeney et al. 1985, 1988; Becnel 1992). There are sound biological reasons to suspect that alternate hosts may be involved in intercolonial transmission of this microsporidian disease in fire ants. Since there are a limited number of arthropods which are symbiotic with fire ants (Wojcik 1990), they were examined for the disease. Observations by Knell et al. (1977) indicated that this disease did not quickly kill the colony. Often very large, apparently healthy colonies with high infection rates of this pathogen were found in the field; however, the ultimate fate of these colonies after a period of time is not known. In fact, we do not really understand fire ant population dynamics in South America very well.

Materials and Methods

Six circular plots, 40 meters in diameter, were set up in unimproved pastures containing cattle and hogs in Saladillo, Buenos Aires Province, Argentina. Each active fire ant colony was plotted on a map by measuring from a central stake using the four compass quadrants. The plots were almost side by side in a series of pastures. The average number of active colonies per hectare was 198 in the six plots. Three of the plots had fairly high disease rates with 35% of the colonies being infected. The other three plots were relatively free of the disease with only 3% of the colonies being infected.

The number of colonies (mounds) in each plot was checked monthly to determine the viability of each colony. Also, if the colony had moved, we needed to know if the diseased ants moved with the colony. Some colony movement was traced with dyes, but mainly it was with observations on new mounds in close proximity to old abandoned colonies or mounds. We also checked for the presence of brood, workers, sexuals, colony health, and presence of disease. Each colony or mound was measured for its height, width and ant activity. A small glass vial was inserted into the mound for less than 10 minutes to collect the ants. Then the vials were removed, capped and returned to the laboratory for microscopic examination for the disease. In the laboratory, the ants were ground with water in a tissue grinder, then a small drop of the ground material was placed on a glass slide, covered by a cover slip and examined by phase-contrast microscopy at 400x for the presence or absence of *Thelohania* spores. Since this sampling was a mixture of many individual ants with both major and minor workers and probably the reserve and forager ants, we could only determine the presence or absence of the disease in the colony; not the percentage or which castes were infected. Jouvenaz (1986) reported that fire ant workers, sexuals, and queens were infected with this disease in Brazil. We later checked individual ants from each caste and the various age classes, as well as any myrmecophiles present in the colony for the disease.

Results and Discussion

The number of active fire ant colonies per hectare in a short grass pasture habitat of the Saladillo area in Argentina was 100 colonies per hectare which was similar to that in the southeastern United States, especially in north central Florida. Adams (1986) reported 60-150 colonies/hectare as being heavy densities of *S. invicta* in Florida and Georgia. There was a 25% reduction in number of active fire ant colonies in the plots from October to March (spring to early fall) in Argentina. This is similar to what is found with *S. invicta* in the Southern U.S.A. (Hays et al. 1982). Of the active colonies, 75% left their original colony site and moved to a new location, usually within a meter of the original site. This also is similar to what has been observed in the U.S. for *S. invicta* and *S. richteri* (Hays et al. 1982).

There was greater loss of colonies with *Thelohania*, than of non-infected colonies (45% versus 25%) during the first year. Once a colony became infected it remained infected whether it moved or not. We observed no colonies losing their infection. However, this is difficult to verify because of colony movement, since we were not always sure of the origin of each new colony. We later examined whole colonies and found that if the queen was infected then all progeny in the colony were infected. In polygyne colonies, it is speculated that the disease might be vertically transmitted transovarially by the infected queen. How it is transmitted horizontally from ant to ant within the colony or to other colonies is a mystery. Probably an intermediate host is involved but we are unsure what it is, although several myrmecophiles have been found infected with the disease. We do know that in the area where 35% of the colonies were initially infected; 80-90% of the colonies were infected four years later and the number of fire ant colonies had decreased to less than 30/hectare. Large fire ant colonies seem to be able to withstand the disease, as noted by Knell et al. (1977) while small colonies appear to die off implying the size of colonies is important. It is probably dependent on when the colony becomes infected and if all the queens are infected. The results of this study with this microsporidian disease to date show that the fire ants move their colonies as frequently in Argentina as they do in the U.S.A. Over 75% of the colonies had moved to a new site within the first six months of this study. Some colonies moved almost monthly, while others remained in the same site for a long time. Since the colonies were numerous and fairly close together, it does not appear that food supplies or mound disturbance caused this movement. There was a total loss in the number of colonies from the study (ca. 25%) during the first year. This occurred whether or not the colonies were infected. When the area was checked two years later, many more diseased colonies were lost. The presence of *T. solenopsae* seems to cause colony mortality, especially of small colonies. However, it may take several years to show a significant decrease. The disease, *T. solenopsae*, as described by Knell et al. (1977) seems to have a wide range of hosts as spores have been observed in many of the other arthropods inhabiting the fire ant mound. Since no intermediate hosts have been identified yet, we do not know how this disease is spread from ant to ant or one colony to another in nature. All the data presently available makes this organism appear very promising as a potential biological control organism. We are continuing our research to determine how this disease can possibly be used in the field to suppress fire ants.

Summary

A number of potential biological control organisms of ants have been identified in the literature. A few of these have been studied in the laboratory. However, except for some limited personal observations and speculations, little has been done in the field to evaluate the impact any have on various ant species or ant densities. We have reported here on laboratory and field studies of two pathogens, *Beauveria bassiana* and *Thelohania solenopsae*, which infect the red imported fire ant, *Solenopsis invicta*. The fungus *Beauveria bassiana*, a strain from Mexico and Texas, was studied in the laboratory and in the field. This is a non-specific pathogen and will affect many insects. Its effectiveness is dependent on the substrate it is mixed with. Control in sterile soils or soils with very low organic matter treated with a 0.5% *B. bassiana* dry mycelium mixture was good with infection of the brood, queens and workers. However, its performance was poor in highly organic or non-sterile soils. The brood and then the queens were most susceptible to the fungus. The workers were the most tolerant of the fungus. *Thelohania solenopsae* was a fairly host specific microsporidian disease. *T. solenopsae* studied in field infected colonies for almost four years showed that it had a great effect on the colony structure and its survival. It has great potential for importation into the U.S.A. as a biological control agent.

Although a number of potential biological control organisms which affect fire ants have been identified in South America, we do not know how each affects the entire colony of the ants. We do know that ants seem to be able to survive most infestations unless the colony is placed under stress. Under stress, the disease seems to quickly take over the colony and the colony collapses. We will continue to work out the life history of a number of pathogens and parasites of fire ants and determine what impact each has on the entire colony. A complex of these, if properly released, should have an adverse effect on the imported fire ant population in the United States.

Resumen

Con el fin de reducir poblaciones de las hormigas de fuego, fueron evaluados en el campo, el hongo *Beuveria bassiana,* y el protozoario, *Thelohania solenopsae*. Se había demostrado en el laboratorio que ambos organismos eran letales a las hormigas de fuego. Cuando las hormigas de fuego eran expuestas a *B. bassiana* combinada en una mezcla de suelo, no se observó mortalidad aunque el hongo parecía estimular el movimiento de la colonia. Si el suelo usado era estéril, el hongo afectaba a las hormigas, especialmente la cría y las reinas. El medio como vermiculita o arena,con un poco, o sin materia organica, no inhibió el hongo. Sinemabrog, las hormigas no pudieron ser controladas al usar mezcla de suelo con *B. bassiana*.

El protozoario, *T. solenopsae,* fué muy efectivo en reducir el numero de las colonias de las hormigas de fuego en el campo. Este organismo parece promisorio como un agente potencial de control biologico de las hormigas de fuego. Se necesita sinembargo, mucha investigación antes que este agente pueda ser introducido en el campo.

References

Adams, C. T. 1986. Agricultural and medical impact of the imported fire ants. Pp. 48-57. In: C. S. Lofgren and R. K. Vander Meer [eds.]. *Fire ant and leaf-cutting ants biology and management.* Westview Press, Boulder, CO.

Allen, G. E. and W. F. Buren. 1974. Microsporidian and fungal diseases of *Solenopsis invicta* Buren in Brazil. *J. New York Entomol. Soc.* 82: 125-130.

Allen, G. E. and A. Silveira-Guido. 1974. Occurrence of microsporidia in *Solenopsis richteri* and *Solenopsis spp.* in Uruguay and Argentina. *Florida Entomol.* 57: 327-329.

Andreadis, T. G. 1985. Experimental transmission of a microsporidian pathogen from mosquitoes to an alternate copepod host. *Proc. Nat. Sci.* 82: 5574-5577.

______. 1988. *Anblyospora connecticus* sp. nov. (Microsporida: Amblyosporidae): Horizontal transmission studies in the mosquito *Aedes cantator* and formal description. *J. Invertebr. Pathol.* 52: 90-101.

Banks, W. A., D. P. Jouvenaz, D. P. Wojcik and C. S. Lofgren. 1985. Observations on fire ants *Solenopsis spp.*, in Mato Grosso, Brazil. *Sociobiology* 11: 143-152.

Becnel, J. J. 1992. Horizontal transmission and subsequent development of *Amblyospora california* (Microporida: Amblyosporidae) in the intermediate and definitive host. *Dis. Aquat. Org.* 13: 17-28.

Bennett, G. W., J. M. Owens and R. M. Corrigann. 1988. Truman's scientific guide to pest control operations. 4th Edition. Edgell Communication, Duluth, MN.

Bieman, D. N. and D. P. Wojcik. 1990. Tracking ants in Florida: The results of a Florida ant survey and a key to common structure- invading ants. PCO, Fla Pest Control Service Mag. 1990 (April) 11-13.

Broome, J. R., P. R. Sikorowski and B. R. Norment. 1976. A mechanism of pathogenicity of *Beauveria bassiana* on larvae of the imported fire ant, *Solenopsis richteri. J. Invertebr. Pathol.* 28: 87-91.

Cherrett, J. M. 1968. Some aspects of the distribution of pest species of leaf cutting ants in the Caribbean. *Proc. Amer. Soc. Hort. Sci. (Trop. Reg.)* 12: 295-310.

_____. 1986. History of the leaf cutting ant problem. Pp. 10-17. In: C.S. Lofgren and R.K. Vander Meer [eds.]. *Fire ant and leaf-cutting ants biology and management.* Westview Press, Boulder, CO.

Ferron, P. 1978. Biological control of insect pests by entomogenous fungi. *Annu. Rev. Ent.* 23: 409-422.

_____. 1981. Pest control by the fungi *Beauveria* and Metarhizium. Pp. 465-482. In: H.D. Buges [ed.]. *Microbial control of pests and plant diseases 1970-1980.* London Academic Press.

Glancey, B. M., R. K. Vander Meer, A. Glover, C. S. Lofgren and S. B. Vinson. 1981. Filtration of microparticles from liquids ingested by the imported fire ant *Solenopsis invicta* Buren. *Insect Soc.* 28: 395-401.

Hays, S. B., P. M. Horton, J. A. Bass and D. Stanley. 1982. Colony movement of imported fire ants. *J. Georgia. Entomol. Soc.* 17: 266-274.

Hölldobler, B. and E. O. Wilson. 1990. *The Ants.* Belknap Press of Harvard University Press Cambridge, Mass.

Jouvenaz, D. P. 1986. Diseases of fire ants: problems and opportunities. Pp. 327-338 In: C.S. Lofgren and R.K. Vander Meer [eds.]. *Fire ants und leaf-cutting ants biology and management.* Westview Press, Boulder, CO.

______. 1990. Approaches to biological control of fire ants in the United States. Pp. 620-627. In: R. K. Vander Meer, K. Jaffe and A. Cedeno [eds.]. *Applied Myrmecology: A world perspective.* Westview Press, Boulder, CO.

Jouvenaz, D. P., G. E. Allen, W. A. Banks and D. P. Wojcik. 1977. A survey for pathogen of fire ants *Solenopsis spp.*, in the Southeastern United States. *Florida Entomol.* 60: 275-279.

Jouvenaz, D. P., W. A. Banks and J. D. Atwood. 1980. Incidence of pathogens in fire ants, *Solenopsis spp.* in Brazil. *Florida Entomol.* 63: 345-346.

Jouvenaz, D. P., M. S. Blum and J. G. MacConnell. 1972. Antibacterial activity of venom alkaloids from the imported fire ant *Solenopsis invicta* Buren. *Antimicrob. Agents Chemother.* 2: 291-293.

Jouvenaz, D. P., C. S. Lofgren, and G. E. Allen. 1981. Transmission and infectivity of spores of *Burenella dimorpha* (Microsporida: Burenellidae) *J. Invert. Path.* 37: 265-268.

Jouvenaz, D. P., C. S. Lofgren and W. A. Banks. 1981. Biological control of imported fire ants: a review of current knowledge. *Bull. Entomol. Soc. Am.* 27: 203-208.

Kermarrec, A., G. Febvay and M. Decharme. 1986. Protection of leaf-cutting ants from biohazards: is there a future for microbiological control? Pp. 339-356. In: C. S. Lofgren and R. K. Vander Meer [eds.]. *Fire ants and leaf-cutting ants biology and management.* Westview Press, Boulder, CO.

Knell, J. D., G. E. Allen and E. I. Hazard. 1977. Light and electron microscope study of *Thelohania solenopsae* n. sp. (Microsporida: Protozoa) in the red imported fire ant, *Solenopsis invicta. J. Invertebr. Pathol.* 29: 192-200.

Lattke, J. E. 1990. Overview. Pp. 71-74. In: R. K. Vander Meer, K. Jaffe and A. Cedeno [eds.]. *Applied myrmecology: a world perspective.* Westview Press, Boulder, CO.

Laumond, C., H. Mauleon and A. Kermarrec. 1979. Donnees nouvelles sur le spectre d'hotes et le parasitisme der nematode entomophage *Neoplectona carpocapsae. Entomophaga* 24: 13-27.

Lofgren, C. S. 1986. History of imported fire ants in the United States. Pp. 36-47. In: C.S. Lofgren and R.K. Vander Meer [eds.]. *Fire ants and leaf-cutting ants: biology and management.* Westview Press, Boulder, CO.

______. 1990. Foreword. Pp. x-xii. In: R. K. Vander Meer, K. Jaffe and A. Cedeno [eds.]. *Applied myrmecology: a world perspective.* Westview Press, Boulder, CO.

Lofgren, C. S., W. A. Banks and B. M. Glancey. 1975. Biology and control of imported fire ants. *Annu. Rev. Entomol.* 20: 1-30.

Lubin, Y. D. 1984. Changes in the native fauna of the Galápagos Islands following invasion by the little red fire ant, *Wasmannia auropunctata. Biol. J. Linn. Soc.* 21: 229-242.

Obin, M. S. and R. K. Vander Meer. 1985. Gaster flagging by fire ants (*Solenopsis* spp): functional significance of venom dispersal behavior. *J. Chem. Ecol.* 11: 1757-1768.

Pollard, G. V. 1982. A review of the distribution, economic importance, and control of leafcutting ants in the Caribbean region with an analysis of current control programmes. Pp. 43-61. In: C. W. D. Brathwaite and G. V. Pollard [eds.]. *Urgent plant pest and disease problems in the Caribbean.* IICA. Misc. Pub.

Porter, S. D., H. G. Fowler and W. P. MacKay. 1990. A comparison of fire ant population densities in North and South America. Pp. 623-624. In: G.K. Veeresh, B. Mallik, C.A. Viraktamth [eds.]. *Social insects and the environment.* Oxford and IBH Publ. Co. Pvt. Ltd., New Delhi, India.

______ . 1992. Fire ant mound densities in the United States and Brazil (Hymenoptera: Formicidae). *J. Econ. Entomol.* 85: 1154-1161.

Smith, E. H. and R. C. Whitman. 1992. NPCA Field Guide to Structural Pests. Natural Pest Control Assoc. Dunn Loring, VA.

Stimac, J. L., S. B. Alves and M. T. V. Camargo. 1989. Controle de *Solenopsis spp.* (Hymenoptera: Formicidae) com *Beauveria bassiana* (Bals.) Vuill. em condições de laboratorio e campo. *An. Soc. Entomol. Brasil* 18: 95-103.

Stimac, J. L., R. M. Pereira, S. B. Alves and L. A. Wood. 1990. Field evaluation of a Brazilian strain of *Beauveria bassiana* for control of the red imported fire ant, *Solenopsis invicta,* in Florida. Proc. 5th Intern. Coll. Invertebr Pathol., Microb. Control.

Sweeney, A. W., E. I. Hazard and M. F. Graham. 1985. Intermediate host for an *Amblyospora* sp. (Microspora) infecting the mosquito, *Culex annulirostris. J. Invertebr. Pathol.* 46: 98-102.

Sweeney, A. W., M. F. Graham and E. I. Hazard. 1988. Life cycle of *Amblyospora dyxenoides* sp. nov. in the mosquito *Culex annulirostris* and the copepod *Mesocyclops albicans. J. Invertebr. Pathol.* 51: 46-57.

Sweetman, H. L. 1958. *The principles of biological control.* W.C. Brown, Dubuque, IA.

Trager, J. C. 1991. A revision of the fire ants, *Solenopsis geminata* group (Hymenoptera: Formicidae, Myrmicinae). *J. New York Entomol. Soc.* 99: 141-198.

Tschinkel, W. R. 1986. The ecological nature of the fire ant: some aspects of colony function and some unanswered questions. Pp. 72-87. In: C. S. Lofgren and R. K. Vander Meer [eds.]. *Fire ants & leaf-cutting ants: biology & management.* Westview Press, Boulder, CO.

Williams, D. F. 1990. Oviposition and growth of the fire ant, *Solenopsis invicta.* Pp. 150-157. In: R. K. Vander Meer, K. Jaffe and A. Cedeno [eds.]. *Applied myremecology: a world perspective.* Westview Press, Boulder, CO.

Williams, D. F. and P. Whelan. 1991. Polygynous colonies of *Solenopsis geminata* (Hymenoptera: Formicidae) in the Galápagos Islands. *Florida Entomol.* 74: 368-371.

Williams, D. F. and M. H. Wilson. 1988. Control of the fire ants, *Ochetomyrmex auropunctata* and *Solenopsis geminata* on the Galápagos Islands. Pp. 61-64. Annual Report 1986-1987, Charles Darwin Research Station.

Wilson, E. O. 1986. The defining traits of fire ants and leaf cutting ants. Pp. 1-9. In: C. S. Lofgren and R.K. Vander Meer [eds.]. *Fire ants and leaf-cutting ants: biology and management.* Westview Press, Boulder, CO.

Wojcik, D. P. 1990. Behavioral interactions of fire ants and their parasites, predators and inquilines. Pp. 329-344. In: R. K. Vander Meer, K. Jaffe and A. Cedeno [eds.]. *Applied myrmecology: a world perspective.* Westview Press, Boulder, CO.

About the Book and Editor

Most of the major problems caused by ants are a result of exotic species that have been introduced into areas where they have escaped any natural control. Some notable examples include imported fire ants in the southern United States, leaf-cutting ants in the tropics, pharaoh ants and Argentine ants in urban environments, big-headed ants in Hawaii, crazy ants in the Seychelles, and, more recently, little fire ants in the Galapagos. It was this last invasion, posing a threat to the delicate ecological balance of the Galapagos Islands, that inspired this volume.

Exotic Ants presents the latest research findings from experts on introduced pest ant species. Discussions include the distribution, biology, ecology, and behavior of several exotic ants and also describe current research on basic and applied topics. Particularly important for Spanish-speaking researchers are the brief summaries in Spanish at the end of each chapter. This volume will prove useful for entomologists, ecologists, ethologists, and agricultural scientists attempting to manage pest ant populations.

David F. Williams is a research entomologist for the Imported Fire Ant and Household Insect Project at the Medical and Veterinary Entomology Research Laboratory, Agricultural Research Service, United States Department of Agriculture.

Contributors

Sandra Abedrabbo
Charles Darwin Research Station
Casilla 17-01-3891
Quito, Ecuador

John W. Beardsley
University of Hawaii at Manoa
Department of Entomology
3050 Maile Way, Room 310
Honolulu, HI 96822 USA

Carlos R. F. Brandão
Museu de Zoologica
Universidade de São Paulo
Caixa Postal 7172
01064 São Paulo, SP, Brazil

Odair C. Bueno
Instituto de Biociencias
UNESP
Departamento de Ecologia
13500 Rio Claro, SP, Brazil

I. M. Cazorla
Divisao de Zoologie
Ceplac/Cepec/Entomologia
Caixa Postal 07
45.600 Itabuna Bahia, Brazil

Daniel Cherix
Musee Zoologique
Palais de Rumine
P. Riponne 6 - Case postale 448
CH-1000 Lausanne, Switzerland

J. M. Cherrett
School of Biological Sciences
University College of North Wales
Bangor, Gwynedd LL57 2UF
Wales, England

P. R. Davis
Agriculture Protection Board of WA
Baron-Hay Court
South Perth, WA 6151, Australia

Alain Dejean
Laboratorie de Zoologie
Faculte des Sciences
Universite de Yaounde
B.P. 812, Yaounde, Cameroon

J. H. C. Delabie
Divisao de Zoologie,
Ceplac/Cepec/Entomologia
Caixa Postal 07
45.600 Itabuna Bahia, Brazil

A. M. V. da Encarnação
Divisao de Zoologie,
Ceplac/Cepec/Entomologia
Caixa Postal 07
45.600 Itabuna Bahia, Brazil

Harold G. Fowler
Instituto de Biociencias
UNESP
Departamento de Ecologia
13500 Rio Claro, SP, Brazil

I. H. Haines
Overseas Development Administration
94 Victoria St.
London, SW1E 5JL, UK

J. B. Haines
Overseas Development Administration
94 Victoria St.
London, SW 1E 5JL, UK

John M. Heraty
Biological Resources Division
Centre for Land & Biological
Resources Research
Agricultural Canada
C.E.F., K.W. Neatby Bldg.
Ottawa, Ontario, K1A OC6, Canada

Klaus Jaffe
Departamento de Biologia de
Organismos
Universidad Simon Bolivar
Apartado 80659
Caracas 1080-A, Venezuela

Gary C. Jahn
U.S. Agency for International
Development
R&D/H/CD
SA-18, Room 1225
Washington, D.C. 20523

Rafael Jusino-Atresino
Texas Tech University
College of Agricultural Sciences
Department of Agronomy,
Horticulture and Entomology
Mail Stop: 2134
Lubbock, Texas 79409-2134 USA

M. Kenne
Universite Paul-Sabatier
Laboratoire d'Entomologie
118 Route de Narbonne
31062 Toulouse Cedex, France

John Lattke
Fundacion Museo de
Ciencias Naturales
Apartado 5883
Caracas 1010-A, Venezuela

J. D. Majer
Curtin University of Technology
School of Biology
GPO Box U 1987
Perth, WA 6001, Australia

Maria Alice de Medeiros
Instituto de Biociencias
UNESP
Departamento de Ecologia
13500 Rio Claro, SP, Brazil

Rolf E. Meier
Glarnischstrasse 152
CH 8708 Mannedorf, Switzerland

Ricardo V. S. Paiva
Museu de Zoologica
Universidade de São Paulo
Caixa Postal 7172
01064 São Paulo, SP, Brazil

Luc Passera
Universite Paul-Sabatier
Laboratorie d'Entomologie
118 Route de Narbonne
31062 Toulouse Cedex, France

Richard S. Patterson
USDA-ARS
Medical and Veterinary Entomology
Research Laboratory
P.O. Box 14565
Gainesville, Florida 32604 USA

Sherman A. Phillips, Jr.
Texas Tech University
College of Agricultural Sciences
Department of Agronomy,
Horticulture and Entomology
Mail Stop: 2134
Lubbock, Texas 79409-2134 USA

Sanford D. Porter
USDA-ARS
Medical and Veterinary Entomology
Research Laboratory
P.O. Box 14565
Gainesville, Florida 32604 USA

Neil J. Reimer
University of Hawaii at Manoa
Department of Entomology
3050 Maile Way, Room 310
Honolulu, Hawaii 96822 USA

J. van Schagen
Agriculture Protection Board of WA
Baron-Hay Court
South Perth, WA 6151, Australia

Marcelo N. Schlindwein
Instituto de Biociencias
UNESP
Departamento de Ecologia
13500 Rio Claro, SP, Brazil

J.-P. Suzzoni
Universite Paul-Sabatier
Laboratoire d'Entomologie
118 Route de Narbonne
31062 Toulouse Cedex, France

Leeanne E. Tennant
Museum of Comparative Zoology
Harvard University
26 Oxford Street
Cambridge, MA, 02138 USA

Patricia Ulloa-Chacon
Universidad del Valle
Departamento de Biologia
AA 25360 Cali, Colombia

Karen M. Vail
USDA-ARS
Medical and Veterinary Entomology
Research Laboratory
P.O. Box 14565
Gainesville, Florida 32604 USA

Ines de la Vega
Catholic University
Quito, Ecuador

S. Bradleigh Vinson
Department of Entomology
Texas A & M University
College Station, Texas 77843 USA

M. A. Widmer
Agriculture Protection Board of WA
Baron-Hay Court
South Perth, WA 6151, Australia

David F. Williams
USDA-ARS
Medical and Veterinary Entomology
Research Laboratory
P.O. Box 14565
Gainesville, Florida 32604 USA

Daniel P. Wojcik
USDA-ARS
Medical and Veterinary Entomology
Research Laboratory
P.O. Box 14565
Gainesville, Florida 32604 USA

Ingeborg Zenner-Polania
Investigacion Basica Agricola
A.A. 151123 El Dorado
Bogota, Colombia

Taxonomic Index

Subject Index